AMERICAN CITIES

A Collection of Essays

Series Editor

NEIL LARRY SHUMSKY

*Virginia Polytechnic Institute
and State University*

A GARLAND SERIES

SERIES CONTENTS

VOLUME

6

TRANSPORTATION AND COMMUNICATION

Edited with introductions by

NEIL LARRY SHUMSKY
*Virginia Polytechnic Institute
and State University*

GARLAND PUBLISHING, Inc.
New York & London
1996

Library of Congress Cataloging-in-Publication Data

American cities : a collection of essays / series editor, Neil Larry
Shumsky.
    p.   cm.
    Includes bibliographical references.
    Contents: v. 1. Urbanization and the growth of cities — v. 2. The
physical city — v. 3. Politics and government — v. 4. The economy —
v. 5. The working class and its culture — v. 6. Transportation
and communication — v. 7. Social structure and social mobility —
v. 8. Institutional life.
    ISBN 0-8153-2191-0 (v. 6 : alk. paper)
    1. Cities and towns—United States. I. Shumsky, Neil L., 1944–
HT123.A6615 1996
307.76'0973 —dc20   95-36145
                                    CIP

Printed on acid-free, 250-year-life paper
Manufactured in the United States of America

CONTENTS

SERIES INTRODUCTION

This collection brings together more than 200 scholarly articles pertaining to the history and development of urban life in the United States during the past two centuries. Less than 100 years ago, the Census Bureau revealed that more than half of all Americans live in urban places; barely fifty years ago, historians began to view cities as distinct places worth studying in their own right, and within the past ten years, these earlier developments were supported by the establishment of the Urban History Association.

Urbanization's rapid occurence during the nineteenth century and its brief period of historical study have led to diverse literature about American cities and their history appearing in a wide variety of publications. Because cities are frequently treated as part of states, articles have appeared in state historical journals. Sometimes, an article has covered a discrete institution and appeared in a journal pertaining only to that institution, for example, the family or the church. At still other times, authors have not consciously stressed the urban aspect of a particular subject, so their work has been published in journals that do not receive widespread attention from urbanists.

This series alleviates some of the problem of locating the varied literature by bringing together articles from journals of all types, organizing them, and making them available together for the first time. The articles in this collection emphasize cities—their growth, politics, economy, and so on—and, until now, many of them were accessible only to those members of a particular field who read the specialized journals of that field, whether it be the South, the Progressive Era, labor, or others. All of this seemingly unrelated literature, however, shares a common focus: cities and urban life. The purpose of this series is to bring out that common theme so that literature is more available, the relationship between the articles more easily seen, and the understandings derived from the articles' juxtaposition more thoroughly developed.

The entire collection is structured to highlight questions and problems that concern historians of cities, as well as to acknowledge distinctions among political, economic, social, cultural, and other varieties of history. Urbanization and many of its attendant processes are analyzed in the first volume, which considers not only the growth of cities but also such topics as the relationship between urbanization and the Westward movement, boosterism and urban rivalry, and company towns, concluding with a consideration of suburbs and their place in American urban culture since the middle of the nineteenth century. The physical development of cities and their infrastructure is considered in Volume 2, which focuses on city planning and its origins in the Rural Cemetery Movement, the City Beautiful Movement, and the role of business in advocating more rational and efficient urban places. Volume 2 also contains articles about essential aspects of the urban infrastructure and the provision of basic services essential for urban survival—water, sewer, and transportation systems. The articles about municipal government contained in the third volume include discussions of how rapid urbanization in the early nineteenth century produced a chain reaction, creating first the need for new political institutions, then the rise of machine politics, and, finally, reform movements that designed, advocated, and implemented new institutional structures such as the commission and city manager forms of government. Volume 3 also includes articles that consider the nature of intergovernmental relations at the end of the twentieth century and the connections between the governments of cities and the governments of the regions surrounding them—localities, states, and the nation.

The selections in Volume 4 of the series concern the development of the urban economy since the early nineteenth century. Three groups of articles, each arranged chronologically, deal with three basic sectors of the economy—trade and commerce (especially retailing), manufacturing and industrialization, and finance. Individual articles address subjects as diverse as merchants and shopping malls, flour milling and scientific management, and the Chicago Board of Trade and redlining. Volume 5 contains articles that are closely related but which concentrate specifically on the changing nature of work in American cities during the past two centuries. While they obviously concern the development of the industrial and post-industrial economies, they also recognize that economic transformations are intimately related to cultural change and that economic and cultural change are inseparable and must be considered together. At the same time, taken as a group, the articles reveal differences in experience between black and white Americans, men and women, and native and foreign-born Americans, necessitating that each of these groups be considered separately. The selections also investigate and illuminate questions about the relationships among these different groups and the kinds of actions they have taken to achieve their goals—political protests, boycotts, strikes, and so on.

The relationship between technology and urban life is the focus of the sixth volume. Developments in communication and transportation—bridges, steamships, railroads, street railways, electric trolleys, automobiles, and highways—played a crucial role in the growth of cities and the nature of urban life as did newspapers, telephones, and telecommunications. The articles in this volume examine the significance of the changing technology.

The final two volumes in the series concentrate on social structure and social/political institutions. Volume 7 looks at social class structure and social mobility. Its articles address questions that have intrigued historians for decades. What has been the class structure of American cities during the past two centuries? How much mobility has been possible? For whom has it been possible? What has been the relationship between social and geographic mobility? Finally, how have all kinds of Americans tried to improve their social status? In Volume 8, the focus is on the institutional life of cities. These articles examine institutions that existed in all societies—rural and urban—at all times, including family, religion, and education. They provide a look at how these institutions have changed and developed in concert with the development of cities in the United States. The final sections in Volume 8 discuss several institutions that are uniquely urban: voluntary associations, vigilance committees, and organized police forces. These articles attempt to consider race and ethnicity, class, gender, and the various experiences of different groups of Americans.

Introduction

Technological change accompanied the rapid urbanization of the United States during the nineteenth century. In fact, the tremendous growth of cities would not have been possible without the enormous variety of new technologies that sprang up everywhere. Techniques to arrange things, implements to manipulate things, utensils to accomplish things, tools to do things, machines to manufacture things—all were essential to urban growth. But nothing affected cities more than the revolution of the transportation system, both within cities and between the city and world beyond.

At the beginning of the nineteenth century, transportation from one place to another depended on foot, animal power, or a combination of wind and water currents. That all changed by the end of the century, as steamships, railroads, horsecars, trolleys, and subways had made raw materials more readily available, expanded the size of markets, and reshaped the internal geography of cities. Automobiles, buses, trucks, and airplanes were on the horizon, and they would have a cumulative impact on cities that can hardly be overstated.

The first new developments—improved bridges and steamships—benefited cities located on waterways by improving their ability to trade with distant places. Before the railroad was introduced, most cities either had coastal sites or were located on rivers for ease of movement and shipping. These sites shared a similar problem—how to get people and goods from one side of the river or harbor to the other. The traditional answer was with small boats, ferries, or elementary bridges if the waterway was narrow enough. In the first half of the nineteenth century, however, architects and engineers began to develop new techniques for building bridges that could cross wider bodies of water, bear heavier loads, and be high enough for ships to pass underneath. First the truss and then the suspension bridge improved connections between cities and their surrounding areas. These developments proved especially significant after the sweeping adoption of the railroad a few years later.

While bridges improved cities' access to their nearby hinter-
lands, steamboats solved the problem of getting upstream. Because most navigable rivers have strong currents, it was impossible for sailing ships, flatboats, keelboats, barges, and similar
vessels to return home once they had travelled downstream. For
farmers in the West, this problem was crushing and seriously
impeded the development of agriculture. After all, there was
little point in tilling hundreds of acres in Ohio, Indiana, Illinois,
or Kentucky, and taking a harvest down the Ohio and Mississippi
Rivers to New Orleans, if they were not able to get home easily.
Powerful steamships, which could overcome overcome the strongest river currents, therefore assured the development of commercial agriculture in the West, transforming cities there into
centers for the national and international distribution of American agricultural products.

A few decades later, steel rails displaced water as the heart of
the transportation system. The rapid creation and expansion of a
national railway network between 1830 and 1860 not only reduced urban dependence on water transit, but also advanced
almost any town or city that had a station, and allowed cities to
develop in the most unlikely spots—places like Denver in the
heart of the Rocky Mountains or Omaha in the middle of the
Great Plains, just to name two. Civic leaders, prominent businessmen, town promoters, and urban boosters soon realized the
potential of railroads. Once these people gained confidence in the
railroads' effectiveness and safety, and understood that cities no
longer depended on waterways (and that water sites no longer
provided an edge), the rush to acquire rail connections was on. It
took place in small towns as well as large cities, spurring some
towns to dream of overtaking nearby river cities. The new railway network presented all cities with access to larger hinter-
lands and transformed the relationships among cities and between cities and rural areas, even regions hundreds of miles
away. The scramble for rail connections altered the precise
placement of cities, policies of state and national governments,
and urban economics and politics.

New transportation technologies also had a major impact on
the internal geography of cities. First the horse-drawn street
railway, then the electric trolley, the subway, the automobile, the
inter-urban, and the bus allowed city residents to be less densely
packed and to live farther from the center than ever before. With
the need to walk to work and elsewhere eliminated, the physical
boundaries of urban areas increased dramatically, and each new
technology increased the potential extent of urban space.

Although the significance of electric streetcars for cities is better understood, it was the horse railway, or street railroad, that first made outlying areas accessible to urban residents. Throughout the latter half of the 1800s, American cities were criss-crossed by rails sunk into streets and over which horses pulled wheeled carriages capable of moving several dozen people at a time. Although horsecars had a limited range, lines could be constructed end to end to cover greater distances, enabling city-dwellers to live farther out. Of course, this form of transportation had certain disadvantages, most notably the physical limitations of the horses, the necessity of caring for the animals, and the constant need to replace the horses—not to mention the pollution they created.

Despite these problems, horsecars were the dominant form of urban public transportation until about 1890 when Frank Sprague, from Richmond, Virginia, discovered a way of providing electric power to streetcars, thus eliminating the need for horses. Within a few years, every major city in the country had a system of trolleys. The streetcar's high speed and low fares encouraged ridership, and opened residential land at distances from city centers that would previously have been unthinkable. The streetcar suburb was born.

While Sprague and other inventors were trying to make street-cars more efficient and less unpleasant, other innovations were also appearing. In cities like New York, located on a series of islands with mainland suburbs, tunnels connecting the heart of the city with its suburbs became an important link in the transportation system. Meanwhile, in Detroit, Michigan, a number of talented inventors were perfecting a different device, the automobile. By 1920, they had so advanced "the machine" that it was rapidly becoming a necessity for America's middle class (in contrast to Europe, where it remained a luxury item for several decades yet to come).

It is hard to imagine any single invention having a greater effect on urban form and structure than the automobile. Cars not only continued the trend toward suburbanization but intensified it, encouraging residents to move farther out and the city itself to sprawl into previously unimaginable tracts of land. The perceived desirability of private transportation and the obsolescence of fixed routes and schedules contributed to the decline of urban mass transit. That, in turn, raised the question of accessibility—without an effective system of public transit, how would people without automobiles get around? Other major problems followed, including traffic congestion, parking, air pollution, and reliance on fossil fuels with its attendant implications for national foreign policy.

Another transportation innovation affecting the growth of cities was the elevator. Rather than moving people along a horizontal plane, the lift (as it is sometimes called) changed a rider's vertical plane by moving up or down. The elevator was a necessary precondition for the development and construction of multistory office buildings because of the limitations stair-climbing imposed on the height of office buildings. With the elevator, downtowns became "canyonized" as skyscrapers lined both sides of many major business streets, walling them off from the sky.

At the same time that cities were dealing with transportation issues, they were also grappling with the problem of how to distribute news to a constantly expanding audience. Like subways and streetcars, some new forms of communication—newspapers, magazines, movies, radio, and, ultimately, television—were intended for mass audiences, while others—the telegraph, telephone, and, later, the computer—were meant primarily for individual use.

In nineteenth-century American cities, journalism's quintessential medium was the newspaper, and "yellow journalism" was its most striking manifestation. Yellow journalism often brings to mind newspapermen like Joseph Pulitzer and William Randolph Hearst, but its first real master, James Gordon Bennett, entered the scene fifty years earlier at the *New York Herald*. As early as the 1830s, Bennett sold his paper cheaply and appealed to a mass audience with sensationalism and human interest stories, especially those covering crime and scandal. Bennett relied on advertising monies in addition to selling newspapers for revenue, and the combination of a large circulation and advertising income freed the press from financial dependence on political factions. By the Civil War, Bennett's style was already being copied in cities across the country.

Yellow journalism wasn't the only journalistic development of the nineteenth century. The suburban and ethnic presses also appeared during that time. Suburban newspapers—those aimed at a specific geographic segment of the urban population—began in the 1870s. They reached the peak of their importance by the end of the century. The ethnic press also began to develop during the 1870s, with the annual arrival of tens of thousands of immigrants after the Civil War. These newspapers not only kept alive immigrants' ties with their homelands but also helped those individuals adapt and acculturate to life in the United States. Ethnic newspapers have found an audience throughout the twentieth century.

While most newspapers served a mass audience located within a single city (even today there are only a handful of national newspapers), the first telecommunications—the telegraph and the telephone—functioned quite differently. In general, they connected two individuals, not a crowd, and those people could be tens, hundreds, or even thousands of miles apart. These new technologies reduced the need for face-to-face contact and basic human interaction. As a result, the need to travel decreased because people could communicate almost instantaneously with distant friends, relatives, associates, or colleagues. Like the new transportation technologies, improved communications systems enabled cities to increase their size and also connected more and more distant cities to each other. In recent years, computers and their global connections have intensified these trends.

THE
WESTERN PENNSYLVANIA
HISTORICAL MAGAZINE

VOLUME 58   OCTOBER 1975   NUMBER 4

A TRINITY OF BRIDGES:
The Smithfield Street Bridge Over
the Monongahela River at Pittsburgh

JAMES D. VAN TRUMP

PITTSBURGH's first river bridge — that over the Monongahela River at what is now Smithfield Street — is, historically speaking, three bridges built successively at the same site by a trinity of famous American bridge engineers, Lewis Wernwag, John A. Roebling, and Gustav Lindenthal, all of whom had been born in Germany and thus had learned their technology from that early and famous fountain of engineering. It was America, however, a new and developing country, that gave them the widest scope for their abilities, and Pittsburgh with its great need for bridges was a special beneficiary of their knowledge, as it was a showcase for their talents.

This essay is a study, therefore, of the three versions of the bridge erected at Smithfield Street as well as a consideration of the development of the technology of bridge construction during the nineteenth century.

From the first settlement at Pittsburgh until 1818, the only means of transportation between the town and the farther banks of the rivers was by canoe or skiff. As the community developed, some kind of ferry service became mandatory, and in 1818 Jones's Ferry operated between the foot of Liberty Street in Pittsburgh to the south bank of the Monongahela. Passengers were carried in skiffs while stock was taken over on flatboats. About 1840 a horse ferry was introduced in which blind horses, as a rule, were used as motive power — they

Mr. Van Trump, architect, historian, and vice-president of the Pittsburgh History and Landmarks Foundation, has written numerous articles on the architecture of Western Pennsylvania.—Editor

1

were made to tramp upon a horizontal wheel, the revolution of which propelled the boat across the stream.

A few years later Captain Erwin established a steam ferry from a site on the south bank of the Ohio slightly below the confluence of the rivers at the Point, but this was never a success.[1] Subsequently the Jones Ferry was abandoned and a steam ferry operated from Saw Mill Run on the south bank of the Ohio to Penn Street in Pittsburgh, and this line was in use until the increasing number of river bridges made it redundant.[2]

Prior to the building of the Monongahela Bridge, all traffic passing from one side of the river to the other at Smithfield Street was carried on a little ferry owned by Enoch Wright of Westmoreland County and Andrew Herd of Allegheny County, who leased the buildings, ferry, and improvements to one Robert Shanhan. Where the ferry landed on the South Side stood Enoch Wright's stone house. After the bridge was constructed, the ferry interests were bought out by the Stock Company.[3] Before the introduction of dams toward the end of the nineteenth century, the streams at slack water were relatively shallow, and numerous islands and sand bars were in evidence. There was, for instance, a long sand bar in the Monongahela at the site of the Smithfield Street Bridge;[4] this river flatland that is shown on the very early maps of Pittsburgh was of sufficient extent that grain could be grown on it at low water. It must be remembered also that there was extensive traffic on all three rivers and the spans of bridges had to be sufficiently high to allow boats to pass beneath them.

The first of Pittsburgh's highway river bridges was the Monongahela (later Smithfield Street) Bridge. Erasmus Wilson, a late nineteenth-century historian of Pittsburgh, has left the best account of its construction and we quote it here :[5]

"In the year 1810 a bill was introduced in the [Pennsylvania]

---

1 J. N. Boucher, *A Century and a Half of Pittsburgh* (New York, 1908), 2, 387. See also C. W. Dahlinger, *Pittsburgh: A Sketch of its Early Social Life* (New York, 1916), 29-30.

2 The route of Jones's Ferry appears on the McGowan map of Pittsburgh of 1852, together with other ferry routes plying the local rivers.

3 Herbert Du Puy, "A Brief History of the Monongahela Bridge, Pittsburgh, Pa.," *Pennsylvania Magazine of History and Biography* 30, No. 2 (1906): 187-88 (hereafter cited as Du Puy, "Monongahela Bridge").

4 A 1795 map of Pittsburgh shows the sand bar, but on the Molineux map of the city of 1830, it has disappeared.

5 Erasmus Wilson, ed., *Standard History of Pittsburg Pennsylvania* (Chicago, 1898), 112-14.

State Legislature providing for the construction of two bridges at Pittsburg — one over the Monongahela and one over the Allegheny, and an estimate of the probable cost of such a structure was made by Judge Findley. It was calculated by him that the 1,200 feet of river would require chains of 1,590 feet and four such chains of inch and a half square iron bar weighing sixty-four pounds to the foot, with some excess, would amount to $8,800; smith work would cost $3,080; a bridge thirty feet wide would require $900 worth of plank; three piers would cost $15,000; other expense, $1,050; right to use certain patents, $1,200; putting together, $1,296; incidentals, $1,000; total $32,326. James O'Hara, William McCandless, David Evans, Ephraim Pentland, Jacob Beltzhoover, Adamson Tannehill, Thomas Cromwell, Thomas Enochs, and Dr. George Stevenson were the commissioners appointed to open books for the subscription of stock in the Monongahela bridge. John Wilkins, James Robinson, Nathaniel Irish, George Shiras, George Robinson, Isaac Craig, James Irvin, John Johnston, and James Riddle, were authorized to open books for the subscription of stock in the Allegheny bridge. Probably owing to the war of 1812, the bridges were not built at that time and in 1816 the law was reenacted, and the Governor, on behalf of the State, was authorized to take 1,600 shares of stock in each bridge. The law specified that one was to be built over the Monongahela at Smithfield Street and one over the Allegheny at St. Clair [now Sixth] Street . . . .

"The last installment of stock for the Monongahela bridge was called for by the treasurer, John Shaw, to be paid May 15, 1818. The first arch was laid on the piers on Saturday, June 20, 1818. [The bridge] was rapidly built, when once begun, and rested on two abutments and seven intermediate piers of stone. It was constructed of wood and iron, with the catenarian curve of arches, the contract price being $110,000. As if to favor the contractor the weather during the fall was excellent.

" 'The beautiful bridge over the Monongahela has nearly reached the northern shore; it will probably be crossed before Christmas. The one over the Allegheny is not so far advanced, but yet enough is done to insure its completion. Pittsburg will then exhibit what no American city or town has ever yet done — two splendid bridges over two mighty streams, within 400 yards of each other.[6]

" 'On Saturday (November 21, 1818) the last arch of the Monongahela [Bridge] being completed, and the whole floored, the

6 *Pittsburgh Gazette*, Nov. 24, 1818, quoted in ibid., 113.

undertakers and builders announced the pleasing event by the discharge of cannon from the middle pier and the display of the United States flag waving over the central arch, having attached to its staff a beautiful banner with appropriate representations. The City Guards and the new company of Washington Guards from Birmingham, heralded on their respective sides of the river, marched across and fired salutes. In the afternoon the workmen sat down to a substantial dinner, at which Mr. Johnston, the meritorious undertaker and superintendent, presided . . . .'

"November 26, 1818, John Shaw, treasurer of the Monongahela Bridge Company called a meeting of the managers to appoint a gatekeeper to receive the toll, as follows: Foot passengers, 2 cents; vehicles of four wheels and six horses, 62½ cents; vehicles of two horses, 25 cents; vehicles of one horse, 20 cents; horse and rider, 6 cents; horse alone, 6 cents; each head of cattle, 3 cents; each head of sheep, 2 cents . . . .

"The State held $40,000 worth of stock in the Monongahela bridge, and was required to assist in repairing the damage caused by the falling [of part] of the span in 1831-2."

Llewellyn Edwards describes the Monongahela Bridge as follows: "The substructure consisted of two abutments and seven piers of stone masonry. The superstructure had eight covered wood truss spans and an overall length of 1500 feet." [7]

Richard Allen also comments on both the Monongahela and Allegheny bridges (the latter not finished until 1819). The Monongahela Bridge was a Burr truss structure, and Allen states that "its outstanding feature was the toll collector's living quarters. He was housed in a small apartment built above the barn-like portal on the Pittsburgh side." [8]

The Burr truss which appears so frequently in the chronicles of early American bridge construction was named for Theodore Burr (1771-1822), a well-known bridge designer of his day. Like his contemporaries, he, for all but very short spans, combined the arch and truss (witness the "catenarian arches" of the Wilson account quoted earlier), but instead of combining them by strengthening the arch by

---

7 Llewellyn Edwards, *A Record of the History and Evolution of Early American Bridges* (Orono, Maine, 1959), 198.
8 Richard S. Allen, *Covered Bridges of the Middle Atlantic States* (Brattleboro, Vermont, 1959), 75. There is also a description of the bridge in "A View of Pittsburgh" in *The Franklin Magazine Almanac for 1820*, 51-52, and in Rebecca Eaton, *Geography of Pennsylvania* (Philadelphia, 1837), 235.

the truss, as did the rest, he strengthened the truss by the arch. His design was in reality merely a series of king posts, and it is safe to say that the majority of wooden covered bridges built in the United States were of the Burr truss design.[9]

The designer of both the Monongahela and the Allegheny bridges was Lewis Wernwag (1769-1843), perhaps the most famous of all early American bridge engineers. Born in Germany, he came to the United States at the age of seventeen and settled in Philadelphia. He specialized in wooden truss spans, his first famous work being a single-span bridge constructed in 1812 over the Schuylkill River at Philadelphia. He later constructed many highway and railroad bridges. A letter from his son John to Samuel Smedley, published in the *Engineering News,* August 13, 1885, includes a list of twenty-nine bridges built by his father during his active career of twenty-seven years.[10] Of all these bridges the Monongahela was among the most famous.

Joseph H. Thompson was selected to build the Monongahela Bridge and a contract was made with him on July 9, 1816, to construct Wernwag's "double-passage wooden-bridge covered from end to end, . . ." The contract price was $110,000.[11]

The Monongahela span gave many years of good service to the developing Pittsburgh region, but it disappeared in ten minutes in a long trailing line of smoke and flame at two o'clock in the afternoon during the Great Fire of April 10, 1845,[12] one of those huge conflagrations that devastated American cities in the nineteenth century.[13] At the time of its destruction it was still the only bridge over the Monongahela at Pittsburgh.

After the fire the old piers and abutments were repaired and on them the now-famous John A. Roebling constructed a new wire cable suspension bridge for a contract price of $55,000. Work began on the

---

9 D. B. Steinman and S. R. Watson, *Bridges and Their Builders* (New York, 1941), 121-22.

10 *Dictionary of American Biography* 10, Pt. 2: 2-3 (cited hereafter as *D.A.B.*). See also Robert Fletcher and J. P. Snow, "A History of the Development of Wooden Bridges," *Proceedings of the American Society of Civil Engineers* 57 (1932).

11 Du Puy, "Monongahela Bridge," 194.

12 Ibid., 198.

13 There is a contemporary oil painting by the Pittsburgh artist, William C. Wall, "The Great Fire of 1845," on display at the Old Post Office Museum, Pittsburgh (on loan from John H. Follansbee), which shows the Monongahela Bridge in flames. Another canvas, attributed to the same painter, and in the possession of the museum, shows the ruins of the city and the bridge just after the fire.

new structure in June 1845, a short time after the fire. D. B. Steinman has given a full and colorful account in his biography of Roebling of the construction of this first of the famous engineer's highway bridges.[14] What began in Pittsburgh culminated in the 1860s in his final master work, the Brooklyn Bridge.

John A. Roebling as a bridge engineer is so well known that any biographical data would seem almost redundant, but some account of his life is necessary here because of his importance in Pittsburgh pontine history.[15] He was born in Mühlhausen, Germany, in 1806, received his engineering education at the Royal Polytechnic Institute in Berlin, and emigrated to America in 1831, settling at Saxonburg in Butler County, some twenty-five miles north of Pittsburgh. This town, which he established, became the chief focus of his early engineering career and here he established his wire rope manufactory which was later moved to Trenton, New Jersey. His wire cables were used first on the inclined planes of the Pennsylvania Canal's Portage Railroad in the mountains of western Pennsylvania; his first important bridge was a suspension aqueduct which he constructed in 1844-45 to carry the Pennsylvania Canal over the Allegheny River at Pittsburgh.[16] As the aqueduct neared completion, the Monongahela Bridge burned, and Roebling almost immediately received the commission to construct the new one. As a result of the fame of these two structures, Roebling now was established as America's foremost bridge engineer. He went on to design such famous structures as (another Pittsburgh work) the second Sixth Street Bridge over the Allegheny River (1858-1860), the Niagara Railway Suspension Bridge (1851-1855), the Cincinnati Bridge over the Ohio (1856-1867), and finally the Brooklyn Bridge which he was never to see finished, since he died as a result of an accident in 1869 when work on the bridge piers had just begun.

John Roebling was more than a competent bridge engineer. He was also a prolific writer and he published his achievements as they appeared. Consequently, the best description of the second Monongahela Bridge is that from his own pen:[17]

14 *The Builders of the Bridge: The Story of John Roebling and His Son* (New York, 1945), 89-100 (hereafter cited as Steinman, *John Roebling*).
15 *D.A.B.* 16: 86-87; Steinman, *John Roebling*; Hamilton Schuyler, *The Roeblings—A Century of Engineers, Bridge Builders, and Industrialists* (Princeton, N. J., 1931); "John A. Roebling," *Engineering News* 10 (May 26, 1883) : 246.
16 John A. Roebling, "The Wire Suspension Aqueduct Over the Allegheny River at Pittsburgh," *Journal of the Franklin Institute,* 3rd ser., 10, (1845) : 306-309. The bridge is illustrated on 307.
17 "The Wire Suspension Bridge Over the Monongahela River at Pittsburgh,"

"The new Suspension Bridge over the Monongahela . . . was commenced in June, 1845, and opened for travel in February, 1846. The piers and abutments of the old wooden structure, which was destroyed by the great fire, required extensive repairs to be fitted for the reception of the new superstructure. The whole length of the work between the abutments, is exactly 1,500 feet, and is divided into eight spans of 188 feet, average distance from centre to centre. The piers are 50 feet long at bottom, 36 feet high, and 11 feet wide on top, battering 1 inch to the foot.

"Two bodies of substantial cut stone masonry, measuring 9 feet square and 3 feet high, are erected on each pier, at a distance of 18 feet apart. On these the bed plates are laid down for the support of the *cast iron towers,* to which the *wire cables* are suspended by means of *pendulums.* Each span being supported by two separate cables, there are therefore, 18 cables suspended to 18 towers.

"The towers are composed of four columns moulded in the form of a two sided or cornered pilaster; they are connected by four lattice panels, secured by screw bolts. The panels up and down stream close the whole side of a tower, but those in the direction of the bridge form an open doorway, which serves for the continuation of sidewalks from one span to the other.

"On top of the pilasters or columns, a massive casting rests, which supports the *pendulum* to which the cables are attached. The upper pin of the pendulum lies in a seat which is formed by the sides and ribs of a square box occupying the centre of the casting. For the purpose of throwing the whole pressure upon the four columns underneath, 12 segments of arches butt against the centre box, and rest with the other end upon the four corners.

"The pendulums are composed of four solid bars of 2 feet 6

The American Railroad Journal 19 (Apr. 4, 1846) : 216; also (June 13, 1846) : 376. The version quoted here is from a reprint in a periodical published at Pittsburgh and edited by Neville B. Craig, *The Olden Time* (June 1846) 1: 286-88 (reprinted 1876 under a Cincinnati imprint). See also Gustav Lindenthal, "The Monongahela Bridge—Rebuilding of the Monongahela Bridge at Pittsburgh," *Engineering News* 10 (July 7, 1883) : 314-15; 11 (Mar. 14, 1884) : 239, 241, 371; "The Suspension Bridge," *American Railroad Journal* 19 (Feb. 21, 1846) : 126; Col. S. M. Wickersham, "The Monongahela Suspension Bridge," *The Scientific American Supplement* 15 (1883) : 6201; "The Monongahela Suspension Bridge at Pittsburgh, Pa.," *Engineering News* 10 (May 26, 1883) : 243-44; *The Iron Age* 31 (June 21, 1883) : 3, 5, 1. Roebling's *American Railroad Journal* article is also illustrated with a lithographed plate of his own drawings for the structure. Steinman, in his biography, also reproduced two Roebling drawings for the bridge—opposite 134.

inches long, from centre to centre of pin, 4 inches by one inch — the pins are three inches in diameter. To the lower pin, the cable of one span is attached directly and the connection formed with the next cable by means of four links of 3 feet 6 inches long and 4 inches by 1¼ inches.

"The opposite cables, as well as the pendulums, are inclined towards each other — the distance between being 27 feet at the top of the towers, and 22 feet at the centre of a span. The pendulums on the abutments, however, occupy a vertical position.

"The two sidewalks are outside of the cables, and 5 feet wide. The roadway is contracted to 20 feet, and separated from the sidewalks by fender rails, which are raised from the floor by means of blocks of 6 inches high, 8 feet apart. The total width of the bridge between the railings is 32 feet.

"The anchor chains which hold the cables of the first and last span, are secured below the ground in the same method which was applied to the [Pennsylvania Canal] aqueduct — their oxidation is guarded against in the same manner.

"The cables are 4½ inches in diameter, and protected by a solid wrapper; they are assisted by stays, made of 1¼ inch round charcoal iron; the suspenders are of the same material, 1½ inch diameter, and placed 4 feet apart.

"The peculiar construction of the Monongahela bridge was planned with the view of obtaining a high degree of stiffness, which is a great desideratum in all suspension bridges; this object has been fully attained. The wind has no effect on this structure, and the vibrations produced by two heavy coal teams, weighing seven tons each, and closely following each other, are no greater than is generally observed on wooden arch and truss bridges of the same span. This bridge is principally used for heavy hauling; a large portion of the coal consumed in the city of Pittsburgh passes over it in four and six horse teams.

"As a heavy load passes over a span, the adjoining pendulums, when closely observed, can be noticed to move correspondingly — the extent of this motion not exceeding one half inch. By this accommodation of the pendulums, all jarring of the cast iron towers is effectually avoided. Another object of the pendulums is to direct the *resultant* of any forces to which the work may be subjected, through the centre of the towers, as well as of the masonry below.

"Two of the piers of the old structure had once given way in consequence of the shaking and pressure of the arch timbers, when subjected to heavy loads. Such an accident can never take place on the new structure, as the piers are only subjected to the quiet and vertical pressure of the towers.

"I do not recommend the application of pendulums in all cases; but in this, it appeared to me the best plan which could be adopted.

"The two towers on each pier are connected by a wooden beam, properly encased and lined by the same mouldings which ornament the top of the castings.

"The lightness and graceful appearance of this structure is somewhat impaired by the heavy proportions of these connections, but I had to resort to it for motives of economy.

"The whole expense of this structure does not exceed $55,000 — a very small sum indeed for such an extensive work.

"A great portion of this work had to be done during the winter, and in cold weather; it was accomplished without any accident, with the exception of one of the workmen who was seized by fits and killed by falling off a pier."

TABLE OF QUANTITIES OF MONONGAHELA BRIDGE

Length of bridge between abutments	1500	feet.
Number of spans	8	
Average width of spans from centre to centre	188	"
Diameter of cables	4½	inch.
Number of wires in each	750	
Weight of superstructure of one span, as far as supported by the cables	70	tons
Tension of cables resulting from it	122	"
Weight of four six horse teams, loaded with 104 bushels of coal each	28	"
Tension resulting from it when at rest	49	"
Weight of 100 head of cattle at 800 lbs.	40	"
Tension resulting from it when at rest	70	"
Aggregate weight of one span as far as supported by the cables, plus 100 cattle at rest	110	"
Tension resulting from it	192	"
Ultimate strength of two cables	860	"
Section of anchor chains	26	inch.
Section of pendulums	63	"

In 1859 an agreement was made with the Pittsburgh and Birmingham Railway Company, a horse car line then being constructed from Pittsburgh to the South Side across the river, to permit the line to cross the bridge at the price of fifteen dollars per car each month. In 1865 the structure acquired gas lights, and the foot toll was reduced to one cent per person.[18] In 1861, a wooden truss bridge was built a little farther upstream at South Tenth Street;[19] Roebling's span was no longer the only bridge crossing the river, and in later years was increasingly referred to as the Smithfield Street Bridge.

The structure during its years of service often was tried sorely — sometimes when it was crowded with people viewing a steamboat race, sudden rushes would be made from one side of the bridge to the other.[20] Such conditions afforded a real test of the designer's foresight in providing various features that assured enduring stability. The bridge continued in service for thirty-five years, carrying the heaviest kind of street traffic, horse cars, steam rollers, and eight-horse teams pulling heavy trucks loaded with iron and machinery. The multiple span arrangement, though quite satisfactory for an aqueduct with its loading constant or uniform in all spans, was under a disadvantage in a suspension bridge carrying variable loading. Despite the system of inclined stays which Roebling had installed, a loaded span sometimes deflected as much as two feet with a corresponding smaller rise of the adjoining spans. Not only the designer but the profession profited by this experience.[21]

Due to the enormous volume of traffic on Roebling's bridge, it began to show signs of strain, and the Board of Managers of the bridge company decided to look into the possibility of providing a new structure. On February 1, 1871, bids were presented to the board, but soon afterward the city of Pittsburgh tried to secure the franchise. This brought out a stockholders' meeting on May 27, 1872, contesting the city's right to such action.[22] This difficulty retarded the new improvement and the Panic of 1873 with its resultant economic woes prevented anything being done, but in the summer of 1880 the board

---

18 Du Puy, "Monongahela Bridge," 202.
19 Henry Joseph White, and M. W. von Bernewitz, *The Bridges of Pittsburgh* (Pittsburgh, 1928), 32.
20 See description of a scull race in *Harper's Weekly* 19 (June 8, 1867): 363-64, with a wood engraving after a sketch by C. S. Reinhart, showing the bridge.
21 Steinman, *John Roebling*, 100.
22 Du Puy, "Monongahela Bridge," 202.

finally decided to demolish the Roebling bridge and construct a new one.[23]

The Board of Managers of the bridge company called to their aid a local engineer, Charles Davis (1837-1907), who submitted a design for another suspension bridge. Davis was one of those American engineers who seemed to have learned their profession "in the field," so to speak, particularly in railway surveying. He had been consulting engineer for Pittsburgh's Point (suspension) Bridge (1875-1877) and in 1881 was elected Engineer of Allegheny County, a position he held until his death.[24]

Work on Davis's bridge began in the summer of 1880. It was to be a suspension bridge having two channel spans of 360 feet each and two shore spans of 180 feet each. Foundations for the channel piers were put in first, and the piers built up to an average height of ten feet each. Because the winter of 1880 was unusually severe further work on the bridge was stopped.[25] None of the drawings for this abortive design seems to have survived.

In February 1881, the bridge company was reorganized, and as a consequence, all work on Davis's bridge stopped and all prior contracts cancelled. The man who now held the controlling interest in the company's stock, David Hostetter, was also largely interested in the Pittsburgh and Lake Erie Railroad and he wished a different type of structure, because he thought it might be possible to run cars from his own line on the south bank to the lines of the Baltimore and Ohio on the north. Consequently, a young German engineer, Gustav Lindenthal, was called in to make a design for a through-truss bridge.[26]

The new bridge engineer was also an immigrant. He was born in 1850 in Brünn, Moravia, Austria-Hungary, and had been educated at the Provincial College of Brünn and at the polytechnical schools of Brünn and Vienna. He worked on railways in Austria and Switzerland before coming to America in 1874; in 1876 he assisted in the construction of buildings for the Centennial Exposition in Philadelphia. In 1881 he established himself in a private engineering practice in Pittsburgh. He was engaged in many railway and bridge projects,

23 Gustav Lindenthal, "Rebuilding of the Monongahela Bridge at Pittsburgh, Pa.," *Transactions of the American Society of Civil Engineers* 11 (Sept. 1883) : 355 (hereafter cited as Lindenthal, "Monongahela Bridge").
24 *Biographical Review, Pittsburg and the Vicinity* (Boston, 1897), 24: 475-77; *Memoirs of Allegheny County, Pennsylvania* (Madison, Wisc., 1904), 1: 37-38; *Pittsburgh Gazette Times*, Feb. 22, 1907.
25 Lindenthal, "Monongahela Bridge," 355.
26 Du Puy, "Monongahela Bridge," 203.

including the reconstruction of bridges on parts of what is now the Erie Railroad, various bridges in and near Pittsburgh, and railway surveys and estimates in Pennsylvania and neighboring states. By the age of forty, he had established a reputation as one of America's great bridge engineers, and certainly the new Smithfield Street project was no small factor in his rise to fame.

In 1890 he set up a consulting office in New York City, devoting most of his time to bridge work. His best-known works are the Queensboro Bridge (1901-1908) over the East River in New York, the Hell Gate Bridge (completed 1917) for the New York Connecting Railroad, and the Sciotoville Bridge (1914-1917) over the Ohio River. In 1902-1903 he served as commissioner of bridges for the city of New York. In this capacity he advocated and established the practice of the association of engineers, in the design of large bridges, with architects whose special interest lay in the esthetics of bridge construction.

As an engineer, his greatest vision never materialized — a bridge over the Hudson River at New York. From 1880 until his death in 1935, he worked on the problem of transportation from New York to the New Jersey side of the river and he constantly urged the adoption of his North River Bridge scheme. However, complications arising from decisions of the United States Army Engineers with reference to clearance defeated final approval of the plan. The long span — 3,100 feet — heavy loading, and the huge costs of this project may be taken as a measure of Lindenthal's vision.

Originality and boldness characterized Lindenthal's designs. He differed from many of his American contemporaries in his frequent choice of more complex structural forms and in some of his views as to working stresses. Like Roebling, he wrote many technical papers and contributed to learned journals, chiefly on bridge design, but his chief monuments were his works. Pittsburgh is fortunate still to possess the first of his great designs which yet functions today, still serving its contiguous land areas and supporting weights that the engineer could not have foreseen when it was designed.[27]

As in the case of Roebling, Lindenthal's is the best account of the construction of the Smithfield Street Bridge.[28] In 1881 "the writer

---

27 *D.A.B.* 21, Supplement 1, 498-99; *Transactions of the American Society of Civil Engineers* 32 (1904); *Who's Who in America* (1934-35); *Who's Who in Engineering* (1931); *Civil Engineering* (Sept. 1935); *Engineering News Record* (Aug. 8, 1935); *Electrical Engineering* (Sept. 1935); *New York Times,* Aug. 1, 1935.
28 Lindenthal, "Monongahela Bridge," 355, ff.

was invited for consultation and to suggest suitable changes in the plans, which should provide for a widening of the bridge by adding another roadway or track, should this ever become necessary in the future. After having submitted such plans, they were accepted and the writer was engaged to carry them out. This plan proposed to utilize the foundations and piers which had been commenced. They were to be built upon, without any offsets, to a width on top of 56 feet.

"As the width of the superstructure may ultimately reach 64 feet, or eight feet wider than the piers, it became necessary to let the side-walks project over the masonry. The present width is 48 feet on the channel spans; the room on the piers for widening the bridge was left on the up-stream side. For the channel spans Pauli trusses[29] were proposed, 25 feet 8 inches apart, centre to centre, and the centre line of the new floor (of 48 feet width) was shifted down-stream 8 feet 2 inches from the centre line of the old bridge. The sidewalk on the up-stream side was proposed to be detachable, so that the floor may be widened and the sidewalks again connected to it.

"For the approaches to the channel spans, plate girder deck spans, on lighter masonry piers, were proposed. This arrangement allows of increasing the width of the bridge by simply adding more plate girders to each span on the piers which are long enough for that purpose. Being a deck bridge, it afforded an unobscured entrance view to the channel spans, the trusses of which were to rest on ornamental towers, giving to the superstructure an architectural appearance of strength and stability.

"The shifting of centre line of new floor 8 feet 2 inches down-stream from the centre line of old bridge allowed of erection of the new superstructure without stopping travel on the old bridge, in a manner described more in detail below.

"The Pauli truss type commended itself for the channel spans in this instance, for several reasons:

"1. The pleasing appearance (for a city bridge) in comparison with the ordinary parallel chord truss.

"2. The fact that the trusses could be made high in the middle (without detriment to their stability in case of high winds),

---

29 These are lenticular, "bow-string," or "fish-belly" through-trusses, named for the famous German engineer Friedrich August von Pauli (1802-1883). For Pauli, see *Zeitschrift für Baukunde* 7 (1884): 379-96; *Allgemeine Deutsch Biographi* 25 (Leipzig, 1887): 251-58; and *Zeitschrift des Vereins Deutscher Ingenieuren* (1865), "den Artikel von Gerber uber die Bereschnung des Bruckentrager nach Paulischen Systems."

thereby reducing the chord strains and chord sections. In connection with the light and slender web-members, it permitted of an economy in the trusses of over 9 per cent, as compared with parallel chord trusses (with inclined end posts) of same height (50 feet). The deflection and vibration of high trusses is small and their rigidity great.

"3. The bottom chord or cable is exposed to the sun's rays as, much as any other truss member; therefore unequal temperature effects in the trusses are avoided. The covered floor construction is independent of the trusses as to temperature effects.

"4. The floor had to be cambered 18 inches in each 360 foot span to agree with the general grade of the new bridge. A straight bottom chord with a rise of 18 inches in 360 feet was undesirable.

"At first it was proposed to build the new structure 15 feet higher at highest point than the old bridge. But the river men, in the interests of navigation, demanded the structure to be at least 20 feet higher, or 57 feet above low water mark, to which the Bridge Company objected, on the ground that the additional 5 feet height would injure travel over the bridge much more, by reason of a steep grade at the Pittsburgh end, than it would benefit navigation.

"There is no statute prescribing the height of bridges over the Monongahela River. The case was taken to a court of equity, and argued there by lawyers pro and contra, resulting in a preliminary injunction against the Bridge Company building the bridge lower than 20 feet. To continue the litigation would have required much time. After a suspension of work at the bridge for 10 months the Bridge Company decided to accede to the demands of the river men.

"The following is a description of the material and methods used in the construction of the bridge:

MASONRY

consists of a gray, hard and durable sandstone, free from admixtures of clay or iron oxide particles. It was quarried near Homewood, Pa. on the Pittsburgh and Lake Erie Railroad where it is found in large blocks of 100 to 500 cubic yards, without any stripping. The masonry is rock-faced, with drafts 1 inch wide all around the face of the stones, which are in courses of alternate headers and stretchers.

"The dimensions of the stones are 24 inches to 16 inches in thickness, 7 feet to 4 feet in length, 3 feet to 11 feet in width, with beds and joints dressed regular and true. The backing for the abutments and wing walls consisted of regular shaped stones, with dressed beds; for the heart of the piers concrete filling was used. It was applied in layers of 12 inches thick. It proved superior in every way to ordinary stone backing. Iron clamps bind the stones in the pier heads in every course.

"The use of spalls was not permitted in any part of the masonry. All spaces between stones were filled with concrete, rammed with iron rammers, making every course absolutely water-tight. Great attention was given to the bond. The stone blocks were laid in alternate header and stretcher courses, which made the coincidence of stone joints in the heart of the pier impossible. In this way each stone is bonded in every direction. The concrete backing, after setting, was very hard and tough; it adhered to the stones with great tenacity, and made the piers monolithic in fact.

"In the execution of the work care was taken to set every stone immediately before setting. When laid in position the stone was settled by repeated blows of a heavy wooden ram. Any stone breaking under this operation was removed.

"The face joints of the finished masonry were cleaned out to a depth of 1 inch, and thoroughly moistened, and caulked with Portland cement and sand mortar, mixed one to one.

"For all face masonry exposed to the weather American Portland cement was used for the mortar; for concrete backing and foundations, Rosendale cement was ordinarily used.

"All cements were required to be so finely ground that 90 per cent of the whole would pass through a sieve of 50 meshes to the lineal inch. Tests as to its tensile strength were conducted on a Fairbanks testing machine with moulded briquettes of pure cement.

"Rosendale cement made of a stiff paste, having been one day in water and one day in the air, at an even average temperature of 70 degrees Fahrenheit (in a room), was tested to show the tensile strength of at least 40 pounds per square inch.

"American Portland cement briquettes, under same conditions, were tested to show a tensile strength of at least 80 pounds per square inch.

"Similar briquettes, after having been four days in water and one day in the air, at the above average temperature, were tested for a tensile strength of 60 pounds per square inch for Rosendale cement, and 150 pounds for American Portland cement.

15

"The concrete used throughout the work was composed of 2 parts of sound broken stone, passing through a 3 inch ring; 2 parts of clean gravel from the size of a pea to 2 inches diameter; 2 parts of washed river sand; 1 part of Rosendale cement of accepted quality.

"For concrete under water 2 parts of cement were used to allow for waste by washing in depositing it under water. With a little care in the operation the loss, however, was insignificant. The stone, gravel and sand were first mixed on a board platform, then the cement added, and the whole mass thoroughly rehandled in a dry state. Water was then added in barely sufficient quantity to reduce the whole mass, by lively and severe shoveling, to a stiff mortar. This was put immediately in place in layers of not over 12 inches thick, and thoroughly rammed with iron rammers about 5 inches square and weighing 36 pounds, until the mass flushed uniformly over the whole surface.

"For depositing the concrete under water for the pier foundations square wooden troughs were used, reaching down to almost the bottom, and the concrete dumped in and raked even with iron rakes having long handles. The running out of the concrete was prevented by sheet piling. When a change in the masonry of Pier No. 4 required the removal of a few stones they were found to form with the concrete backing one solid mass, which had to be rent asunder with steel wedges and sledge-hammer, and would sometimes break through the stone rather than through the concrete.

"Openings or slots for one car track were left in the new abutments and piers to accommodate travel on the old bridge.

"The pier posts of the channel spans on the down-stream side have their bearing near to the pier ends, and to prevent cracking of the channel piers or uneven settlement after the superstructure should be in place, riveted iron anchors were walled into the top of piers Nos. 2, 3 and 4.

"The coping on the piers, consisting of two projecting courses of cut stone, was nearly all in place for a grade 15 feet higher than the old bridge, at the time of the dispute with the river men. When the height of the piers was increased to suit a grade 20 feet higher than the old bridge, the additional masonry was built on top of the coping in the form of pedestals of cut stone.

"After the erection of the superstructure had so far progressed that travel could be turned on to one track on the new bridge, the old bridge was abandoned, and the taps and openings in the masonry of the new abutments and piers successively walled in and closed. In this

The Wernwag bridge over the Monongahela, built in 1818, and destroyed by fire in 1845. Picture painted by Leander McCandless.

This is John A. Roebling's Monongahela suspension bridge, built in 1846. View is looking from downtown towards Mt. Washington.

17

North portal of Lindenthal's Smithfield Street Bridge. Photograph shows the original wrought-iron portal.

One of the present portals of the Smithfield Street Bridge.

The Smithfield Street Bridge as it appears today.

View of the through-trusses of the Smithfield Street Bridge.

wise it was possible to complete the masonry work without stopping travel on the old or new bridge.

SUPERSTRUCTURE

"The roadway is at present 22 feet 10 inches wide in the clear, and two sidewalks each 10 feet in the clear. The full width of the bridge on the deck spans of approaches is 43 feet 6 inches, and on the channel spans, which are through spans, 48 feet.

"The bridge can be widened out, if ever required, to 64 feet. This made it necessary to erect the present superstructure nearer to the down-stream end of the piers. It detracts much from the appearance of the bridge, which is unsymmetrical at present.

"It was important not to stop travel during the rebuilding of the bridge. Passengers and freights from and to the Pittsburgh and Lake Erie Railroad must pass over it. Besides, there is a heavy traffic in coke, iron and other mill material, which would have been compelled to take a long, roundabout way. The construction of the superstructure had to be arranged to allow of the erection first of one track and then of the other.

"If the new bridge had really been built 15 feet instead of 20 feet higher than the old one there would not have been left height enough near the ends of the channel spans for teams to pass under on the old bridge. It was therefore intended to erect the channel spans about 5 feet higher than their proper grade, and to complete the floor and tracks of the same.

"The pier posts would have temporarily rested on sand jacks, by means of which both spans, weighing about 1600 tons when completed, could have been simultaneously lowered in a few hours to their proper grade. One track and sidewalk on the plate girder approaches on the down-stream side would have been meanwhile prepared for use. In this way travel would have been interrupted only for one day. But this operation became unnecessary when the new grade was raised 20 feet above the old bridge.

CHANNEL SPANS

"It was found that *the use of steel, in the trusses at least*, would prove economical as compared with wrought iron. The saving based on the prices at that time was over $21,600.

"The Pauli trusses were designed with an uneven number of panels, namely 13, in order to get two tangential points of attachment for each truss to the floor-construction, thereby securing greater longitudinal and transverse rigidity of the entire bridge frame. Roller bearings for pier posts were avoided; the middle posts, supporting two truss ends each, have a fixed and square bearing on heavy pedestal castings on the pier. Each end post has a bearing on a 6 inch steel pin in a cast-iron pedestal on which it can rock. It is probable that very little movement takes place on account of friction on the pin, and that the posts would bend or spring. The resulting bending moments on the end posts have been considered in proportioning them.

"The projected length, 27 feet 7-$\frac{5}{8}$ inches, of all panels being alike, it follows that the lengths of chords in a curved line are unlike, and if the curve were a circle or a parabola, then the angles formed by the straight chord sections would also all be unlike.

"For practical reasons it is desirable to have these angles all alike, so as to have only one template for the beveled joints. This condition would prescribe the character of the curve, in this instance a sine-curve. The difference in curvature between a sine-curve and arc of a circle was found to be small (2$\frac{1}{2}$ inches). The difference in the bevel points was inappreciable (3/64 inch). Therefore a true arc line was then assumed for the chords to facilitate other calculations.

"The vertical web-members are in tension from the dead load or from a uniformly distributed live load. They will sustain compression strains only from an uneven distributed load. Near the centre of truss they are long and slender, requiring intermediate bracing, which was placed at half the truss height for the entire length of trusses.

"The suspenders from trusses to floor, which were all stiffened to prevent vibration, were not made adjustable; their exact lengths were calculated to give the required camber of 18 inches to the floor construction. The truss camber was obtained by shortening the lower and lengthening the upper chord members 3/16 inch, so that after erection it amounted to 2 inches.

"All diagonal bars were made adjustable and single; they are strained from partial loads only. The trusses were adjusted to their proper shape by means of these ties, which received a slight initial strain.

"The top and bottom *chords, pier-posts, diagonal-ties,* and *pins* are of steel; *all other parts are of wrought-iron with steel rivets.* The calculated sections of the vertical web-members for steel were so light that for practical reasons they were all made of *wrought-iron* and of the same section.

QUALITY OF STEEL USED

"Every heat of steel was tested and its quality determined before any more work was done to it.

"For the compression members and pins, the steel was required to stand the following tests on specimen bars $\frac{3}{8}$ inch diameter:

Elastic limit: 50 to 55,000 pounds per square inch.

Ultimate strength: 80 to 90,000 pounds per square inch.

Elongation in 8 inches: Minimum 12 per cent.

Reduction of area at fracture: Minimum 20 per cent.

Cold bending: 180 degrees around its own diameter without crack.

Cold punching of holes in flat $3x\frac{1}{4}$ inch bars; 3/16 inch from the edge without crack or distention of metal.

"All specimens and shapes were required to be finished at nearly the same heat, as it was observed that rods finished at a lower heat would give higher tension results than samples of same steel finished at a higher heat.

"The Andrew Kloman firm in Pittsburgh had contracted to procure the steel and to furnish the steel shapes.

"The intention was to use Bessemer steel for the compression members; a large lot of Bessemer steel was tested, but few samples were found to stand the required tests. The difficulty seemed to consist in controlling the uniformity of the steel within close limits for quality and strength. After a while the attempt was given up and *open hearth steel* was substituted. No trouble was then experienced in getting a uniform grade of steel of prescribed quality.

"The top chord sections consist of four leaves, which were originally designed to be each a 20 inch steel plate with 4x4 inch angles for flanges. In ordering the steel it was discovered that enough plates of that width could not be procured in the required time. Therefore, the chord sections were changed to 10 inches and 12 inches steel plates, with 4x4 inch angles, composed as shown in the drawings [not reproduced here].

"Notwithstanding the great care used, the finished plates and angles were by no means a uniform product. According as they in rolling were finished at a higher or lower heat, they would have different degrees of hardness. Steel plates and angles finished at a lower heat had a smooth surface, and the noise of punching them resembled pistol-shots, while plates finished at a higher heat had a rougher surface, and there was hardly more resistance to punching than in wrought-iron.

"The specifications for riveted steel work provided that the punched rivet-holes, ¾ inch diameter, should in the assembled parts be enlarged to 1 inch diameter by reaming. The time for the delivery of the steel work growing short, the question was considered whether the reaming of the holes could be avoided, to hasten the completion of the work at the shops. Messrs. Kellogg and Maurice, in Athens, Pa., had the contract for this part of the work.

"To that end the following experiments were made:

"Ten specimens were cut from the same steel plate ¼ inch thick; one specimen was tested to ascertain the tensile strength of the steel in the specimen. The nine other specimens, all alike in form, were prepared as shown in sketch [not reproduced], for the purpose of ascertaining the effects of punching holes, of punching and reaming, and of drilling. The tests were expected to show the amount of reaming required, and whether any annealing effects from the hot rivet on the injured steel around the punched hole could be observed.

			Strain per Square Inch Pounds
1. Plain specimen 3x¼"........	Without holes..........	Broke with	89 730
2. 2 holes punched 1" dia.	No rivets in holes....	Broke in punched	
3 holes punched ¼" dia. ... reamed to 1"		hole with	72 000
3. Prepared as No. 2.............	No rivets in holes....	Broke in punched hole with	63 870
4. Prepared as Nos. 2 & 3.....	Rivets in all holes....	Broke through punched	
5. 2 holes punched 1" dia.		hole with	85 000
3 holes punched ¼" dia. ... reamed to 1"..................	No rivets................	Broke in punched hole with	71 000
6. Prepared as No. 5.............	No rivets.................	Broke in punched hole with	55 200
7. Prepared as Nos. 5 and 6..	Rivets in all holes....	Broke in punched hole with	83 320
8. All holes drilled 1" dia.	No rivets.................	Broke in hole with....	79 330
9. 2 holes punched ¼".......... reamed to 1"...................			
3 holes punched ¼"........	No rivets.................	Broke in hole with....	64 400
10. Prepared as No. 9.............	Rivets in all holes....	Broke in hole with....	83 32 -

"The conclusion from these tests was that the injured steel (of the quality used in this instance) around the punched hole was in part restored by annealing in contact with the hot rivets, the size of which was large in proportion with thickness of steel plates and angles as used in the chords.

"The reaming of the punched holes to a greater extent than to make the rivet holes smooth and straight was therefore dispensed with, and a reduction in the price for the finished work agreed upon.

"The same quality of steel as for the compression members was used for them; they were forged from solid steel billets, and turned to size. No appreciable difference in the hardness of the metal in the pins was observed.

"For *tension members* and rivets, the steel was required to stand the following tests on specimen bars ⅝ inch diameter:

Elastic limit: 45 to 40,000 pounds per square inch. — Yield.

Ultimate strength: 70 to 80,000 pounds per square inch.

Elongation in 3 inches: Minimum 18 per cent.

Reduction of area at fracture: Minimum 30 per cent.

Cold bending: to a loop 360° around its own diameter, without crack.

Cold punching in 3x¾ inch bars of 1 inch rivet holes: ¼ inch from the edge without crack or distention of metal.

Open hearth steel of the above and uniform quality was obtained without trouble.

"The eye-bars were made by the Kloman process, i.e., the bars were rolled from billets between reversible and adjustable rolls, in such manner as to leave the ends thicker than the bar. The ends were then spread and forged to the proper shape of the eye, under a steam hammer. The heaviest steel bars for this bridge were 28 feet 6½ inches long, centre to centre of eyes, and 1-13/16 inches thick. All steel billets and all steel bars required very close inspection for flaws, the detection of which was sometimes difficult.

"It has been stated that for the detection of flaws in steel or iron, a magnetic needle had been used with success, though the manner of its use the writer has not heard stated. A device for the certain discovery of flaws in steel bars is certainly needed. Where the solid metal sections are proportioned very economically to the work they have to do, flaws are a source of great danger, especially in attenuated steel structures; flaws in wrought-iron are more likely to happen in the direction of the fibre, but in steel they can as well happen crosswise to the direction of the tension strain as any other way.

"Three steel bars 9 feet long between centres of eyes, and 4 inches x 1-1/16 inches in section were tested to ascertain the effect, if any, of annealing the finished bars. The results were as follows:

	Bar A Annealed		Bar B Not annealed		Bar C Not annealed	
	Eye	Eye	Eye	Eye	Eye	Eye
Diameter of eye	9"	9"	9¼"	9¼"	10"	9"
Least cross-section of eye	5.72"	5.82"	5.83"	5.72"	6.80"	5.77"
Excess of metal in eye over bar	7.9%	3.6%	3%	3.2%	3%	3%
Elongation of pin-hole	0.4"	0.72"	0.45"	0.44"	0.38"	0.4"
Average section of bar	4.42"		4.22"		4.33"	
Average reduced area after test	3.90"		3.97"		3.80"	
Reduction in percents	10%		5%		10%	
Reduction at fracture	43.87%		37.5%		37.55%	
Elongation of whole bar	10.5%		10.3%		11.1%	
Elongation for 12 inches near fracture	24.6%		23.2%		22.1%	
Elastic limit per square inch	43,140 pounds		45,360 pounds		40,940 pounds	
Ultimate strength	74,310　"		78,186　"		73,760　"	

Pin-hole in one eye of bar C was bored ¼ inch out of centre line of bar, and accounts for its lower ultimate and elastic limit.

A specimen from the same heat of steel, of which the above bars were made, showed on a ¾ inch round—

Elastic limit .... 46,389 pounds per square inch
Ultimate limit .... 78,898　"　　"　　"　　"
Elongation in 8 inches 18.0%
Reduction 30.2%

"The net section of the heads through the pin-holes for all eye-bars being at least 50 percent more than the bars, and the good effects from annealing being doubtful in the above tests, it was thought not necessary to anneal the steel bars.

"For steel rivets the above quality of tension steel proved very suitable. The rivets were tough and tenacious.

"It was, however, observed that the manufactured rivet-heads would easily break off with few blows, the fracture in each instance showing a fine granulated appearance.

"Rivet-heads, however, made by hand or riveting machine were very tough, and could not be broken off; they had to be cut off.

"The cause for the brittle rivet-heads was supposed to be the upsetting by blows, in forming the head at a high heat in dies, producing sharp corners under the rivet-head and around the rivet-stem.

PLATE GIRDER SPANS

"There are six plate girders in each span beneath the flooring, namely, one girder under each rail and one girder under each sidewalk, which is detachable on the up-stream side.

"This arrangement was chosen to admit of the erection first of the new down-stream street track, which came to lie sideways of, but on a higher grade than, the down-stream track of the old bridge. To this track travel was confined during erection. Plate girders were chosen for this reason that for the limited depth of floor (for a grade 15 feet higher than old bridge, as at first contemplated), it gave a more rigid construction than open girders of low depth. It was also more convenient to work into them, and *get rid of a lot of wrought-iron which was on hand,* and was left over from orders for the suspension bridge originally intended to be built.

"Could the writer have foreseen that the new grade would be 20 feet higher than the old bridge, the deck spans of the approaches would have been made of two open girders of greater depth, in a manner that would have admitted of finishing both tracks on them at the same time. This would have been also more economical. As it was, the plate-girders were nearly finished when the change in grade was made.

"For all wrought-iron work in the bridge the quality of iron was required to be equal to that of standard bridge iron.

"Steel rivets were used for all wrought-iron bridge-members and girders.

REMOVAL OF OLD AND ERECTION OF NEW BRIDGE

"The new north abutment wall was located 40 feet back of the old one. In preparing the foundation for the same it was necessary to remove the anchorage of the old cables, and to construct temporarily two anchor chains attached to the second pair of old towers. Previous to this wrought-iron anchors had been imbedded into the foundation of new pier No. 1, which had been built up to obtain the requisite weight for the temporary anchorage.

"These anchor chains were composed of steel eye-bars, which were on hand from the intended suspension bridge. Each chain was made adjustable in length by means of a transverse screw rod, and four sets of eye-bars, forming a funicular machine. The chain could thereby be shortened with comparatively little power. To the cable the chain bars were attached by means of two wrought-iron plates. Between these plates were cast-iron friction clutches holding the cable, and pressed and held together with bolts passing through the plates. These were attached to the cable as near to the towers as possible. To prevent slipping of the clutches on the cable, the wire wrapping was removed, and spikes driven through the cable wires behind the clutches.

"The transfer of the anchorage was done without mishap while travel as usual was going over the old bridge on both tracks. The pull per anchor chain was at times 160 tons.

"Under the first north span of the old bridge, false works had been built, which, after the transfer of the anchorage, supported the old roadway, and at the same time served for the erection of the iron girders for the new bridge. No other part of the old bridge was removed till after the erection of the new channel spans.

"In the false works for the latter an opening 100 feet wide near the Pittsburgh end was left for navigation, and temporarily bridged over with wooden Howe trusses. The false works were further so arranged as to clear one track on the old bridge, on which the team-travel moved in squads in alternate directions.

"To prevent accidents from anything falling from above on pedestrians or teams below, the false works were covered with a platform of planks, which were afterwards used for the new floor. The upper staging was built up on the outside of, and to half the height of, the trusses to be erected; at that height a traveling derrick, 30 feet high, moved on a track of iron rails. All material for the channel spans was lifted (by a hoisting engine near the south Pittsburgh end) to the platform, on which a temporary track was laid, and all material transferred on push-cars.

"With another hoisting engine, conveniently located on the up-stream end of pier No. 3, the material for both spans could be handled and put in place without moving the engines.

"The Pittsburgh span was erected first. After the pier posts were put into position, the bottom chords and connecting web-members were put in place. The top chord sections, weighing from 7 to 9 tons, were picked up and placed on the verticals, one after the other, from each end in each truss. For closing the top chord, the two middle chord-sections were raised at one end till they met, and then sprung into line by pulling down these ends towards the bottom chord with block and tackle acting as a funicular machine.

"The false works of the Pittsburgh spans had settled more than was anticipated. Before it was possible to close the top chords the different panel points had to be jacked up 2 to 6 inches. No such trouble was experienced with the other span.

"During the erection of the channel spans no little anxiety existed at the possibility of an accident from some heavy weight dropping to the platform and breaking through to the constantly crowded old bridge below. Fortunately, the work was completed without such accident, but there were two casualties, which both resulted luckily. One man fell from a height of 80 feet into the river, but was picked up and next morning was at work. Another man fell from a height of 50 feet into shallow water; he was able to report for work after two days.

"The *iron* floor construction was suspended to the trusses after these were swung.

"The detail of connections in the Pauli trusses being simple, the erection of the steel and *iron* work went off smoothly, and with no more expense than in parallel chord trusses. It commenced in the middle of September, 1882, and was completed December 31st, 1882.

"To the new *iron* floor construction the old bridge floor was temporarily suspended with *iron* plate-girders at the south end of the bridge. These down-stream towers were removed first, together with the cable they supported. The old bridge floor where it was not suspended from the new bridge was held up on wooden trestles.

"Three plate-girders in each span, supporting the down-stream track and sidewalk, equal to half the width of the new bridge, were put into position, and the paving for one car-track finished for the entire length of the bridge, without interrupting travel on the old bridge below.

"Temporary wooden trestle approaches, with plank floors for one track, were built at both abutments, because the filling in would have interfered with travel on the old bridge. All this work was much retarded by a stormy and severe winter. Travel was turned over the new bridge on the down-stream track on March 19th, 1883.[30]

"During a high water, February 22nd, 1883, a heavy mass of ice came down the river on a swift current, and tore away a part of the false works supporting the old bridge in a place where it was not suspended from the new one. The old bridge was then in danger of falling into the river; but by promptly suspending the old floor to the new one, first with ropes and chains, and then with iron rods, the old bridge, after one and a half day's interruption, was again safe. This was the only interruption of travel throughout the whole work.

"After travel was turned on the new bridge, the gaps and open-

---

30 The opening of the bridge was noticed in the following: *The Scientific American* 49 (Sept. 22, 1883) : 175-80. There is a large full-page wood engraving of the north portal, with an insert showing a general view of the bridge; "Monongahela-Brucke, zu Pittsburgh in der Verlangerung von Smithfield Street," *Zeitschrift für Baukunde* 11 (1884) : 258-60; "The Rebuilding of the Monongahela Bridge at Pittsburgh, Pa.," *Engineering News* 11 (May 24, 1884) : 251-53, 265-66.

ings in the abutments and piers were walled in, as stated before. The remaining old towers, cables and bridge floor were removed, and the up-stream half of the plate-girder approaches completed.

"This was done by placing in position the remaining three plate-girders in each span, and the iron columns (supporting the girders near the abutments). At the same time the erection of the hand-railing and of the ornamental cast-iron tower progressed. The adjustment nuts of the diagonal ties in the channel spans were covered with ornamental castings, which prevent tampering with the sleeve-nuts.

"The filling in and regrading of the approaches at both ends, and the building of the toll-houses and bridge office, were completed simultaneously with the superstructure.

THE FLOORING OF ROADWAY AND SIDEWALK

"This consists of preserved wood, namely, gum-wood and white pine, preserved by the zinc-tannin process. On both the roadway and sidewalks the bottom planking distributes the weight on the *iron* and girders, so that the top sheeting or paving forms merely the wearing surface.

"To the top of *iron* floor girders are bolted wooden bolsters, to which is spiked the bottom cross-planking, 3 inches thick for the roadway, and 2 inches thick for the sidewalks.

"No provision is made to carry off the water sideways. The grade of the bridge is sufficient to carry off all surface water lengthwise. Besides, the durability of the preserved gum-wood is increased by keeping the floor moist (by sprinkling during the dry season).

"The space between the track rails and in the middle of bridge is paved with preserved gum-wood blocks, 3 inches thick and 3 inches high, laid with $\frac{1}{4}$ inch strips between.

"Every paving block is fastened down to the bottom planking with diagonal spikes. The paving blocks for the tracks rest on a 1 inch longitudinal sheeting of preserved white pine, which serves to distribute lengthwise any uneven pressure to the cross-planking beneath. The joints between paving blocks were filled with a hot mixture of tar, pitch, rosin, lard, lime and sand in such proportions as to run freely from the ladle.

"The space between the tracks and sidewalks is covered with a lengthwise top planking 3 inches thick.

"The sidewalks are 9 inches higher than the roadway. The wearing surface consists of white pine, 1 inch thick, on the bottom planking of gum-wood, 2 inches thick. In the curb are openings 30 inches long, and on the average 3 feet apart for cleaning the roadway of mud and snow. Under the sidewalk, on the down-stream side, extends a box with a movable cover, the entire length of the bridge. This contains the water and gas pipes and telegraph cables. Every 150 feet are covered openings for hose attachment, provided for sprinkling the floor and for use during a fire.

"A small fire occurred in April, 1883. It originated on the old bridge, and scorched the floor of the new bridge near the southern end. It showed the necessity of guarding against fire on the new bridge. All wooden flooring is to be protected by a paint of quicklime and glue water, and all crevices and joints in the wooden floor to be filled with it.

"For the preservation of the lumber by the zinc-tannin process, the specifications stipulated that steaming in the curing cylinder should continue at 18 pounds pressure for four and a half hours; the vacuum should not be more than two pounds per square inch. Gum-wood should absorb 25 percent, and white pine 12½ percent of the antiseptic solution under a pressure of 30 pounds per square inch. The solution is to be 5 parts (in bulk) of chloride of zinc to 95 parts of water. The lumber was to be left in it till each cubic foot of gum-wood had absorbed one and a half gallons, and each cubic foot of white pine 1.05 gallons of the antiseptic. After this a solution of tannin was forced into the cylinder, and the lumber kept immersed in it for 3 hours, under 80 pounds pressure.

"Borings from the end of a stick which were analyzed, contained 0.370 percent metallic-zinc in weight, equivalent to 0.789 percent of zinc chloride. This was rather a high showing, as 0.5 percent of zinc chloride was all that was expected.

"Borings taken from the middle of a stick 26 feet long, 12 inches by 6 inches, were found, on analysis, to contain 0.125 percent of zinc chloride, or only one-quarter of the amount intended to be injected; it is doubtful whether in long sticks the desired percentage can be attained, without very materially increasing the strength of the solution, which again would probably increase the percentage at the ends to such an extent as to render the lumber brittle after a while.

"The borings were from freshly treated lumber. It is probable that the percentage is gradually increased to a limited extent in the heart of a long stick, owing to interchange of the solution by capillary attraction along the grain of the wood. The real antiseptic substance is the zinc chloride, while the tannin serves only to increase the adhesion of the precipitate to the wood fibres.

"Care was required in the inspection of the lumber before treatment. Sap, loose knots, cracks, windshakes, are of course as much a defect in treated as in untreated lumber. Any unsound, weak or soft wood will not be improved by the treatment, which aims merely to make the lumber more durable, by preventing rotting. It may be added that lumber, treated by the zinc-tannin process, will not lose anything in its value as a combustible, as the experience with the fire at Monongahela Bridge proved.

"The track rails on the bridge are 12 inches wide, and composed of flat bars 7x¾ inches, 2 feet long under the joints, and ¾ inch round spikes, with conical heads, countersunk to the full thickness of the rail (¾ inch), so that the spikes may hold down the rail, no matter how thin it may have worn.

THE ORNAMENTAL TOWERS

are built of cast-iron, the roofs being of wrought-iron; they support merely their own weight; they incase the steel posts, which, to the eye, would seem very slender supports, and would appear out of proportion in comparison with the heavy piers and high trusses. The end posts can rock inside of the towers, which are not in any way connected with them. Where the trusses pass through the towers, room is left for expansion from temperature changes.

"The architecture of the towers is so planned, and the composing parts so arranged, that the portals may be widened out to suit the entrance to a wider bridge, should it be required.

PAINTING

"Besides painting the metal with raw linseed oil at the mills, and iron oxide paint at the bridge shops, two coats of white lead paint were applied to the erected steel and iron work. The white lead paint was used without any dryer, and mixed with boiled linseed oil only. All joints and crevices where water might collect, were puttied all around and raw linseed oil poured in, as much as they would hold.

"As the erection took place mostly in inclement weather, the shop paint came off in many places by dragging the pieces through slush and mud, which, especially in Pittsburgh, rusts iron rapidly.

"Rusty places were coated with a thin lime paste, which, after drying, was scrubbed off with wire brushes and freshly painted.

"All iron work under the flooring has been painted brown, all iron and steel work above the flooring is blue. The towers have a stone color.

LOADS AND UNIT STRAINS

Beginning from the north end, there are:

1. One 40 foot span, six equal plate-girders, proportioned for a live load of 10,800 pounds per lineal foot of bridge.

2. One 81 foot span, six equal plate-girders, proportioned for a live load of 9,000 pounds per lineal foot of bridge.

3. One 87 foot span, six equal plate-girders, proportioned for a live load of 9,000 pounds per lineal foot of bridge.

4 & 5. Two channel spans, 360 feet each, two equal Pauli trusses of steel and floor construction of iron, proportioned for a live load of 4,500 pounds per lineal foot of bridge and in addition a concentrated load of 40 tons on a 20 foot wheel base for each track; of these loads the sidewalks were assumed to carry 100 pounds per square foot.

6. One span, 88 feet 3 inches, six equal plate-girders.

7. One span, 84 feet 9 inches, six equal plate-girders.

8 & 9. Two spans, 60 feet each, six equal plate-girders in each.

All of these plate-girder spans proportioned for 9,000 pounds live load per lineal foot of bridge.

The wind truss and lateral bracing under the floor is proportioned for a wind force of 400 pounds per lineal foot of bridge.

The above live loads, in addition to the load of the superstructure in the different spans, produce no greater strains per square inch of useful metal areas than:

IRON

8,000 pounds in compression flanges of all plate-girders, floor beams, stringers, etc.

9,000 pounds in tension flanges of all plate-girders, floor beams, stringers, etc.

8,000 pounds tension in suspenders and hangers of channel spans.

4,000 pounds shear in iron web-plates.

12,500 pounds bearing strain on iron in rivet and pin-holes.

FOR STEEL

9,800 pounds to 13,200 pounds in compression members.

15,000 pounds in steel eye-bars.

10,000 pounds shear on steel rivets and steel pins.

20,000 fibre strain on steel rivets and pins from bending-moment.

18,000 pounds bearing strain on steel in rivet and pin-holes.

The following quantities of material were consumed in the construction of the Monongahela Bridge:

	Lumber, feet B.M.	594,000
Foundations	Piles, lineal feet	10,800
	Concrete, cubic yards	1,280
	Iron, tons	32
Stone masonry, cubic yards		10,500
	Iron, tons	1,070
	Steel, tons	740
	Cast-iron of towers, pedestals, etc., tons	196
	Preserved lumber for floor, feet B.M.	358,000
	Steel rails, tons	134
	Hand-railing, 2,980 lineal feet, pounds	120,200
	Filling, cubic yards	10,000
Approaches	Sidewalk pavements, square yards	1,400
	Street pavements, ” ”	2,200

The total cost of construction amounts to about $460,000."

In 1890-1891 the bridge was widened by utilizing the provisions built into the original bridge.[31] Lindenthal was both designer and contractor for this change. A third truss was added to each span on the upstream side of the bridge, which increased the width of the structure by twenty feet, eight inches and provided a second roadway. When this was done the street railway tracks ran on each side of the center line, but twenty-one years later the upstream trusses were moved four-and-a-half feet to the eastward, and the additional width made it possible to put both the electric car tracks on that half of the bridge and devote the opposite roadway to other vehicular traffic. Sidewalks eleven feet wide projected beyond the truss work. The floor system beneath the car tracks was also modified in 1911, but the other half remained essentially the same as in 1883.

Between 1911 and 1925 the elaborate wrought-iron bridge portals were removed and much simpler gateways of cast steel, designed by Stanley Roush, substituted. Here we can see the old idea of the monumental bridge portal in the process of disappearing, but even these later portals were highly ornamental.[32]

In 1895 the city of Pittsburgh determined to secure title to the bridge and open it to the public, an action which was in accordance with the trend of the times. After the appointment of viewers and the taking of testimony on both sides, the commissioner's report

31 *Engineering News* 39 (Apr. 11, 1912) : 676-80.
32 "Decorating a City Bridge With Structural Steel Portals," *Engineering News* 42 (Dec. 3, 1915) : 1086.

was filed in court, and no exception being taken, the city assumed complete ownership of the corporation through purchase of the outstanding stock.[33] The purchase price of the bridge was $1,000,000.[34]

After 1911-1915 there were few changes in the bridge, but by the early 1930s it began to be in need of repair. In order to lighten the load on the structure, it was proposed to install an aluminum deck on the vehicular roadway, and this was carried out in 1934. According to *The Engineer* of London: "This is, as far as we have been able to ascertain, the largest bridge undertaking in aluminum that has yet been carried out. It has afforded engineers an opportunity to gain experience in the use of aluminum on a large scale. Regarded as an experiment in bridge building we suggest that its importance cannot be overrated." [35]

In the early 1960s the bridge once again was exhibiting signs of wear and stress. The *Pittsburgh Post-Gazette* of April 10, 1964, announced that "approximately $700,000 will be spent next year to rehabilitate the Smithfield Street Bridge . . . weight limits have been placed on the bridge."

The *Pittsburgh Press* for May 7, 1967, stated that the bridge would close in June to permit the installation of a new deck: "the contract for the repairs had been awarded to the Mosites Construction Company in July, 1966, and since that time substructure work has been completed as well as the fabrication of new aluminum deck panels. A polyester non-skid coating is to be applied to the panels." The *Post-Gazette* of November 16, 1967, announced that the bridge "closed since last June will reopen today at three o'clock. The total cost of repairing the structure was $712,615."

In 1970 the Pittsburgh History and Landmarks Foundation placed one of its historic landmark plaques on the structure, and on May 28, 1974, the bridge was named an official city landmark by the City Planning Commission under the city's landmarks ordinance.[36]

Perhaps the most famous double lenticular truss span in the world is the Saltash Railway Bridge spanning the River Tamar in Cornwall, England, which was designed by the great engineer Isam-

33 Du Puy, "Monongahela Bridge," 204.
34 *Pittsburgh Bulletin, 32* : 24 (Apr. 18, 1896).
35 "Aluminum Floor for an Old Bridge," *The Engineer* (London) 157 (July 27, 1934) : 91.
36 *Pittsburgh Press,* May 29, 1974.

bard Kingdom Brunel and built in 1857-1859 just before his death.[37] Perhaps the Smithfield Street Bridge deserves no lesser fame. Impressive as is the Brunel bridge, the former is the more graceful and beautiful.

According to David Plowden: "The Smithfeld Street Bridge was the first and largest bridge in the New World to employ the Pauli system of lenticular trusses, it remains the only example of this type in America." [38]

The Monongahela Bridge at Smithfield Street is now within a few years of attaining its centenary, that magical state which should ensure its veneration by all who care about our technological monuments. Rumors of demolition still trouble the local air, but our great bridge, now the oldest on our three rivers at Pittsburgh, as well as the oldest steel through-truss span in America—if it cannot continue to bear increasing burdens—should be, like the great bridges at Wheeling and Cincinnati, eased into an honorable quasi retirement. Lindenthal's splendid spans have served long and well; they are now an almost indissoluble part of the cityscape. It is profoundly to be hoped that this tough, but graceful structure will, as it begins its second century, enter upon a new period of usefulness.

---

37 Eric de Mare, *The Bridges of Britain* (London, 1954), 180-81; see also David Plowden, *Bridges* (New York, 1974), 66.
38 Plowden, *Bridges,* 167.

Steamships: A Hundred Years Ago

BY A. A. DORNFELD

*Chicago's first great population explosion occurred immediately
after its incorporation as a town.
If you think that all those people got here by covered wagon, read on.*

CHICAGOANS ONCE BOASTED that their city was the greatest railroad center in the world. It still is, although nowadays civic pride is more likely to be manifested by a declaration that O'Hare is the busiest airport anywhere.

The truth is that transportation, of whatever mode, has always been of prime importance to this city, situated as it is at the hub of the nation. So it is merely a slight exaggeration to assert that the history of Chicago is the history of its transportation.

Back in the third and fourth decades of the previous century, when the inconsequential white settlement at the foot of Lake Michigan was making its first giant stride toward becoming the nation's second city, none of today's sophisticated modes of transportation was available. The predecessors of our expressways—a sprawl of miry roads—were a discouragement rather than an aid to travel. The transport plane was still a century away. No railroad linked Chicago to the highly populated East until 1852, nineteen years after its first incorporation, as a town. Despite these handicaps, the growth of the community was swift, so explosively swift that it is hard to grasp today.

A hint is given by Mark Twain, who in the late-nineteenth century needed a telling phrase to emphasize the phenomenal expansion Berlin was experiencing at that time. He called it "the Chicago of Europe."

Chicago had only 28 registered voters in 1833, when it became a town. Twenty years later, the population had swelled to 60,662. How did all those newcomers travel? A few trudged over the muddy roads, an even smaller number rode horses, covered wagons fetched in some, a growing fleet of slow and erratic sailing ships brought others. Add a handful of Indians and pioneers who had been born here, and the sum falls far short of accounting for the population boom which took place before the Michigan Central Railroad reached far enough westward.

One steam locomotive, the *Pioneer,* had indeed been puffing in and out of Chicago since 1848. But this small engine, now owned by the Chicago Historical Society, ran on tracks reaching only toward Galena, Illinois. The locomotive itself had been brought here on the deck of a ship, rather than on rails. And it was also on ships—mostly steamships—that the first swarms of immigrants came to Chicago.

So heavy was the travel on ships that Knut Gjerset, a recent historian, declares that five thousand persons embarked on a single day in 1839 from Buffalo for the upper Great Lakes, which includes Lake Michigan. It seems safe to assume that many, perhaps most, of these migrants landed here. In that year, one steamship line was already running eight scheduled trips a week between Detroit and Chicago.

It was 1821 when the first steamer reached Chicago. Before that, an occasional sailing ship, an even more occasional bateau manned by Canadians, and a few Indian canoes carried whatever waterborne commerce the settlement knew. Chicago's first steamship visitor, the *Walk-in-the-Water,* was the third ever launched in the

A. A. Dornfeld, veteran newspaperman and student of maritime history, is a frequent contributor to *Chicago History.*

The *Walk-in-the-Water*, the first steamship to dock in Chicago, sketched from a drawing on the ship's bill of lading.

Great Lakes. Fitted with masts and sails, it was so weakly engined that it needed the assistance of a "horned breeze" to stem the current when passing through a river. The "breeze" consisted of several yoke of oxen which walked along on shore and pulled the ship onward by a rope.

A round trip from Detroit to Chicago took the *Walk-in-the-Water* thirteen days, including a layover at Green Bay. Not fast, certainly, but a good deal faster than a schooner. It took one sailing ship three months to make the same trip.

Steam traffic boomed in 1832 when the government chartered three steamships to bring troops from the East to suppress Chief Black Hawk, who had taken to the warpath to protest ill-treatment by whites. There were many casualties on one of those ships. Cholera attacked the soldiers on the *Thompson* as it steamed toward Chicago from Detroit, causing forty-four deaths, and several others died after the *Thompson* anchored outside the Chicago River bar, where the bodies were dropped overside. So gruesome was the sight that the vessel was moved to a different anchorage before the survivors were ferried ashore in small boats. Another steamship had far better luck. The *William Penn* arrived with four companies of soldiers, without losing a man to cholera.

The winning of that war freed settlers in Illinois and Wisconsin from fear of further reprisals by Indians. All that open farmland was a potent factor in luring strangers to those parts. But there were other factors as well: In 1825, the Erie Canal had been opened, providing handy and fairly brisk transit for migrants and their goods from the seaboard states to the lake ships at Buffalo.

The early steamships, commonly called steamboats by the men who built and navigated them, were small by our standards. The *Walk-in-the-Water* was 140 feet long, 31 feet wide, and had a tonnage of 330. All of the early steamships were built of wood, which deteriorated rapidly.

In a book written some years ago, a veteran Great Lakes shipmaster, Capt. H. C. Inches, said that although framed and planked of finest oak, these ships were "at their best for only 15 years. After that it usually cost all the ship could clear to keep it seaworthy. One winter the ship would need repair on the port bow; soon afterward on the starboard quarter. And then a new deck." Captain Inches also stated that the engines often outlasted two wooden hulls.

Testimony on the quality of travel on early steamers is divided. The ships were not particularly reliable nor speedy, according to many accounts written by people who used them. Their wooden hulls always leaked a little, and often they leaked a good deal. Rainfall and spray from boarding waves could seep down into the passenger quarters, which were usually divided into the Ladies' Cabin and the Gentlemen's Cabin. One traveler told of a storm during which the leaks forced the ladies to take refuge in the Gentlemen's Cabin. Even there the downpour was so great that umbrellas were raised. In other emergencies, passengers crawled into their bunks to stay dry. These bunks, stacked one above the other, were a relatively safe haven—except for the top ones.

Other accounts give quite a different picture of early steamship travel on the Great Lakes. One woman passenger wrote that the Ladies' Cabin was "the very climax of comfort and convenience." Another passenger, the colorful New York State lobbyist Thurlow Weed, said that the steamship *Empire* on which he traveled to Chicago was in every respect comparable to the Hudson River steamboats, which had achieved wide fame for luxury. Expansively, Weed also declared that on the shores of the Great Lakes there was country capable of supporting a quarter of a million people within half a century—a wild understatement, as it turned out.

Other accounts of the splendors to be found on the fancier passenger steamships mention French wallpaper, marble fountains, oil paint-

Chicago Historical Society
Its owners boasted that this "Queen of the Lakes," which operated out of Traverse City, Michigan, was "fitted expressly for pleasure parties and sportsmen."

ings, tall mirrors, barbershops, and nurseries for infants. Each new steamship in its turn became known as the "Queen of the Lakes," only to be surpassed in elegance when the next was launched.

Ironically, the Michigan Central Railroad, which along with several other eastern roads eventually gobbled up the bulk of the passenger trade, built two of the most pretentious of these ornate vessels. The largest was the *Mayflower*, known as the fastest ship on fresh water. It had eighty-five individual staterooms and could carry over three hundred other passengers elsewhere in its capacious hull. A sister ship, the *Atlantic*, was almost as large. But in their last years, the *Atlantic* and *Mayflower*, shorn of their luxurious appointments and even their engines, became motionless docks because the railroads had taken over their function as passenger carriers.

To our eyes, the very earliest lake steamers appear only a little less grotesque than a drawing which survives of one of the pioneer steamboats built by John Fitch on the Delaware River in the late 1700s. This was propelled by

a bank of steam-activated paddles arranged along each side, rather resembling a Carthaginian galley of classic times. Many lake steamers had reinforcing girders which arched from front to back, high above the deck and cabin houses, lending them a somewhat bug-like profile, a bit like the Volkswagen's. The girders served the same purpose as the thick ropes which stretched fore and aft over posts set in Egyptian boats on the Nile centuries and centuries ago: they kept the vessel from sagging too badly.

An equally picturesque feature of the early steamers was the "walking beam," a long, diamond-shaped structure set well above the deck and supported by a pair of "crutches." In use, the beam teeter-tottered on its central pivot as the engine down in the hold pushed and pulled at one end of it by means of a long rod. At the other end of the beam, another rod rotated a crank on a shaft running clear across the ship. The side wheels were mounted on this shaft.

Chicago History 151

41

Sometimes the image of a galloping horse or crowing rooster was fastened to the center of the walking beam to pitch forward and backward with it in stately rhythm.

Steam whistles were not introduced for many years, and captains commonly announced their intention to arrive or depart by firing a cannon. The *Illinois,* a well-known early steamship, boasted an historic piece of artillery dredged up from the mud near the mouth of the Chicago River—a relic of Fort Dearborn, which had been burned down by Indians during the War of 1812.

Navigating the lakes involved many of the problems of ocean navigation, and a few special ones. Storms could sink ships, shoals could wreck them, and early freeze-ups of northern harbors could immobilize them long before the scheduled end of their operating season. Just as sailing-ship masters had a particular dread of lee shores against which adverse winds could crush their vessels, so steamship crews had their own special specters of fires and boiler explosion to haunt them. The art of safely installing a fiery furnace inside a highly flammable wooden hull which kept rocking about endlessly was not always well understood in the early days of steam engineer-

ing. One widely accepted practice was to mount the furnace and boiler on short iron legs and then place them inside an iron platform with upturned edges. Functioning somewhat like an oversized roasting pan, the platform was supposed to catch flaming debris from an open furnace door before it could reach the timbers below and start a fire. Passenger steamers carried "cabin watchmen" to patrol the decks at all hours and sound an alarm at the first sight or scent of fire.

Engine breakdowns, which were fairly common in the early days, caused headaches to captain and crew but seldom entailed tragic consequences. Disabled steamers could usually make it to port under the sailing rig they often carried—or thumb a tow from another ship, if one happened along. But boiler explosions, caused by defective material or careless water tending, were calamitous. Such explosions were relatively rare in the early years when engines operated at boiler pressures as low as ten pounds per square inch, but they became more common as machinery grew more complex and steam pressures mounted. Explosions finally dwindled after the middle of the last century, when the United States Steamboat Inspection Service was formed to test the bursting point of boilers and the competence of masters, mates, and engineers.

Still, early steamboat mariners on the Great Lakes were spared some problems which beset their oceangoing counterparts, such as carrying large amounts of fuel to feed their inefficient engines on month-long voyages. The early steamers used a great deal of wood: for example, the *Empire,* which won Thurlow Weed's approval, burned over eight hundred tons of wood steaming from Buffalo to Chicago, just over a thousand miles. On the Great Lakes, wood was available at dozens of ports separated only by a day or two of steaming. And when wood became scarce, coal was substituted. It, too, could be procured readily on the lakes, though not as cheaply.

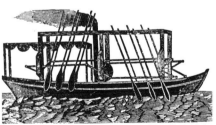

Chicago Historical Society
John Fitch's Delaware River steamboat, built in the late 1700s.

The reinforced girders stretching above the deck of this passenger steamer look decorative, but their function was to keep the ship from sagging. Its owners, the Buffalo, Chicago & Lake Huron Railway, promised three-day passage between Buffalo and Chicago.

Food was also less of a problem. Tough salt beef and weevily biscuits were staples on ocean-going ships, but the menus of the lake ships, which made frequent stops at ports situated in one of the world's most fertile regions, were limited only by their stewards' willingness to disburse cash and their chefs' talents. Most lake vessels, even freighters and sailing ships, set a good and ample table.

Another headache which caused engineers on ocean steamships much concern—the shortage of fresh water for the boilers—simply did not exist on the lakes. Seagoing ships were obliged to recycle their fresh water endlessly and often had to replenish their boilers with brine. The *Savannah,* the first ship equipped with an engine capable of making the Atlantic crossing, consistently used salt water in its machinery—and salt corrodes iron, causes foaming of the boiler water, and brings in a whole train of other troubles. The lakers, for their part, floated on an endless supply of unpolluted water.

Early steam engines had an unhappy way of refusing to start up again if the engineer on duty absentmindedly stopped them "on dead center" —that is, when the connecting rod was in direct line with the cylinder bore and the crank arm was directly above or below the cylinder. This is somewhat analogous to a bicycle with one pedal missing when the other pedal stops directly below the rider's seat. No amount of shoving on that pedal will start the bicycle because the thrust is in line with the pedal arm.

Dead centering could be embarrassing if it occurred when a captain was docking his ship against an adverse current with a strong wind blowing. To get the engine moving again, the mechanical staff had to move the crank arm a little past dead center—usually by muscle power, a time-consuming process. Eventually, improved "moving gear" eased the problem although it did not eliminate it.

Side wheels were the only means of steam propulsion on the lakes for many years. Stern wheels, used on the placid Mississippi, were not practical on the violent waters of Lake Michigan. There were many experiments with screw propellers, but it wasn't until 1841 that the first propeller-driven craft appeared on the lakes. This ship was the twin-screw *Vandalia,* whose propellers were of the sort devised by Swedish-born inventor John Ericsson—better known as the builder of the iron-clad *Monitor* which battled the Confederate ship *Merrimac* to a draw in history's first clash of iron ships. Those were not the first iron war vessels ever built, by the way: twenty years earlier, two such vessels had been launched on the Great Lakes. In 1843, the Canadians constructed the *Mohawk* on the shores of Lake Erie and, at the same time, this nation was building the *Michigan*—later renamed *Wolverine*—on the opposite side of the lake. Neither ship ever saw combat.

Even as the *Vandalia* was being launched, the Great Lakes steamships were losing out—at first to the sailing ships which they were eventually to supersede. By 1843, the city's pressing need for building materials and manufactured goods of all kinds, combined with the demand for Midwest farm products in the East, had created a boom in sailing ships. Early steam vessels were not well adapted for heavy trade. Their engines were expensive and their engineers commanded high salaries. The wind, however, was free, and there was a steady influx of European immigrants, many of them Scandinavians skilled in handling sail and ready to work for modest wages. As a result, sailing ships began to eclipse steamers in numbers and tonnage, and they maintained their advantage for almost four decades.

Both Chicago's need for imported material and the ability of its adjacent countryside to raise and send up food needed in the East in-

creased dramatically in 1848, when the Illinois & Michigan Canal was opened. A flood of farm products began to pour from downstate Illinois into Chicago, most of it was destined for transshipment. In 1855, when the Soo Canal linked Lake Superior to Lakes Huron and Michigan, the region exported so vast a freightage of mineral wealth that the United States leaped from fifth to first place in world steel production in a span of only twenty years. That same year, 1855, there were 1,150 sailing craft on the Great Lakes but only 238 steam vessels. Those were lean years for the steamers, but they survived.

In fact, one class of small steamer became very common. Tug boats, heavily timbered and heavily engined, were needed in ever-growing numbers for the growing fleets of sailing ships which had to be forced through narrow rivers and into cramped harbors. As the sailing ships proliferated, so did the tugs. Low-sided and usually less than a hundred feet long, the tugs did not look particularly seaworthy; nevertheless, they regularly steamed out into the open lakes to succor ships beset by gales—for a fee, of course. During calms, tugs sometimes puffed as far north as Wilmette to tow in immobilized schooners. At one time, there were almost a hundred tugs based in the Chicago River.

By 1869, sailing ships were having things pretty much their own way, and their continued prosperity seemed assured. They outnumbered steam vessels by 1,752 to 636 on the lakes, and their tonnage totaled 277,893 versus 146,237 for steamers. The wind ships carried most of the long-haul bulk cargoes which constituted the backbone of the freight traffic—metallic ore, lumber, and grain. Steamers of that time, their decks cluttered with engine houses and passenger cabins, were less than ideally arranged to carry such commodities. To them was left the handling of packages, the excursion passenger business, and a residue of the regular passenger trade which the railroads had not succeeded in gobbling up.

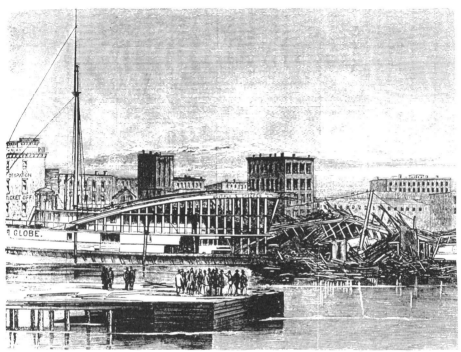

The fate of the *Globe*, which exploded at its wharf in Chicago in 1860.

But in that year of 1869, unseen influences were at work. Technological improvements had been quietly going on for sometime—not only the screw propeller introduced on the *Vandalia,* but also the steady development of the reliability and efficiency of the steam engines themselves.

The first steam engines had had only one cylinder, but later designs were fitted with two or three. In multicylinder steam engines, the weary steam, after shoving one piston ahead in the first or "high pressure" cylinder, is piped to a second. There more energy is squeezed out of it, against another piston head. After that it is piped to a third cylinder or to a condenser. The condenser reconverts it to a watery state for another trip through the boiler and the engine. The fuel efficiency made possible by using several cylinders was largely responsible for the eventual victory of steamships over schooners.

Chicago History 155

45

The *Medusa Challenger*, a modern freight steamer whose
uncluttered lines hark back to 1869, when the
R. J. Hackett was launched.

And in 1852, when shipping by steam was still in the commercial doldrums, the first iron freight vessel, the steamship *Merchant*, was launched. The durability of the wrought-iron hull was an important factor in the later resurgence of the steamship on the Great Lakes. To illustrate, the *Wolverine* stayed in commission until 1923. The time-worn hull finally reached the scrap pile in 1949, more than a century after it was built and long after the introduction of steel hulls. Another prime example of wrought iron's lasting powers is tied up in Chicago's front yard today. Built eighty-eight years ago as an excursion steamer, the wrought-iron hull of the *Florida* now serves as the base for the Columbia Yacht Club on the lakefront, south of Navy Pier. The ancient vessel has been cropped of its side wheels, smokestack, and pilothouse, but it still floats on the waters of Lake Michigan as jauntily as ever.

All these were portents of the future—if anyone was noticing. Then, in 1869, the first true

Great Lakes'-type steam freighter was launched. Although only 211 feet long and built of wood, the *R. J. Hackett* had a continuous stretch of uncluttered deck amidships—easily accessible for the loading of bulk cargo, just like the freight schooners of its day. With its engine in the extreme stern and the pilothouse perched up front, practically over the bow, the *Hackett* originated a profile which has remained standard on the lakes to this day.

The following year, 1870, the number of sailing craft on the Great Lakes diminished for the first time since such ships were counted. Within two decades their construction practically ceased. In the decade from 1870 to 1880, Knut Gjerset remarks, "It became apparent that the sailing ship had no future on the lakes."

After 1880 the future belonged to the steamship.

At least until the diesel engine arrived.

Making a Railroad:
The Political Economy of the
Ithaca and Owego, 1828-1842

JAMES A. HIJIYA

The story of a railroad between two central New York villages reveals that "metropolitan mercantilism," which historians have associated with the larger cities, also influenced smaller communities. Mr. Hijiya is a graduate student in history at Cornell University, Ithaca.

I N THE middle period of American history, railroads initiated a new phase of the "transportation revolution" which in a few decades would harness the continent with iron track. Between 1830 and 1840 railroad mileage in the United States vaulted from 73 to 3,328—almost doubling the total in Europe. By 1850 America boasted 8,879 miles, by 1860 an astonishing 30,636.[1]

Historians have found two basic economic motives for Americans' energetic construction of railroads. One, of course, was an entrepreneurial desire to turn a profit in this risky but potentially lucrative new field of investment. The other motive, perhaps even more important, was not confined to the relatively small group of people with money to invest, but was active among all the populace. George Rogers Taylor has explained it as "metropolitan mercantilism."

It is the peculiarity of any great improvement in transportation that it may bring tremendous indirect benefits, that the multiplier effect of a relatively small investment in a railroad may have the result of greatly increasing the productivity of a whole area. The incidence of these gains varies depending on circumstances, but in general farmers, manufacturers and mine owners receive higher prices, consumers

---

[1] George Rogers Taylor, *The Transportation Revolution* (New York, 1951), pp. 74, 79.

get more for their money, real-estate values mount, favorably located middlemen and bankers augment their profits, and government revenues rise.[2]

If the indirect benefit of increased productivity was a carrot enticing Americans to build railroads, there was a stick spurring them along as well—the threat of commercial isolation and subsequent stagnation of business if they did not do so. Fearing for their continued prosperity, or even survival, cities and towns had to improve their lines of communication to prevent rival communities from monopolizing trade. Competition among cities had always existed, Taylor says, but the advent of the railroad intensified the struggle, for a rail line could be laid where a canal could not be dug. "Thus there developed a sort of metropolitan mercantilism in which railroads, rather than merchant fleets, were the chief weapons of warfare."[3] Heated inter-city competition, then, set off an explosion in railroad building.

To cover their land with rails, Americans depended heavily on public funds. Federal, state and local governments sometimes built lines themselves, sometimes bought stock in railroad companies, sometimes granted companies loans of money or public credit. This governmental support was necessary, for few entrepreneurs could accumulate enough capital to build a railroad. Only in England—which had a large supply of money for investment, experience in the use of the corporate form, and dense settlement which would provide substantial traffic as soon as a railway was completed—could private investors carry out transportation improvements unassisted. In the United States, as well as continental Europe, these ventures required governmental aid.[4]

For almost seventy years historians have remarked on the readiness of Americans—famous for their self-reliance—to turn to the government for help in making internal improvements. In 1902 G. S. Callender first noted this pragmatic attitude toward free enterprise. Lacking capital but wanting railroads desperately, "it was inevitable that the American people should turn to the only means in their power to pro-

---

[2] Ibid., p. 86.
[3] Ibid., p. 98.
[4] Carter Goodrich, *Government Promotion of American Canals and Railroads 1800-1890* (New York, 1960), p. 6.

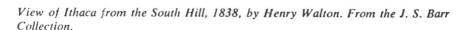

View of Ithaca from the South Hill, 1838, by Henry Walton. From the J. S. Barr Collection.

vide for their needs."[5] More recently Nathan Miller has shown how the people of New York State, while not abandoning individual initiative, were willing to spend public funds on privately owned and operated projects that benefited all.[6]

In demonstrating that metropolitan mercantilism propelled railroad building and that Americans willingly provided public funding for private enterprise, historians have focused on major improvements. Julius Rubin, for example, has detailed the commercial rivalry among New York, Philadelphia, Baltimore, and Boston and the resultant creation of such grand projects as the Erie Canal, the Pennsylvania Mainline, the Baltimore and Ohio Railroad, and the Boston and Albany Railroad.[7] In discussing governmental assistance to railroads, Callender restricted himself to "the more important works" which required several millions in capital.

However, most early railroads were short lines built for less than a million dollars. New York State had dozens of such railways, which received grants or loans from more than 300 cities and towns of all sizes. As for the state government, in only one case—that of the New York and Erie—did it provide more than a million dollars in assistance.[8] Thus, impor-

[5] G. S. Callender, "The Early Transportation and Banking Enterprises of the States in Relation to the Growth of Corporations," *Quarterly Journal of Economics, XVII* (November 1902), 153.

[6] Nathan Miller, *The Enterprise of a Free People* (Ithaca, 1962), p. xv.

[7] Julius Rubin, *Canal or Railroad?* (Philadelphia, 1961).

[8] Harry H. Pierce, *Railroads of New York* (Cambridge, 1953), tables on pp. 178-193.

tant questions arise: How much did "metropolitan" mercantilism affect small towns? Did the people of these communities see public funding as necessary for the building of minor railroads, and therefore willingly accept governmental intervention in the economy?

An answer to these questions is suggested in the case of one early railroad between two small New York towns, the Ithaca and Owego. Chartered in 1828 and granted a loan from the state ten years later, the I & O may offer further insight into motives for the great spurt of railroad building in mid-nineteenth-century America and may help fill out the picture of Americans' attitude toward public support of private enterprise.

THE *Ithaca Journal* of April 5, 1826, broke custom by relegating advertisements to the inside pages and displaying news on the front. The big story was an unwelcome one: Ithaca had lost the great canal race. For a year Ithacans had petitioned the state legislature to construct a canal from Cayuga Lake at Ithaca to the Susquehanna River at Owego, twenty-nine miles south in Tioga County.[9] Since the state in 1824 had authorized an extension of the nearly completed Erie Canal to Cayuga Lake, a Cayuga-to-Susquehanna canal would be the shortest possible connection between the canal system of New York State and the river system of Pennsylvania and Maryland. Boats laden with cargo and passengers could float from the Great Lakes to Chesapeake Bay—bringing growth and prosperity to the little town at the head of Cayuga's waters.

But Ithaca was not the only community with such ambitions. To the west, Newtown (renamed Elmira in 1828) pleaded for thirty-one miles of canal to connect Seneca Lake— also joined to the Erie system—with the Chemung River, a tributary of the Susquehanna; and to the east, Binghamton began its long campaign for a ninety-mile canal along the valley of the Chenango River from the Erie to the Susquehanna. In 1825 state surveyors examined the three proposed routes, and in March 1826 they presented to the legislature their estimates of cost. David Thomas, experienced in the survey of the Erie Canal, figured the Cayuga-Susquehanna Canal at $320,000 plus 10 percent for contingencies. James Geddes, re-

---

[9] *Ithaca Chronicle*, Feb. 1, 1825; *Ithaca Journal*, March 2, 1825; *Owego Gazette*, Jan. 24, 1826.

nowned for his work on the Erie, estimated the Chenango Canal at an exorbitant $715,478, but the Chemung at a mere $239,118.[10]

It seemed clear that Newtown would win the contest (the Chemung Canal was approved in 1829), but the people of Binghamton were not disheartened. Though their route was apparently the one least feasible, they persisted in making new surveys and petitioning the legislature until, in 1833, the Chenango Canal finally was authorized and construction begun (but never completed).

The people of Ithaca, in contrast, gave up on their canal quickly—but not without a snarl. The *Journal* of April 5 printed an acrimonious letter from "Xenophon": "Far be it from me to insinuate that the Engineer [Geddes] has intentionally stated any thing untrue; but I apprehend that the feasting, toasting and flattery he received, so far turned his head that he credited and has given currency to every idle tale told by persons deeply interested to deceive him." On April 19, immediately after the Cayuga-Susquehanna Canal disappeared from the realm of likely realization, the town's other newspaper, the *Chronicle,* suggested a startling alternative:

> After all which has been said, a canal is not exactly the communication between the lakes and the river, which is wanted. A cheaper communication, which can be used all seasons of the year—particularly in winter, so that property could be transported to the banks of the river to be in readiness for the spring freshets—such as a *rail road,* for instance—would possess many important advantages over a canal.

However, Ithacans did not yet see any need for such an audacious innovation; the idea of a railroad languished.

In 1825, four years after its incorporation as a town, Ithaca produced exportable surpluses of whiskey, lumber, flour, and optimism. The town was growing fast: from 859 residents in 1820 to 1,268 in 1823 to 1,548 in 1825.[11] Its two newspapers regularly published censuses by the state, the county, and the town, and pointed with pride and exhilaration to Ithaca's growth from month to month (more than 200 people between January and August 1825).[12] In that year Tompkins

---

[10] *Journal,* March 22 and April 5, 1826.
[11] H. A. Manning, *Ithaca Chronological Events* (Springfield, 1939).
[12] *Chronicle,* Aug. 31, 1825; *Journal,* Jan. 18, 1826.

County contained 32,908 people; 51 grist mills, 145 saw mills, 34 fulling mills, 48 carding machines, 4 woolen factories, 1 cotton and woolen factory, 29 distilleries, and 27 asheries.[13]

Probably the main source of growth was commerce. Cayuga Lake provided water transportation north, and the opening of a turnpike road between Ithaca and Owego in 1811 made land communication to the south possible, though relatively expensive and slow. One of the most important products passing through Ithaca was gypsum, a lime compound used in the manufacture of fertilizer. Before 1812 gypsum plaster had been shipped from Canada to the east coast of the United States, but the war had forced American manufacturers to find a new source—the lime-laden Finger Lakes region. As many as 800 teams were said to have crossed the turnpike to Owego in a single day carrying plaster to the Susquehanna, whence it would be conveyed by barge to the coast.[14] With the coming of peace, the plaster trade diminished, but remained substantial. Ithacans hoped a cheaper, faster communication to Owego would revive this important trade.

Besides plaster, an estimated 5,000 barrels of salt taken from the waters of Onondaga Lake was hauled each year down Cayuga Lake, through Ithaca, and over the turnpike. From the Ithaca region itself came lumber, grain, whiskey, cloth, potash, pork, livestock, and produce to be sent south on the turnpike or north to the Erie Canal and tidewater.[15]

Business was good, but the people of Ithaca and Owego were not averse to having it get better. The Cayuga-Susquehanna Canal, they thought, would eliminate the troublesome land portage between the lake and the river, making transportation quicker and less expensive, and monopolizing trade between the Erie and Susquehanna waterways. When in 1826 their pet project seemed doomed, they foresaw the demise of their towns as the Chemung Canal would divert traffic west.

Two weeks after it announced the surveyors' discouraging reports, the *Ithaca Journal* brought more bad news: "The state road bill, as will appear from the Legislative proceedings, has been finally defeated, by a postponement to a day beyond the

[13] Proceedings under the Charter, in *An Act to Incorporate the Ithaca and Owego Rail Road Company* (probably published in Ithaca by Mack & Andrus in 1831), pp. 6-7; *Journal*, Jan. 12, 1831.

[14] Anonymous [Henry B. Peirce], *History of Tioga, Chemung, Tompkins, and Schuyler Counties* (Philadelphia, 1879), p. 408.

[15] *Journal*, Dec. 26, 1827.

rising of the Legislature." The road bill, a favorite hope of New York State's "southern tier" of counties along the Pennsylvania border, had called for the construction by the state of two highways—one from Gerry, Chautauqua County, by the heads of Seneca and Cayuga lakes to the head of the Delaware River, and the other from Gerry, through Elmira, Owego, and Binghamton to Liberty, Sullivan County. The *Journal* blamed the bill's defeat on the sectional jealousy of New York City and the northern counties, which had benefited from the Erie Canal and stood to lose from the creation of a competing east-west arterial.[16]

In the summer of 1826 Ithacans organized committees to petition the legislature to approve the road at its next session. The *Journal* on June 14 declared, "It stands those in hand who feel interested in the success of the state road, to use every effort to promote its interest, and to see that no member shall be returned to the next Legislature from this section of the State, who does not consider this an object of primary consequence." At Ithaca's Independence Day celebration in 1826 the town fathers drank the toasts *"The State Road—An improvement second to none but the grand canal"* and *"The State of New York—First in population, first in wealth, first in improvements. Every section equally entitled to its resources."*[17]

But in February 1827 the state assembly voted down the road bill for the last time, and the people of Ithaca were in despair. The state had refused to build them a road to the east and west, or a canal to the north and south. It seemed that their town might become isolated, a backwater, never fulfilling its magnificent potential as a commercial emporium. To prevent this catastrophe, they were ready for desperate measures —"a *rail road*, for instance."

A railroad south from Ithaca had been discussed for some time. On April 20, 1825, the *Journal,* after mentioning a series of articles on railroads in the *Baltimore American,* mused, "It has frequently occurred to us that a *rail road* from this place to the Susquehanna, would be cheaper, of more easy construction and of greater advantage, than the contemplated canals. Our present goal, however, is to excite the attention of those who are qualified to investigate this interesting subject of in-

---

[16] *Journal,* April 19, 1826.
[17] *Chronicle,* July 12, 1826.

quiry." On July 27 a correspondent, "B," maintained that railroads were preferable to canals in economy and practicability of construction, cost of conveyance, safety, and certainty of usability. The editor of the *Journal*, while less enthusiastic, was not hostile to a railroad: "We do not think that present experience, and other *exigencies*, public and private, will admit the accomplishment of so great a design as is advocated by our correspondent B. A discussion of the subject, however, will elicit useful information upon an important branch of internal improvements little understood among us, and which must eventually lead to beneficial practical results."

On December 7 a notice appeared in the *Journal* announcing that application would be made to the legislature for a Cayuga and Susquehannah Rail Road Company with capital of $500,000 to construct a line from the head of Cayuga Lake to the Susquehanna River at Athens, Pennsylvania. Apparently, however, no action was taken. The *Chronicle* also brought attention to railroads in filler articles and lengthy feature stories, including one of December 21, 1825, which began with "1. The friction is always in inverse ratio to the diameter of the wheels" and ended with "42. The labor and consequent expense of transporting on a rail way, when compared with a canal, is as 20 to 32 dollars."

Still, the railroad remained merely talk until certain *"exigencies"* changed; that is, until hopes for a canal from Ithaca to Owego were destroyed. Even then, plans for a railway took more than a year to materialize, and the impetus for this great design came not from Ithaca, but from Owego. On September 26, 1827, the citizens of Tioga County convened at Goodman's Hotel in Owego and resolved in favor of a railroad:

Whereas the object of opening a communication between the waters of the Erie Canal and the Susquehanna River, has for a number of years occupied much of the public attention in this State,—and whereas we are deeply impressed with a sense of the important benefits that its accomplishment would confer upon a widely extended and fertile country, and whereas we are well satisfied that the grand object in view, would be as well if not better effected, by the construction of a RAIL ROAD than that of a canal: . . . *Resolved,* That we will bring the subject before the Legislature of this State, at their next session, and adopt such measures in relation thereto, as shall appear most likely to ensure success to the application.[18]

[18] *Journal,* Oct. 3, 1827.

Following this lead, Ithacans on October 31 held a similar meeting which resolved "that it cordially adopts the sentiments expressed by the resolutions" passed at Owego;[19] and on November 20, in Owego, committees from both communities agreed to petition the legislature to incorporate a company, capitalized at $150,000, to construct a railroad from Cayuga Lake at Ithaca to the Susquehannah at Owego.

In its article on the meeting the *Journal* for November 28 reported that, "Having full confidence in the practicability and advantage of the undertaking, it is deemed advisable to apply, not for legislative bounty, but simply for an act of incorporation for a company, which shall complete the work upon its own responsibility."

Having spied a deliverance from economic isolation, the Ithaca leaders promoted their new project with enthusiasm. On December 5 the *Journal* compared railroads to turnpikes and canals: A horse could pull eight times as much weight on

[19] *Journal,* Nov. 7, 1827.

The locomotive pictured here at Little Falls, about 1838, was similar to that used on the Ithaca & Owego line. W. H. Bartlett in Willis, American Scenery.

a railroad as on a turnpike; three to eight times as much as on a canal, depending on the speed maintained. On December 26—when it was clear that if the state built any canal it would be the Chemung, not the Cayuga-Susquehanna—the *Journal* realized how impractical canal building really was:

> The connexion of the Erie Canal with the waters of the Susquehannah, for purposes of commercial intercourse, has hitherto been deemed an object of great publick importance. Several routes for canals, embracing the object, have been proposed; but the great elevations to be overcome, compared with the distance, renders the communication by canal difficult and expensive. These circumstances, startling to the prudence and economy of our Legislature, have hitherto prevented any public appropriation for a canal. Against the proposed connexion by Railway, however, these objections do not exist.

In the same issue the *Journal* estimated the cost of building the railroad at $112,000, annual income at $9,540, cost of superintendence and repairs at $1,700, and profits at $7,840—a neat 7 percent on capital invested. Those calculations, the *Journal* pointed out, did not take into account the shipment of coal and iron from the Lackawanna Valley, which would certainly account for even more revenue, once communication there was established.

The propaganda campaign proved successful. An Act to Incorporate the Ithaca and Owego Rail Road Company passed the assembly 92-2 and the senate 26-0, becoming law on January 28, 1828. In May Lieutenant W. H. Swift of the U. S. Corps of Engineers began a survey of the two suggested routes between Ithaca and Owego—both of which had previously been considered for the Cayuga-Susquehanna Canal. Judging the one along Six Mile Creek and the east branch of Catatonk Creek more feasible than that along Cayuga Inlet Valley and the west branch of Catatonk Creek, through the village of Spencer, he estimated cost for construction at $177,028 and annual maintenance at $21,189.[20] On December 24 the *Journal* reported subscription for all $150,000 of the company's stock, and on February 10, 1829, the first board of directors was elected.

IF THE DESIRE to keep Owego and Ithaca growing, or at least to keep them from disappearing, was the predominant motive behind the construction of the I & O, this motive could take two forms. On the one hand, it might appeal to men's public

[20] *Journal*, Aug. 20 and Sept. 3, 1828.

spirit, prompting them to action to save their communities. On the other, it might touch their self-interest, for the economic welfare of many depended directly on commerce. These twin motives seldom occurred in isolation but usually combined in an individual, who could be both a philanthropist and a "robber baron."

The men generally credited with organizing community support for the I & O were James Pumpelly of Owego and Ebenezer Mack of Ithaca. Pumpelly, according to local lore, came to Owego in 1802 with capital of five shillings, which he promptly divided with a penniless friend. But from that time on, he was on the make. Trained as a surveyor, he married a widow who had inherited extensive real estate from both her father and her husband. Shortly thereafter Pumpelly opened a land office, and thereby made his fortune. "As is usually the case," records a historian of Owego, "many purchasers failed to make their payments in full and forfeited what they had already paid, allowing the land to go back into Mr. Pumpelly's possession, to be sold again."[21] Pumpelly expanded his investments to fields besides real estate. He was a partner in a lumber manufacturing company, president of the Bank of Owego, a director of the Cayuga Steamboat Company, treasurer and president of the Ithaca and Owego Turnpike Company, and president of the Susquehanna Steam Navigation Company.

It is not surprising that such a man of business would have an interest—a very direct interest—in the construction of a railroad to facilitate trade. Pumpelly was chairman of both the September 26 and November 20 meetings in Owego, was elected to the first board of directors of the I & O, served for a time as president, and bought $1,800 worth of stock.

Ebenezer Mack, who had acquired printing experience on the *Owego Gazette* before coming to Ithaca in 1816, edited and published the *Ithaca Journal* until 1833. In time he acquired a bookstore, a book publishing office, a bindery, and the village postmastership. Like Pumpelly he wore many hats. He sold state lottery tickets as well as policies for the Equitable Fire Insurance Company of New York. But unlike his counter-

[21] L. W. Kingman, *Early Owego* (Owego, 1907), p. 146. Most of this article's biographical data on Owegans are from this book. The information on Ithacans comes mainly from Peirce; Thomas W. Burns, *Initial Ithacans* (Ithaca, 1904); newspaper articles, and obituaries.

James Pumpelly. From Kingman,
Early Owego.

part in Owego, Mack had the kind of enterprises less likely to benefit immediately from the Ithaca and Owego Rail Road. While general prosperity would of course help his business, the transportation improvement would not reward him as directly as it would a shipper such as Pumpelly.

Mack probably boosted internal improvements largely out of the same spirit that made him a trustee of Ithaca Academy, a colonel in the militia, an organizer of St. John's Episcopal Church, a village trustee, a founder of the circulating library, an assemblyman, and a state senator. His ethic of public service was expressed in his biography of Lafayette:

> In private life, Lafayette was a model of the social and domestic virtues, as he was in publick of disinterested patriotism and unbending integrity. Lafayette was devoid of that ambition which seeks personal aggrandizement; but he gloried in the acts he had performed for the liberty and happiness of mankind.[22]

In November 1827 Mack served on the committee petitioning the legislature to incorporate the I & O; his newspaper plumped hard for it; in February 1829 he was elected a board member and its first secretary.

While Pumpelly and Mack may have been the fathers of the Ithaca and Owego, the railroad also had many godfathers. In the *Journal* for November 28, 1827, notice was served that an application for the company's charter would be made to the next legislature. This notice was signed by six men: three from Owego—Charles Pumpelly (brother of James), John R. Drake, and Jonathan Platt—and three from Ithaca—Jeremiah S. Beebe, Francis A. Bloodgood, and Luther Gere.

[22] Ebenezer Mack, *The Life of Gilbert Motier de Lafayette* (Ithaca, 1841), p. 369.

Charles Pumpelly was a director of the turnpike company, a merchant, and a sawmill owner, who shipped lumber, salt and plaster down the Susquehanna. When the I & O was incorporated, he was one of three commissioners named in its charter to receive subscriptions to the capital stock.

Drake, a large land owner, merchant, and mill owner (a partner of James Pumpelly in lumber manufacturing), was said to have more rafts and arks on the river than any other Owego shipper. He was a delegate to the Tioga County meeting of September and the combined Tioga and Tompkins meeting of November 20, and he served on both the committee to petition the legislature and the general corresponding committee.

Platt, owner of a general store, a saw mill, a distillery, and an iron foundry, dealt extensively in lumber and grain. He became the first agent for the railroad in Owego.

Beebe was a merchant, manufacturer, and contractor (he built the Fall Creek tunnel and the Clinton House in Ithaca). Besides serving as a director of the company from 1833 to 1842, including several terms as president, he was one of the major contractors for the road.[23]

Bloodgood was a large property owner and a director of the Cayuga Steamboat Company. He also functioned as financial agent in Ithaca for Simeon De Witt, surveyor general of New York, "Proprietor of Ithaca," and the biggest investor in the railroad. Bloodgood served at least four terms as I & O president.

Gere was a partner of Beebe in the Fall Creek Manufacturing Company and a director of the steamboat company. He was one of the three commissioners appointed in 1828 to take subscriptions for stock.

In addition to these eight men, at least three others were important in founding the I & O. Andrew De Witt Bruyn, chair-

[23] Kingman, *Early Owego*, p. 446, mentions only Beebe and his brother Alvah, a member of the first I & O board of directors, as builders of the road. However, Hardy Campbell Lee, *A History of the Railroads in Tompkins County* (Ithaca, 1947), p. 10, reports a contractor named Merrill working on the road in 1832 and complaining that the company had not employed enough engineers. Robert J. Casey and W. A. S. Douglas, *The Lackawanna Story* (New York, 1951), p. 57, records Simeon De Witt building the road in 1832. This source, however, probably can be discounted. In the same paragraph in which it asserts that DeWitt resigned as state surveyor general in order to construct the road, it says that in 1836 he bought "one of those new-fangled railroad engines." This is unlikely, since he died in 1834.

man of the Ithaca railroad meeting and a delegate to the second one in Owego, was the third commissioner appointed to sell stock. A successful businessman, dealing in real estate and manufacturing properties, he was also a full-time public servant—county surrogate; judge of court of common pleas; village trustee, supervisor, treasurer, and president; school commissioner; fire department trustee; congressman.

Simeon De Witt and his son Richard Varick De Witt were the biggest stockholders in the company, with Richard controlling 3,691 of the 8,319 shares sold by 1842.[24] Both were directors of the company, and Richard was perennial treasurer. However, living in Albany, and with other enterprises to attend to, they seem not to have taken active roles in the organization and operation of the road.

Most of these eleven movers behind the railroad—particularly a merchant like Drake or a contractor like Beebe—stood to make money if the road were built and, conversely, to lose if it were not. Real estate speculators such as James Pumpelly, Bruyn, and Simeon De Witt might expect the railroad to raise the value of their lands. These were practical businessmen with, as the *Journal* had reported after the second railroad convention in Owego, "full confidence in the practicability and advantage of the undertaking."

There were also some promoters, like Mack, who apparently were motivated largely by a disinterested concern for the prosperity of their communities. In either case, whether acting out of a public-spirited drive to save their towns or a more personal desire to save their shirts— or, as probably was most common, both—these eleven men saw the railroad as a means to avoid commercial isolation and decline. They must have agreed with the *Chronicle* of April 20, 1830: "This important work . . . will secure to Ithaca all the advantages of its favorable location, which might be jeoparded [*sic*] by the construction of other works of improvement at other points, if this should not be executed."

The Ithaca and Owego Rail Road originated as a substitute for a canal. Though in the early 1820s Ithacans had shown

---

[24] New York State Assembly Document No. 144, March 23, 1842. In compliance with a resolution of Feb. 21, the I & O submitted the names of stockholders and the number of shares held by each. Of 9,000 shares (face value $50 each), 3,691 were held by the DeWitts, 3,275 in New York City, 513 in Owego, 198 in Ithaca, and 844 elsewhere or in places unspecified; 489 were unsold.

interest in the strange new mode of transportation, the news-paper writer "B" (Bloodgood? Beebe? Bruyn?) had been alone in advocating the building of a railroad, until after hope for a canal to Owego had perished. The railroad was a last resort.

The cause of this desperation was a commercial crisis. If Ithaca did not reach out and monopolize the Erie-Susque-hanna trade, Elmira or Binghamton would. On a miniature scale, this competition mimicked that among big cities, which Julius Rubin has described:

> The [Erie] Canal immediately took over from the turnpikes a part of the westward trade and, when the immigrants it transported had built up the northern midwest, it carried their agricultural products back to the east. For the first time, east and west were linked by a direct two-way trade. New York, already in the lead in other fields because of its geographic advantages and superior enterprise, seemed now in a position to dominate the trade of the northern midwest and to extend that domination southward. Her major rivals—Boston, Phil-adelphia, and Baltimore—were left in the rear facing a most difficult problem.

The three cities, owing to varying attitudes toward innova-tion among their leaders, chose three different courses. Un-imaginative Philadelphia immediately began building a canal, but, finding it impossible to construct a waterway over the mountains, ended up with an inefficient mongrel Pennsylvania Mainline consisting of alternating sections of canal and rail-road. Baltimore, led by an energetic and bold group of busi-nessmen, quickly built, under private auspices, the Baltimore and Ohio, the nation's first unbroken rail link from the coast to Ohio. Boston, playing it safe, waited a decade while the in-novation proved itself practical and profitable, before under-taking construction of a railroad across the Berkshires.[25]

The mercantile warfare among these cities was re-enacted only a few years later in south-central New York. Among small towns the rivalry, though resulting in less grand projects, was no less acute. Elmira, with its Chemung Canal, threatened to monopolize trade. Binghamton responded with the unsuc-cessful Chenango Canal. Ithaca and Owego built a railroad. Thus, the I & O, as much as the B & O, was a creature of metropolitan mercantilism.

[25] Rubin, *Canal or Railroad?*, p. 6.

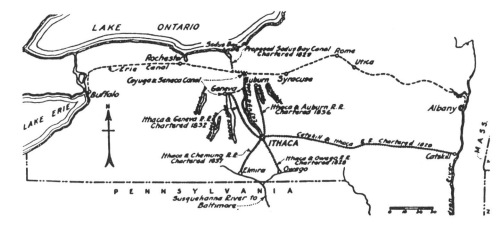

Transportation dreams of Ithaca, 1827-1837. From Abt, Ithaca.

ONCE THE I & O had entered the dreams of Ithacans, railroad fever became epidemic. By 1836 charters had been issued for separate railways connecting Ithaca with Chemung, Geneva, Auburn, and Catskill.[26] Ithaca was to be the hub of a railroad network—as Chicago later became. In the lyceum, orators debated the question, "Ought Rail Ways to supersede use of Canals?"[27] The newspapers churned out technical explanations of acceleration and inertia, reports on successful or contemplated railways, statistics on dividends from the Baltimore and Ohio.[28] The *Journal* for December 9, 1829, told how the locomotive would open "a new era for the most extraordinary improvements in travelling ever made":

A London paper referring to this subject observes:—"The engine of Braithwaite and Erickson, moved at the astonishing speed of 28 miles an hour. It seemed, indeed, says a spectator, to fly—presenting one of the most sublime spectacles of human ingenuity and human daring the world ever beheld. It actually made one giddy to look at it, and filled thousands with lively fear for the safety of the individuals who were on it, and who seemed not to run along the earth, but to fly on the wings of the wind."

Ithacans' grandest dream was the Sodus Bay Canal. According to this plan, a trench to be dug from Sodus Bay on Lake Ontario (due north of Seneca Lake) to an enlarged Erie Canal would enable ships laden with products of the old Northwest

[26] Lee, *Railroads of Tompkins County,* p. 8.
[27] *Journal,* June 16, 1830.
[28] For example, *Chronicle,* May 5, Dec. 22, 1830, Jan. 26, Feb. 16, 1831; *Owego Free Press,* Jan. 14, 1829.

to sail direct from the Great Lakes to Ithaca, from which a radiating web of railroads would distribute their cargoes all over the east. The *Journal* of August 17, 1836, reported all $800,000 of the Sodus Bay Canal stock taken up shortly after the books were opened in Geneva. Within two or three years, the story asserted, the project would be executed. In November 1838—even after the Panic of 1837, which dried up sources of capital and brought many construction projects to a sudden halt—Henry Walton painted his "View from South Hill" with sailing ships anchored off Cayuga Inlet, and noted that "Vessels are introduced upon the lake to shew the result of the Sodus Canal being made according to plan."[29]

Two other projects were particularly important to persons interested in the Ithaca and Owego Rail Road. One was the North Branch Canal, approved by the Pennsylvania legislature early in 1828, which was to follow the north branch of the Susquehanna River from Northumberland to the New York State line. The *Owego Gazette* of March 4, 1828, assured its readers the canal would secure "immense trade" by bringing the Lackawanna Valley coal and iron traffic to Owego.

The second important project was a railroad from Owego to New York. Founders of the I & O took leading roles in conventions at Binghamton (December 15, 1831) and Owego (December 20) which called for the construction of such a railway.[30] December 1831 also saw the publication of a "Petition for a Rail Road from the Termination of the *Ithaca and Owego Rail Road* at Owego, to the city of New York," which contained an essay on self-reliance and public profligacy:

> The subscribers, inhabitants of the county of   respectfully represent: That they have witnessed with satisfaction, and with feelings of emulation, the spirit of improvement which now so generally prevails among the citizens of this state. The magnificent work, connecting the western lakes with the Atlantick ocean, which has been constructed by the government, with the sanction of the people of this state, has surpassed, in its facilities and its revenues, the most sanguine expectations. Hence have arisen local claims and applications to the legislature, for the construction at the publick expense of similar works in various sections of the state—some of which were within the means and entitled to the favourable consideration of the Legislature; but taken in the aggregate, would involve an amount of expenditure far

---

[29] The painting is at the De Witt Historical Society of Tompkins County Museum.
[30] *Chronicle*, Dec, 28, 1831.

exceeding the present resources of the government, except by a resort
to a system of taxation which could not be expected to meet the ap-
probation of the people . . .

Your memorialists conceive it to be both the duty and sound policy
of a free government, to encourage that spirit of emulation and individual enterprise which is calculated to impress the people with a
dependence upon their own industry and resources, rather than upon
executive power, or legislative bounty.[31]

In 1832, to the joy of the people of Ithaca and Owego, the
legislature chartered the New York and Lake Erie Railroad
Company, which was to run from New York through Owego
and on across the state. Among the company's incorporators
were three Owegans: John R. Drake, Stephen B. Leonard
(editor of the *Gazette*), and J. H. Avery, all of whom had
helped found the I & O.[32] But construction, not begun till the

[31] A copy of the petition is in the De Witt Historical Society manuscript
collection, box 1-6-6.
[32] *Journal*, March 21, 1832.

*Route of the I & O (dotted line) illustrating the link, at Owego, with Erie Rail-
road (dotted line) and the Susquehanna River. From Pierce and Hurd,* History
of Tioga, Chemung, Tompkins, Schuyler Counties.

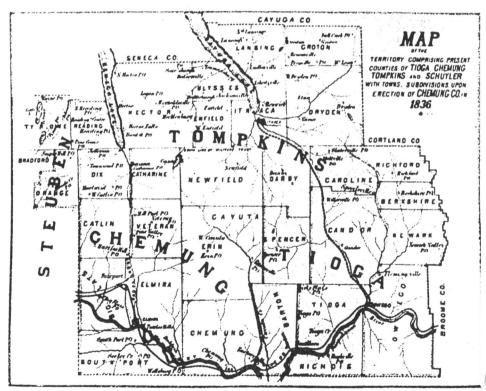

Stephen B. Leonard. From Kingman, Early
Owego.

fall of 1835, was slow. Shortage of capital and the initial reluctance of state legislators to subsidize the project (although
they eventually would dole out over six million dollars) prevented it from reaching Owego till 1849. In 1835 the *Ithaca
Chronicle,* a Whig paper, had urged the state to build and
operate the road, and by 1838 the Democratic *Ithaca Journal*
had come around to the same view. The *Journal* urged every
man in the county to sign a memorial to the legislature which
said the petitioners believed that—

such an artery of commercial communication, traversing the whole
territorial limits of the commonwealth, and deeply affecting the present
and future welfare of its inhabitants, should be kept at all times within the sole and undivided control of the public authorities, and that
the commonweal would be much better subserved by constructing this
great thoroughfare at the public expense, and adopting such rules for
its regulation as should reduce, to the lowest reasonable limit, the expense of transportation, than by granting its revenues to an incorporated body, who would claim remuneration for the risk assumed in its
construction, by extracting a large and constantly increasing tribute
from the internal commerce of the state.[33]

The Democratic *Owego Gazette* on February 13, 1840,
urged the legislature to make the New York and Erie a state

[33] *Chronicle,* April 1, 1835; *Journal,* Jan. 3, 1838.

65

work, saying, "The Friends of the Rail Road, ask for no further aid to the Company.—They have lost all confidence in the Company and want nothing more to do with it."

Between 1831 and 1840 Ithacans and Owegans had renounced that "spirit of emulation and individual enterprise which is calculated to impress the people with a dependence upon their own resources" and called instead for the "sole and undivided control of the public authorities." The history of the New York and Erie foreshadows that of its smaller neighbor, the Ithaca and Owego.

Four years elapsed between incorporation of the I & O and the beginning of construction. New surveys were taken; the stock was re-distributed twice.³⁴ But enthusiasm still bubbled high. After the second transfer of stock, the *Journal* of April 27, 1831, reprinted the report of the transaction from the *Albany Argus*:

> One hundred thousand dollars of this stock was offered in New York to the public on the 13th inst. by Messrs. Bucknor, Woolsey & Pumpelly, as attornies for the proprietors. This rail road has the confidence of the most judicious persons on Wall st. No newspaper puffing was resorted to; but public opinion was left to itself. The road will benefit three states materially, and make Ithaca perhaps the greatest place of trade between Albany and Buffalo.

In September 1831 the *Journal* advertised for bids on materials for building the line, and in the following February construction began. The *Journal* for March 7, 1832, said that "nothing but the difficulty of procuring a requisite number of experienced contractors, and a sufficient force of laborers, is likely to prevent the entire completion of the road within the ensuing season."

Meanwhile, the directors had realized that the $150,000 capital allowed by the charter was insufficient, and appealed to the legislature for an additional $150,000. In the assembly the Committee on Rail Roads, of which Horace Mack (brother of Ebenezer and another director of the railroad) was a member, saw no reason to deny the request, and the assembly on February 17, 1832, concurred, 95-0. On March 31 the senate approved the doubling in capital stock, 25-0.

By July 1833 a mile of track had been laid, from the inlet

---

³⁴ The first was reported by the *Chronicle* on April 20, 1830: "A more general distribution of stock has recently been effected, and large amounts have been applied for, which could not be obtained."

of Cayuga Lake to near the foot of South Hill. On Independence Day the company offered free rides, and more than a thousand citizens took part. Henceforth the public-relations-conscious I & O made two excursion runs a week.[35] In Owego, too, enthusiasm reached a peak. The village trustees authorized laying the track through the public park and streets to the town square. Farmer Elizur Talcot, following the lead of his townsman James Pumpelly, donated to the company a right of way through his land.[36]

But financial realities were catching up again. On February 26, 1833, John Randel Jr., chief company engineer, estimated the entire road could be made ready for use for $299,475—conveniently under the $300,000 limit on the capital of the company. However, to build the road on a "more permanent plan," substituting stone fixtures for wood, an additional $83,854 would be required. Even Randel's lower estimate exceeded by more than a hundred thousand dollars the estimate of 1828 by Lieutenant Swift. Swift had figured on using bridges over ravines, rather than making embankments and cuttings. In June 1832 the railroad's special Committee to Inspect Accounts and Plans stated that Swift's plan would be too liable to accident and would not permit the use of a locomotive. The committee therefore recommended the building of a more solid, though costlier, road.[37]

Still, in 1833 the promoters of the railroad saw no reason for dismay. In his report to the stockholders on March 14, president Bloodgood said that on the subject of prospects for dividends, "we have full confidence in the assurance that if the calculations for the future can be regulated by the experience of the past, they are flattering." He estimated revenues for 1834 at $91,125; expenses, "at a liberal calculation," at $20,000; leaving $71,125—more than enough to meet all obligations.[38]

On April 7, 1834, a single track had been completed between Ithaca and Owego, and a train of forty-nine cars—four

[35] *Journal*, July 10, 1833.

[36] Kingman, *Early Owego*, p. 447. De Witt, Bloodgood, and Gere also donated land through which the road was to pass, according to the Engineer's Report of Feb. 26, 1833, in *Report of the President and Directors to the Stockholders* (Ithaca, 1833), p. 47.

[37] Engineer's Report, p. 81; Report of the Committee to Inspect Accounts and Plans, June 5, 1832, in *Report of the President and Directors*, p. 105.

[38] *Report of the President and Directors*, March 14, 1833, pp. 3, 6-7.

I & O handbill. From Kingman, Early Owego.

filled with passengers and 45 with salt and plaster—made the twenty-nine-mile run to Owego in three hours, excluding stops. The *Chronicle* of April 9 rejoiced:

> This road forms the first direct communication between the Erie Canal and the Susquehannah river . . . accomplished by individual enterprise, and individual funds, economically expended, and forming a durable track which can be maintained in repair at a comparatively trifling cost. It forms, too, the first link in a communication by Rail Road with the city of New York, through the southern counties, the sequestered portion of our state which has hitherto been called upon to join in contributing, by government patronage, to the prosperity of more favored portions without enjoying an adequate return for favors bestowed.

On May 3, after votes of 91-0 in the assembly and 22-0 in the senate, where Ebenezer Mack sat on the Committee on Rail Roads, the limit of the company's capital was increased another $150,000.

Throughout 1835, 1836, and 1837 all mention of the I & O in the newspapers was favorable—though considerably less frequent than in previous years. The *Chronicle* for July 15, 1835, said business already done on the road that season, together with existing contracts, would equal the highest estimates previously made. The *Journal* of August 17, 1836, said all problems would soon be overcome:

> It is true the road is at present much out of repair and little or no system observed on the line, all which however results from the fact that the road never was completed and put in order, and from the great number of persons now busily employed in completing and repairing it—of course the cars are frequently detained for the rails to be temporarily placed. We understand and also observed that the work is now rapidly progressing and will soon be completed in a permanent and substantial manner, and when so, all must be aware of its great importance to the manufacturing and commercial interests of Ithaca.

The *Chronicle* on January 25, 1837, reported a 25 percent yearly increase in business on the line since it opened, and predicted a still greater rise once a steam locomotive were put into use. "The stock of the road," the story concluded optimistically, "will soon be valuable."

Unfortunately the outlook was not as bright as the newspapers claimed. Company stock was selling at only seventy-five cents on the dollar.[39] From 1834 to 1837 annual revenues averaged only a little over $20,000—somewhat less than $91,125 predicted by Bloodgood in 1833.[40] Moreover, the cost of construction was higher than expected—$570,000 by March 1838—and to pay for it, the company had resorted to relatively high-interest (7 percent) loans from private individuals.

In 1838 the I & O went to the state with hat in hand—but it was not the first in line. As early as 1827 the state had granted the Delaware and Hudson a loan of $500,000; after 1836 the New York and Erie was seeking public money constantly.[41] Precedent for state aid to privately owned railroads had been set, and the I & O's case seemed as urgent as any. When in 1838 the Assembly Committee on Rail Roads looked over documents submitted by the I & O, chairman John M. Holley of Wayne County reported in favor of state assistance. Much of the grading, two inclined planes, and all bridges, cul-

---

[39] Assembly Document No. 144, 1842. Of 9,000 fifty-dollar shares in 1842, 3,529 had been sold at $50; 4,790 at $37.50; and 489 were not sold. The company also had issued 192 certificates entitling the bearer to a share.

[40] Assembly Document No. 138, March 23, 1842. In pursuance of a resolution of Feb. 3, the I & O submitted data on expenditures and receipts from 1831 to 1841. This is the most complete record of the company's finances that has survived.

[41] Pierce, *Railroads of New York*, p. 13.

verts and viaducts for a double track had been completed, he noted. However, more rail still needed to be laid, and locomotives and engine houses purchased. Since the opening of the road, salt, plaster, flour, and other merchandise had been taken down the Susquehanna, and lumber and coal brought up to the Erie Canal. Tolls on the Erie from that lumber alone had totalled $9,010 a year, and "no portion of this lumber, it is believed, would have reached the Erie but for the existence of this road." The I & O, thus, was indirectly serving the state.

Concerning the company's financing, Holley reported that $105,000 had been obtained from the sale of capital stock, and $300,000 from the mortgage of the road to individual lenders. In addition, the company owed $165,000 to contractors, laborers, and small creditors. The object of the company now, he said, was to extinguish the $300,000 mortgage by assigning it to the state, with a 5 percent interest loan instead of the current 7 percent. Unless this were done, the lenders would foreclose the mortgage, and "long and bitter litigation" would shut down the road, destroying its usefulness.[42]

After being amended, the aid bill was passed by an unrecorded vote in the assembly, approved 16-11 in the senate, and signed into law by Governor William Marcy on April 18, 1838. The amended bill instructed the comptroller to issue the I & O special certificates of state stock to the amount of either $250,000 or half the sum the company had expended, not exceeding $300,000 ($287,700 was eventually loaned under this act). The stock, bearing 4½ percent interest, was to be auctioned off, and the credit of the state pledged for payment of its interest and principal. The I & O was to provide for the redemption of stock and the payment of interest. Should it fail to do so, the comptroller would be empowered to auction off the road to the highest bidder or to buy the road for the use of the state, subject to the disposition of the legislature.[43]

On December 15 the stock went on sale in New York. However, when it became apparent that no one would buy it except at discount of at least 7 to 10 percent, William G. Bucknor, an agent for the I & O, bought it up for the company. Two weeks later the railroad memorialized the legislature to

[42] Assembly Document No. 237, March 6, 1838.
[43] An Act to Aid in Construction of the Ithaca and Owego Rail Road, in *Charters, Laws and Leases of the Cayuga and Susquehanna Railroad Company 1828-86*, compiled by Fred F. Chambers (New York, 1887).

take back the stock and re-issue it at 5 percent interest, making it more salable.[44] The assembly approved the measure on January 23, 1839, but on February 18 the senate defeated it 15-12.

The I & O had to sell its state stock cheap, and most of it was bought by European financiers.[45] However, the railroad did eventually get its compensation. In May 1840 the legislature granted an additional loan of $28,000 in state stock at 6 percent interest, by a vote of 50-49 in the assembly and 13-10 (after reconsideration) in the senate.

For the time being, at least, the Ithaca and Owego Rail Road was saved, but the rescue attracted little attention in the newspapers. The *Journal* on April 25, 1838, informed the public of the loan in only nine words. On May 9 the *Chronicle* published the act verbatim but made no comment. Apparently railroad fever had cooled down. In any case, no one complained of the road's failure to pull through on just "individual enterprise, and individual funds, economically expended."

In seeking legislative bounty after having proclaimed self-reliance, the I & O promoters were merely coming full circle. They had originally wanted the state to build a canal; they started the railroad with private funds only because they had "full confidence in the practicability and advantage of the undertaking" (since, they thought, it could be built for half the cost of a canal). Despite their rhetoric of individual enterprise, the promoters were ready when the time came to go to Albany with an open palm.

Thus the history of the Ithaca and Owego illustrates that pragmatic attitude toward the role of government in the economy described by Callender and Miller. The small transportation improvement needed capital as desperately as the large. For the American in the small town, as well as for the one in the big city, nothing in his political ideology restrained him from seeking governmental support for private industry—not when his survival seemed to be at stake.

Even after receiving state aid, the railroad did not earn significant profits. From April 1839 to January 1840 the road took in $24,244 and paid out $13,598, leaving $10,646—not

[44] A draft of the memorial is in Cornell University's Regional History Manuscript Collection, box 2594, folder 5.

[45] Comptroller's Report to the Legislature, Assembly Document No. 10, Jan. 11, 1843. Of $315,000 from the loans of 1838 and 1840, $221,000 was held in Europe, $82,000 in New York State, and $12,000 in other states.

even enough to pay the interest on the loans. To some extent this failure was due to the inefficiency of the company. For example, during several months of the year no tolls were collected, though traffic was "considerable."[46] But another reason for the I & O's failure was the inability of other communication lines to connect with it. The New York and Erie had not yet reached Owego, and Pennsylvania had suspended work on the North Branch Canal.[47] Without these links to New York and the Lackawanna Valley, the Ithaca and Owego's hoped-for boom in traffic and revenues never occurred.

Though out of money and unable to pay its debts, the railroad kept plugging along. In the same month that it received its second state loan, it announced a round trip fare of seventy-five cents, and half price to passengers attending any public meeting.[48] Fourth of July 1840 saw the introduction of the line's first locomotive—a great occasion joyfully remembered by an old-timer:

A gala day was announced, a free ride was offered to all the world from Ithaca to Owego, and return. It was called a grand celebration; such it was. Our train of sixteen flat cars stopped at every crossing for passengers. We made the round trip under Conductor Hatch with only one accident; John Haviland was crowded off or fell off the train and was killed.[49]

October 1840 brought the first publication of words critical of the Ithaca and Owego. Levi Hubbell, a former Democrat and director of the company, was running for the state legislature as a Whig (he won), and the Democratic *Journal* did everything possible to keep the traitor out of office. On October 21 it denounced him as "a dissolute man, in whose company the wives of bosom friends are not secure from insult." A week later it claimed that his intervention in 1838 and May 1840 had procured the state loans to the I & O, and suggested that he now ran for the legislature to secure the golden lining in the company's pocket.

The *Journal's* assailing of the railroad, motivated not by

---

[46] Letter from Daniel Bishop to Hon. Demos Hubbard Jr. of Committee on Rail Roads, March 16, 1840. Copy of letter in De Witt Historical Society manuscript collection, box 1-6-6.

[47] William Heidt Jr., *Simeon De Witt* (Ithaca, 1968), p. 31.

[48] *Chronicle*, May 27, 1840.

[49] Alvin Merrill, "The First Passenger Railway in America," manuscript probably written before 1902, located in De Witt Historical Society manuscript collection, box 1-6-6.

philosophical opposition to state aid to private enterprise but by the desire to defeat Levi Hubbell and the Whigs, introduced something new into the history of the I & O. Previously—as long as things were going well—politics had been kept out. Of the founders, about half were Jackson men—Bloodgood (its first president), the De Witts, Mack, Drake, and Bruyn—and about half anti-Jacksonians—the Pumpellys, Platt, Gere, and Beebe (its last president). The first board of directors consisted mostly of Jacksonians, but in the following decade it came to be dominated by Whigs, partly because some Jacksonians, like Hubbell, Bloodgood, and Drake, became Whigs after 1836.[50] Still, until 1840 neither the Democratic *Journal* nor the Whig *Chronicle* had had anything but praise for the railroad and its management.

In 1841 the Ithaca and Owego failed to pay the interest on its state-guaranteed stock, and in October of that year Comptroller Azariah Flagg foreclosed the company and advertised it for sale to the highest bidder. Henry Yates and Archibald McIntyre, business partners from Albany, had long been interested in the I & O: Yates had invested $11,475, McIntyre $6,037.[51] With the help of the company's officers they arranged for the transfer of the road.[52]

On May 20 the Ithaca and Owego Rail Road, which had cost $590,039 to bring into being, was bought at auction by McIntyre for $4,500. (The rolling stock went later for an additional $13,000.)

Begun with unanimous enthusiastic optimism, the Ithaca and Owego Rail Road reached its end amid a flurry of petty recrimination. Democrats blamed the line's demise on its Whig directors and the Whig state administration; Whigs scolded the Democrats for lending money to a shaky enterprise and for selling the road at a ridiculously low price. But from the standpoint of political-economic ideology, all this bickering was pointless.[53] Both the Whigs and the Democrats believed the

[50] Newspapers, particularly in election time, record the party affiliations of the founders. For example, *Chronicle*, Nov. 16, 1825, Oct. 11, 1826, Oct. 3, 1832, Oct. 30, 1833, July 9, 1834, Sept. 29, 1841; *Journal*, Oct. 26, 1825, Jan. 1, April 16, 1828, Sept. 26, 1832, Feb. 14, July 10, 1833, April 30, 1834, Nov. 9, 1842.

[51] Assembly Document No. 144, 1842.

[52] Copies of correspondence among company secretary Daniel Bishop, Yates, and Bucknor are in the Cornell Regional History manuscript collection, box 2594.

[53] Only once was expression given to an opposition to railroads as such.

railroad necessary for Ithaca's development. Neither side opposed governmental aid to railroads on principle. As late as May 1840, when the I & O received its second state loan, no one in Ithaca raised a complaint. Only in October, when the Democrats saw the opportunity to make a partisan issue of the loan, was there any word spoken against it. Such retrospective resistance does not attest to an unshakeable devotion to unassisted, unfettered free private enterprise.

Facing the problem of what to do with a bankrupt railroad, the Whigs and the Democrats offered divergent analyses of alternatives. The Whigs saw the choice of selling it immediately at a low price or retaining it until a higher one could be had, and preferred the latter. The Democrats said the state could sell the road immediately or become its proprietor, and chose the former. Neither group wanted the state to take over the railroad permanently.

This reluctance to surrender lasting control of the road to the state contrasted with Ithacans' advocacy of state ownership of the New York and Erie. While the bigger project, "traversing the whole territorial limits of the commonwealth, and deeply affecting the present and future welfare of its inhabitants," had required "the sole and undivided control of the public authorities," the I & O apparently needed only a temporary assist from the state. Desiring as much governmental regulation of the economy as necessary, but no more than necessary, Ithacans demonstrated again the flexibility of their brand of capitalism.

On April 18, 1843, the state legislature chartered McIntyre's reorganized company, named the Cayuga and Susquehannah. The company had capital of only $18,000; but it had a railway already built, and no loans to repay. (The state, having lent its credit to the Ithaca and Owego, had to make good on the $315,700 in principal and $336,114 in interest.)[54] In 1848 Pennsylvania mining entrepreneur George Scranton bought the Cayuga and Susquehannah for $50,000. Beginning in 1849 the new owner tore up the rail of the old line and laid down a new route to Owego. In 1851 he united the southern

---

The *Tompkins Volunteer* of May 17, 1842, reported that the Homo Patha Life Preserving Philanthropic Society of Ithaca had denounced the manufacture and use of railroads and all other potentially deadly merchandise, including steamboats, gunpowder, narrow shoes, baker's bread, and ladies' stays.

[54] Pierce, *Railroads of New York*, p. 193.

end of the C & S to the Liggett's Gap Rail Road (soon to be part of the Delaware, Lackawanna and Western), and achieved his goal—a route to northern markets for anthracite from his mines in the Lackawanna Valley. The C & S's net receipts reached $41,813 by 1851, and kept on rising. By 1854, when the Delaware, Lackawanna and Western issued its first annual report, Ithaca was "the principal coal depot of the company."[55]

Thus reincarnated, the Ithaca and Owego Rail Road was finally a success. Though Ithaca never became a Chicago, nor even "the greatest place of trade between Albany and Buffalo," it at least survived. The DeWitts of Albany may have lost $150,000, and the State of New York four times as much, but Ithaca and Owego remained competitive in a world of metropolitan mercantilism.

Just as the Ithaca and Owego replicated major railroads as an innovative improvement forced into existence by commercial competition, so did it resemble its big brothers in requiring the nurture of the state and in stimulating popular advocacy of governmental aid. The generalizations of historians Taylor, Rubin, Callender, Miller and Goodrich concerning big cities and grand projects hold equally true in regard to Ithaca, Owego and their railroad.

[55] D.L.&W. report quoted in Lee, *Railroads of Tompkins County*, p. 17.

Dallas and Its First Railroad

By Leon J. Rosenberg and Grant M. Davis

Theories of city location frequently attribute the growth of an area to multiple geographic and economic factors that focus on temporal-spatial dimensions. Given this perspective, the development of a major north Texas city would appear to have been both logical and inevitable. A fortuitous combination of soil, climate, and water facilitated the production of wheat, cotton, and cattle in the region. Later, the discovery of oil became a significant factor. However, the actions of individual men likewise have contributed to the development of cities. Local historians emphasize this factor, often to the exclusion of others, and isolate specific actions by community leaders to explain the development of a city.[1]

Certainly, Dallas, Texas, the modern metropolis and transportation hub of the southwest, owes its growth to numerous advantages, political, economic, and geographic. But, unlike a city such as St. Louis, which grew up at a precise site primarily because of the location of waterways, Dallas does not possess significantly better geographic characteristics than any other locale in north Texas. The city's environmental setting in the rich blacklands of north Texas does not adequately explain its growth, since dozens of other towns, including nearby Mesquite and Corsicana, had equal access to the same natural assets.

This study began with a single question: To what extent were railroad routes influenced by the precise location of Dallas? The question is simple, but potential answers are numerous and complex. Any analysis of city location and development must recognize that there exists an intricate interplay of economic, social, and ecological relationships.[2] The purpose of this study is not to provide a total socio-economic history of Dallas. Rather it attempts to focus on one aspect in the history of the city, the circumstances under which its first south to north and east to west railroads were induced to intersect at a specific location in north Texas.

Dallas experienced an inauspicious beginning when, in 1843, John Neely Bryant built a log cabin. Just over a dozen years later, in February 1856, the town was incorporated by an act of the Texas legislature. Although one Alex Simon was advertising "cheap prices for dry goods, groceries, hardware, ready-made clothing, pants, and coats," Dallas then had a population of only 430.[3]

On July 8, 1860, a fire destroyed almost all buildings in Dallas. Most were subsequently rebuilt, but Dallas as yet displayed no

34

indications that it would later expand and become a major metropolis. Although it was relatively unaffected by the Civil War, by 1865 the aggregate assessed value of city lots was only $78,000 and the assessed value of all real estate in Dallas County was only $1,182,000.[4]

The despair that permeated the defeated Confederate States of America during the Reconstruction Period following the war naturally pervaded Texas. But, paradoxically, the feeling of hopelessness in the South contributed significantly to the growth of north Texas. Many former slaveowners apparently felt that without slave labor they would be unable to grow cotton profitably. At that time the principal agricultural crop in north Texas was wheat, which required relatively less productive manpower per acre than cotton. Consequently, a considerable number of families moved from the older southeastern states to the north Texas area.[5]

During the late 1860s, Dallas ceased to be the peaceful settlement that it had been during the antebellum period. There was a steady influx of new settlers and the development of a brisk trade with other north Texas towns. Regular wagon trains connected with Decatur and Weatherford, both west of Dallas. Photographs taken during the period, moreover, show Dallas' business streets lined with frame buildings. Many had wooden awnings sheltering plank-floored porches, which for the most part served as sidewalks. Numerous hitching posts were located along the streets—themselves merely lanes of swirling hot dust during the summer and channels of black mud during rainy periods—but thronged with horsemen, wagons, and livestock. Though the actual population of Dallas in 1871 was estimated at only 700 to 800 people,[6] its proportional growth was substantial and it was becoming a bustling frontier town. However, without railroads its potential for continued growth was clearly limited.

The Railroads Before the Civil War

The growth of Dallas, and of Texas, should be considered from a national perspective. The discovery of gold in California, in 1848, had impressed upon the people of the United States the need for a railroad connecting the Atlantic seaboard with the Pacific Ocean. In this regard, most historians have recognized that perhaps the most thorough explorations made by Americans during the nineteenth century were those pertaining to the transcontinental railroad.[7] Texans were naturally anxious to have a line that would traverse their state, so in 1850 the Texas legislature approved a resolution to provide financial assistance to such a railroad.[8] During the 1850s, under the supervision of the Secretary of War, the Army conducted

35

studies to ascertain which of five proposed transcontinental routes should be developed.[9] Proponents of a southern route urged a line from Memphis, extending through Arkansas and Texas to El Paso, and terminating in San Diego, California.[10]

On December 17, 1853, the Texas legislature granted a charter to the Memphis and Pacific Railroad. The line was to originate at Fulton, Arkansas, on the Red River, and follow a route across Texas along a line as near the 32nd parallel as practical. It was to cross the Trinity, Brazos, and Colorado rivers, and proceed westward to El Paso. The Texas legislature subsequently changed the name of the proposed route to the Memphis, El Paso, and Pacific.[11]

This was not the only rail line proposed to cross Texas. By 1854 there were at least four other chartered companies authorized to construct railroads from its eastern boundary to the Rio Grande. These included the Texas Western Railroad Company, the Texas and Louisiana Railroad Company, and the New Orleans, Texas, and Pacific Railroad Company.[12] During 1856, the charter of the Texas Western was amended substantially, and its corporate name changed to the Southern Pacific Railroad Company.[13]

In 1857 construction began on the Memphis, El Paso, and Pacific, with groundbreaking taking place near Texarkana, at Moore's Ferry landing on the Red River. However, high water conditions required an extension of the roadbed to Jefferson, Texas, so that necessary supplies could be more easily delivered by Red River steamboats. Fifty-two miles of roadbed were graded and five miles of track were laid by 1861. Construction was discontinued at the outbreak of the Civil War.[14]

Construction also started on the Southern Pacific Railroad in 1857. The roadbed started near Marshall, Texas, at Caddo Lake, on the state's eastern boundary. Some twenty-seven miles of rail were laid prior to 1861.[15] The Memphis, El Paso, and Pacific and the Southern Pacific had simultaneously begun construction on parallel rail lines across Texas, and were engaged in a race from its eastern border westward toward El Paso.

During the same year, 1857, construction of the Houston and Texas Central Railroad commenced from the Gulf Coast northward. Its objective was to serve the north Texas region and extend into the Indian Territory, now the state of Oklahoma. Tracks had reached Millican, seventy-five miles northwest of Houston, by 1861, when the outbreak of the Civil War resulted in the cessation of building.[16]

The Railroads After the Civil War

Capital was scarce in Texas during the years immediately following the Civil War, but a widespread demand for cotton existed.

36

The Houston and Texas Central resumed hauling cotton from south-central Texas to the Gulf Coast, where it was transloaded aboard vessels for shipment to the industrial northeast and to British ports. Consequently, the road earned sufficient revenue to finance an ambitious expansion program.

However, the railroad was adversely affected by a crisis in the Texas cotton plantation system that occurred in 1868 and 1869. Two major factors were involved: heavy federal taxes on cotton, and the collapse of the plantation system. These problems in the cotton economy spurred the railroad to push further northward toward the wheat growing section of north Texas. In 1867, the railroad moved to Bryan and then to Hearne. By 1870, it had continued northward to Bremond and Kosse.[17] It reached Groesbeck, eighty miles south of Dallas, in 1871.[18]

To their consternation, community leaders of Dallas discovered that the route of the Houston and Texas Central was projected to pass north through the village of Mesquite, eight miles east of Dallas. Acting to encourage railroad officials to build through Dallas, these leaders raised $5,000, secured title to 116 acres of land, and arranged for a free right-of-way for the railroad through their town.[19] They did not get exactly what they wanted from the railroad. However, negotiations did result in an agreement that the tracks would be constructed seven miles west of the original plan, or one mile east of the Dallas business center. The contractual arrangement concluding the agreement was signed in December 1871.[20]

Simultaneously, statewide interest existed in resuming rail construction across Texas, from east to west. On May 24, 1871, the Texas legislature passed an act designed to encourage the speedy construction of a railroad through Texas to the Pacific Ocean.[21] Dallas business leaders wanted their town to be on that line. Not satisfied to be served only by a north-south railroad, they were convinced that the east to west line would have to be attracted if the town's role as the principal business city of north Texas was to be assured.

Sim Duncan, bookkeeper of a Dallas mercentile firm, Clark & Bryan, knew that the charter of the Memphis, El Paso, and Pacific would route it through Corsicana, near the 32nd parallel. He also was aware that the survey of the Southern Pacific, entering Texas about fifty miles south of its competitor, provided a route near or through Corsicana. Duncan and W. J. Clark, president of Clark & Bryan, devised a scheme that they anticipated would bring the railroads through Dallas instead of Corsicana. Both the Memphis, El Paso, and Pacific and the Southern Pacific had completed only a few miles of construction across Texas. Duncan and Clark's plan called for a third

37

railroad, whose trackage would be roughly midway between the two lines already under construction.

Civil and business leaders located along the line of the proposed railroad were invited to a meeting in Dallas during the summer of 1871. Those attending passed resolutions and agreed to attempt to secure a charter for a new railroad, the Jefferson and Dallas. In the meanwhile, contacts were made with the Memphis, El Paso, and Pacific, urging the line to go through Dallas. Leaders of the proposed Jefferson and Dallas Railroad stipulated that Dallas would abandon its plans for a new parallel route, and instead attempt to secure a charter for a line to Wichita.

Simultaneously, John W. Lane, owner and editor of the Dallas *Herald*, and a state representative of Dallas, Tarrant (Fort Worth), and Collins counties, agreed to help the Southern Pacific have its charter amended, the ostensible reason being to change its name to Texas and Pacific Railroad.[22]

The Southern Pacific had been granted a new charter by Congress on March 3, 1871. The same legislation provided that the name of the Memphis, El Paso, and Pacific be changed to Southern Transcontinental Railroad. The Southern Transcontinental was authorized to extend its trackage—which had already been built from Fulton, Arkansas, to Texarkana—westward along the Red River. Its proposed route through Texas would run parallel to the tracks of the Southern Pacific for over three hundred miles. Further, the legislation provided that the two lines merge some 170 miles west of Dallas, near Schackelford County, Texas.[23]

In accordance with his agreement to help the Southern Pacific have the name of its line changed to the Texas and Pacific, Lane introduced an act in the Texas legislature. The act was supported by the other members of the legislature who represented Dallas, Tarrant, and Collins counties, Senator Samuel Evans and Representative A. F. Leonard.[24] The company would be permitted to build from Longview, its existing terminus in the eastern part of the state, westward along the 32nd degree of latitude, by "such route as the company may deem most practical."

Just a mile south of the Dallas of 1871, in what today is Dallas City Park, a trickle of water exists known locally as Browder Springs. Lane and other civic strategists apparently decided that this could provide just the obsure cause they needed. Lane inserted a specific restriction: the railroad would have to be built ". . .crossing the Central Railroad within one mile of Browder Springs." The Dallasites employed two tactics that are still favorites. First, they suggested their bill be attached as a rider to a legislative proposal that was

38

already moving quickly toward passage. Second, the maneuver was timed in such a fashion that the bill appeared for final consideration shortly before the legislature was scheduled to adjourn.

It retrospect, it is difficult to understand why no one at the state capitol in Austin, with the notable exception of the Dallas and Fort Worth group, apparently knew or bothered to determine the exact location of Browder Springs. One possible explanation is that the Dallas, Tarrant, and Collins legislators had handled the political phase in such a clever manner. The Southern Pacific Railroad management did not question the phrase, nor did the representatives of Corsicana, which was the site of the critical intersection with the Houston and Texas Central Railroad.

On November 21, 1871, the Texas legislature passed the bill without a single dissenting vote.[25] It included the rider which perhaps as much as any single factor is responsible for the status of modern Dallas. The bill was presented to the governor for his approval, and when it was not returned by him in the prescribed time it became law without his signature.[26] Undoubtedly, a considerable uproar must have ensued when certain people discovered that Browder Springs was only one mile from Dallas, but by then the legislature had adjourned.

Perhaps the management of the Southern Pacific had been outsmarted. Still, the railroad actually had the last word. The Dallasites' plan had been designed to draw the line right through the town. But the company indicated that it would indeed build one mile south, claiming this would be cheaper. In order to encourage the railroad to modify its plan, Dallas business leaders quickly subscribed to a $100,000 city bond issue for the railroad, proving they had money as well as tricks up their sleeves. They also agreed to provide the railroad with a broad right-of-way through town.[27]

Four months later, on March 30, 1872, the Southern Pacific, and the former Memphis, El Paso, and Pacific, by then known as the Southern Transcontinental, executed deeds of consolidation to form the Texas and Pacific Railroad.[28] Plans were modified to eliminate the parallel construction across north Texas. The Texas and Pacific agreed to build through Dallas, and the plans for a Jefferson, Dallas, and Wichita Railroad were changed to provide simply for a Dallas and Wichita Railroad. The first three railroads that were to intersect in Dallas cost approximately 150 acres of land, $5,000 cash, and $200,000 in bonds.[29] The total includes an additional $100,000 in Dallas city bonds that Dallas voters approved and donated to the Texas and Pacific in April 1875.[30]

39

The pending arrival of the railroads provided an immediate stimulus to Dallas growth. People quickly converted survey parcels of land into plots of blocks, lots, and streets. Some probably visualized the spires and turret stones of a great city, and heard the rumble and roar of many railroad trains. By 1872, the town had a population estimated at 1500, approximately twice as large as the previous year.[31]

The Houston and Texas Central tracks were constructed on a right-of-way now occupied by Dallas Central Expressway, and the area seven miles east was left to become the suburb of Mesquite. The first train steamed into Dallas on July 16, 1872, and was enthusiastically welcomed by a crowd estimated between 5,000 and 6,000 persons.[32] Construction of the Texas and Pacific continued with trackage extending west from Longview and east from Dallas. The connecting rails were joined near Grand Saline on August 16, 1873.[33] Dallasites had won a complete victory, just as planned. The Texas and Pacific tracks were laid one mile north of the innocuous Browder Springs, right through the heart of Dallas.[34] Shortly thereafter, the main line of the Texas and Pacific was opened and the inaugural train steamed into Dallas, having traveled the 124 mile route from Longview.[35] Completion of the tracks signified the first direct connection between north Texas and the east.

During the same year, 1873, the Houston and Texas Central, whose tracks to the north were located almost a mile east of downtown Dallas, constructed the Marilla Street spur track, leading to the downtown district. On Wood Street, the carrier built a freight terminal, adjacent to the business district and to the depot of the Texas and Pacific.[36]

Like Reconstruction, the Panic of 1873 also benefitted the young city. Economic conditions forced a temporary cessation of the westward construction of the Texas and Pacific, so Dallas enjoyed the distinction of being its western terminus for almost three years.[37] Railroad transportation contributed to the development of a number of wholesale concerns. Also, the town became the shipping point for raw materials moving from regions north, south, and west of Dallas to the large consumer markets in the east.[38]

Because of the railroad and its accompanying telegraph line, commodity buyers swarmed into the city. By then Dallas had become the center of the state's finest cotton producing area as well as maintaining its position in the wheat trade.[39] Furthermore, Dallas became an outfitting point for railroad passengers who came to town intending to continue an overland trek westward. Wagon yards were filled with emigrant wagons bound for west and southwest Texas.[40]

40

Although it is perhaps difficult to understand, even in the early 1870s Dallas possessed a group of businessmen who mutually felt a strong civic loyalty. They were willing to commit their time and money to ventures for the good of the young town. Its leaders, merchants, bankers, and salesmen, were solid, sober, middle-class citizens who believed in homes, churches, and schools. They were drawn to Dallas by a desire to make money and get ahead.

The character of the city's leadership, possessed of a remarkable degree of local patriotism, is reflected in the manuverings over railroad routes. The Houston and Texas Central would probably have gone through Mesquite if Dallasites had not taken action to bring it through their town. The Texas and Pacific would likely have passed through Corsicana. But the point is that Dallasites *did* bring effective action to bear on behalf of their town, thereby assuring its place as the crossroads of the state's first east-west and south-north railroad lines.

FOOTNOTES

1. Charles N. Glaab, *Kansas City and The Railroads* (Madison, Wisconsin: The State Historical Society of Wisconsin, 1962), p. 1.
2. Glaab, *Kansas City,* pp. 1-2.
3. Works Projects Administration, American Guide Series, *Dallas Guide and History* (Dallas, reprint 1970), p. 228.
4. Phillip Lindsley, *A History of Greater Dallas and Vicinity* (Chicago: The Lewis Publishing Company, 1909), p. 71.
5. John William Rogers, *The Lusty Texans of Dallas* (New York: E. P. Dutton and Co., 1951), p. 101.
6. Lindsley, *History of Greater Dallas,* p. 369.
7. A. Theodore Brown, *Frontier Community: Kansas City to 1870* (Columbia, Missouri: University of Missouri Press, 1963), p. 5.
8. Alexander Deussen, *The Beginnings of the Texas Railroad System* (Austin: Transactions Texas Academy of Science, 1906), p. 44.
9. Brown, *Frontier Community,* p. 5.
10. *Minutes and Proceedings of the Memphis Convention, Assembled October 23, 1849* (Memphis, 1850), pp. 21-23, 29-30.
11. Lindsley, *History of Greater Dallas,* p. 370.
12. Deussen, *Beginnings of the Texas Railroad System,* p. 44.
13. Act of August 16, 1856, *The Laws of Texas 1822-1897* (Austin: The Grammel Book Company, 1898), Vol. 4, p. 622.
14. *Texas Almanac for 1868* (Galveston: A. H. Belo & Company, 1869), pp. 136, 137.
15. *Idem.*
16. *A Memorial and Bibliographical History of Dallas County, Texas* (Chicago: Lewis Publishing Company, 1892), p. 820.
17. *The Kosse Cyclone,* June 24, 1933.

41

18. *The Groesbeck Journal*, May 15, 1936.
19. George Jackson, *Sixty Years in Texas* (Dallas: Wilkinson Printing Co., 1908), p. 231.
20. Lindsley, *History of Greater Dallas*, p. 370.
21. Grammel, *Laws of Texas*, Vol. 7, p. 202.
22. Lindsley, *History of Greater Dallas*, pp. 372, 373.
23. S. G. Reed, *A History of Texas Railroads*, (Houston: The Clair Publishing Co., 1971), pp. 151, 358.
24. Texas House of Representatives, *Members of the Legislature of the State of Texas from 1846 to 1939*, 46th Texas Legislature, June 15, 1939, p. 62.
25. Lindsley, *History of Greater Dallas*, p. 373.
26. *Laws of Texas*, Vol. 7, pp. 203-206.
27. Muir, Andrew Forrest, "The Thirty-Second Parallel Pacific Road in Texas to 1872," (Ph.D. diss., University of Texas, 1949), p. 214.
28. Reed, *A History of Texas Railroads*, p. 363.
29. Lindsley, *History of Greater Dallas*, pp. 375, 376.
30. Charles S. Potts, "Railroad Transportation in Texas", *University of Texas Bulletin*, No. 119 (March, 1909), p. 67.
31. Jacksc.*, Sixty Years in Texas*, p. 231.
32. Nelson A. Hutto, *The Dallas Story* (Dallas: Willima Noel Sewell, 1953), p. 28.
33. Samuel Bertram McAllister, "The Building of the Texas and Pacific Railroad," (M.A. thesis, University of Texas, 1926), p. 47.
34. R. E. Butterfield and C. M. Rundlett, *Directory of Dallas for the Year 1875* (St. Louis: St. Louis Democrat Lithograph and Publishing Co., 1875), p. 50.
35. Potts, "Railroad Transportation," p. 5.
36. Reed, *A History of Texas Railroads*, p. 209.
37. Works Projects Administration, *Dallas Guide and History*, p. 19.
38. *Ibid.*, p. 231.
39. Rogers, *The Lusty Texans of Dallas*, p. 108.
40. Works Projects Administration, *Dallas Guide and History*, p. 20.

42

Railroad Promotion and Economic Expansion at Council Bluffs, Iowa, 1857-1869

Sidney Halma

PRIOR TO 1857 THE UNITED STATES ENJOYED increasing prosperity. Gold discoveries in the Far West preceded the gold rush of 1849 and encouraged migration from eastern cities. Western movement accelerated demands for transportation, resulting in rapid railroad expansion, speculation in real estate, and a general scramble for quick fortunes. Speculation was particularly common in outfitting centers that engaged in the lucrative business of supplying emigrants. Overexpansion led to financial panic in 1857.

The first sign of an impending financial crash in one of these outfitting centers, Council Bluffs, was the closing of the Benton banking house in September. A prominent citizen, Dexter C. Bloomer, recalled the event: "All my money was in Benton's bank which failed."[1] A number of other residents suffered similar losses in 1857. Benton made earnest and persistent efforts to meet his obligations but was unable to repay and eventually lost his homestead. The collapse of one after another bank in nearby Nebraska rendered their bills worthless.[2]

Of the four banking firms in Council Bluffs at that time, all but the Officer & Pusey firm dealt extensively with Nebraska currency, which circulated locally in large quantities. Prior to 1855, the circulating medium throughout the Council Bluffs region had been confined primarily to gold or silver specie. Eagles and double eagles were abundant; bank bills were the exception rather

[1] D. C. Bloomer, "Commonplace Book," Iowa Department of History and Archives.

[2] D. C. Bloomer, "Notes on the Early History of Pottawattamie County, *Annals of Iowa,* First Series, X (July, 1872), 185-186.

than the rule. The Iowa Constitution of 1846 prohibited banks from issuing paper money.[3] A number of easterners and Iowans incorporated "wildcat" banks across the Missouri River in the Territory of Nebraska. This paper money rapidly filled the vacuum caused by the drain of specie at Council Bluffs.

The Council Bluffs banking and real estate firm of Baldwin and Dodge likewise was seriously hurt by the collapse of banks and decline in land values. Town lots depreciated in value; prices fell lower and lower. Lots in Council Bluffs selling in 1856-7 for $3,000-$4,000 could hardly be sold for $750-$1,000 in 1860.[4] Money became very scarce. Council Bluffs residents were literally without money for some time. People resorted to trade and barter and wore their old clothes. Merchants issued pieces of paste-board which were good for five, ten, and fifty cents in merchandise. Farmers went ragged and burned corn for fuel.

The full impact of the Panic of 1857 was not felt in Council Bluffs because of the yearly emigrant trade. Emigrants continued to move across Iowa on their way west and because it was an outfitting center, the town continued to reap the benefit of active trading. Merchandise sold freely for cash, and the farmers found a good market for corn and wheat.[5]

Fortunately in 1858 the announcement that gold had been discovered in the Colorado region vaguely known as Pike's Peak once more brought large numbers of outfitters to Council Bluffs. The news of the discovery of gold in Cherry Creek near Pike's Peak was at once circulated by travelers and newspapers. The *Iowa Weekly Citizen* of Des Moines was one of the first newspapers to publicize the Cherry Creek discovery. The newspaper reported that two men with "inferior implements washed out $600 in one week. . . ."[6] The *Council Bluffs Nonpareil* followed three days later with the banner:

Pike's Peak Gold Diggins! Eureka! Eureka! Gold Mines Within 500 miles! The Best Route Thither! The Yellow Fever Spreading Rapidly! The Only Antidote—Pick and Shovel!!![7]

[3]Ruth A. Gallaher, "Money in Pioneer Iowa, 1838-1865," *Iowa Journal of History and Politics,* XXXII (January, 1934), 20-21.
[4]Bloomer, *op. cit.,* X (July, 1872).
[5]*Ibid.,* pp. 191-192.
[6]Des Moines *Iowa Weekly Citizen,* September 8, 1858.
[7]Council Bluffs *Nonpareil,* September 11, 1858.

Newspapers undoubtedly exaggerated these reports about the gold discoveries because "the Panic of 1857 was depressing business and a gold rush would restore prosperity to pioneer towns where miners bought supplies."[8]

Attempting to substantiate rumors about the gold rush, the Council Bluffs newspaper published accounts stating that the gold discovery had been officially confirmed by James W. Denver, Governor of the Kansas Territory, in a letter to the Secretary of the Interior.[9] For the next several years much space in the Council Bluffs newspaper was devoted to articles on the new gold regions and the best routes to them. Council Bluffs, Nebraska City, and St. Joseph competed for the outfitting business. The press in each town advertised that the best and most direct route to the gold fields lay through its limits. The *Council Bluffs Nonpareil* published an elaborate map of the road from Council Bluffs to the Cherry Creek gold mines. Several Council Bluffs' residents responded to the "gold fever." Samuel Curtis, a prominent citizen, was among those who joined the throng headed for the gold fields. Curtis served as a correspondent for the *Nonpareil,* which published his letters reporting favorable conditions.[10]

By the spring of 1859, reports to the contrary alleged that the entire rush was a "humbug" and that it was ridiculous to believe a fortune could be made quickly. The *Burlington Hawkeye,* published in a rival outfitting center, pronounced the whole story about the discovery of gold on Cherry Creek a wicked deception and fraud.[11] Disillusioned gold seekers influenced many to return. Hundreds of returning emigrants recrossed the Missouri, highly indignant at the merchants and newspapermen of Council Bluffs and rival towns, and charged that they had been misled. Threats of vengeance were sometimes heard, leading the press and businessmen to fear for their safety and property.[12] *The Council Bluffs Bugle* attempted to learn the truth of the so-called

[8] Ray Allen Billington, *Westward Expansion* (New York: Macmillan, 1967), pp. 619-620.

[9] Council Bluffs *Nonpareil,* October 9, 1858.

[10] *Ibid.,* September 18, October 16, 1858. Bloomer, *op. cit.,* X (July, 1872), 192.

[11] Bloomer, *op. cit.,* X (July, 1872), 193-194.

[12] Nathan P. Dodge, "Early Emigration Through and to Council Bluffs," *Annals of Iowa,* Third Series, XVIII (January, 1932), 167-169.

gold discovery, but stated that no returnees had been as far as the mines.

> We are fully persuaded that the return stampede has been caused by speculation beyond Fort Kearney, those who have turned the Emigration back—bought their outfits for almost nothing and are making a big speculation out of their frauds, falsehoods and lying. . . .

> Emigrants are told all kinds of stories. Some say that the reports of the Mines are gotten up by persons in the frontier towns for the purpose of selling outfits . . . such assertions, as far as we are concerned, are unqualifiedly false. . . .[13]

The reports in the newspapers became so conflicting that three eastern editors, Horace Greeley, Henry Villard, and Albert D. Richardson, went to the region to report the facts for their newspapers. On June 9, 1859, these three newspapermen signed a widely publicized statement from Gregory's Diggings which stated that, while there was gold in this region, mining was a business that required "capital, experience, energy, endurance. . . ." The report enumerated the successes of several individuals in that area and stated that many others had left too hastily.[14] The report, coming in the midst of the 1859 gold rush, may have discouraged some potential miners, nevertheless, Council Bluffs saw hordes of hopeful '59ers.[15] Approximately 15,000 people passed through in 1859, some of whom probably bought provisions at Omaha.

The Council Bluffs press made further attempts to destroy the apparent myth of the humbug by sending a correspondent to the Colorado gold fields to verify the findings. William H. Kinsman, a correspondent for the *Nonpareil,* walked the entire distance from Council Bluffs to Cherry Creek. He reached the mining region in early June, and sent back a record of his observations. His optimistic reports produced further excitement in Council Bluffs and points east.[16]

The renewed emigration brought a revival of trade at Council Bluffs. By March of 1859 a "steady tide of hoofs and horns and covered wagons" was passing through the streets, and leaving

[13]Council Bluffs *Bugle,* May 18, 1859.
[14]Des Moines Iowa *Weekly Citizen,* June 29, 1859.
[15]Council Bluffs *Weekly Nonpareil,* June 11, 1859.
[16]Bloomer, *op. cit.,* X(April, 1872), 193-194.

daily for Colorado. Every stage from the east brought passengers, and five or six steamers unloaded from fifty to seventy-five "Peakers" weekly. By April these numbers had grown to hundreds daily—most of them from the upper Mississippi Valley. One prospector observed that Council Bluffs was "not a very inviting looking place it is situated between two bluffs one main street only and a very few good buildings. Streets and ravines crowded with teams and emigrants."[17] While some camped nearby, quite a few companies disregarded the inclement weather and advice of "those who knew" and departed immediately.

Although the rush was over in June, the effect of this banner year upon Council Bluffs was noteworthy. Several large hotels catered to the emigrants; outfitters increased in numbers, eleven advertising in one newspaper; a "horse-railroad" was organized to run the three miles from the city to the landing; and a pork-packing plant was established.

The decade of the sixties brought another great westward movement. Farmers destined for the Far West were generally better equipped than the emigrants who had headed for Colorado. Nevertheless, Council Bluffs merchants did a brisk business, since their advertising was designed to create a felt-need for new products. Publicity agents promised the public an abundance of reasonably priced goods:

> . . . articles can be purchased here cheaper than at any other town or city in Iowa. We have mills in the city that can manufacture from two to three hundred sacks of flour daily, and the mills in the county within four miles to the city can make as many more. There is an abundance of wheat to be manufactured—enough to supply all the wants of the country and emigration, and a large surplus for shipping.[18]

Publicity centered around the superiority of the Mormon Trail, recommended as the "natural highway" to the mines. The estimated 25,000 people who had passed over the Mormon Trail previously could not be wrong. In 1860, according to emigrant guidebooks, the trail boasted a shorter route, abundant supplies of wood and water, a well-protected and settled route up to Ft.

---

[17]Kenneth F. Millsap, (ed.), "Romanzo Kingman's Pike Peak Journal, 1859," *Iowa Journal of History and Politics* (January, 1950), 67-69.
[18]Council Bluffs *Weekly Bugle,* February 29, 1860.

Kearney, and easily-crossed rivers.[19] A map of the region between the Missouri and the Rockies, which showed Council Bluffs closer to the mines than it actually was, ran for weeks in the *Bugle.* A detailed "Table of Distances" informed the emigrant of every station, ferry, and bridge, and of wood and water along the entire route. In a more spectacular vein, a traveling artist exhibited 10,000-feet of canvas paintings of Council Bluffs, Omaha, the overland trail and the mines in both towns before taking them east on a tour.[20]

Inducements such as these encourgaged the arrival of increasing numbers of emigrants. By April and May some fifty wagons were leaving for the prairies daily, and each week some 1,000 emigrants arrived by steamer, wagon, or stage.[21] The two steam ferries, each capable of carrying twelve teams per trip and making from twenty to thirty trips each day, were kept busy crossing the "Big Muddy!" Livestock in great droves accompanied many wagons bound for California, Oregon, and the Rocky Mountain territory.

The emigrant of the sixties made a somewhat different impression than his counterpart of the late fifties. He was, noted an observer, generally "of the wealthy class, and for sobriety, morality and general good behavior is not surpassed by the resident population. . . . Out of the vast crowd which has passed through, we have not seen but one man who was any worse for liquor."[22]

Although Council Bluffs employed publicity agents, a local editor denied sending "runners" to the states to the east, as he alleged other outfitting towns did. He claimed that this was unnecessary because the emigrants generally followed the most accessible routes such as the North Platte Route. Statistics reported by the company owning the two steam ferry boats supported this claim. For the week ending April 24, 1860, 514 emigrants were counted. The total for the six weeks ending May 26 was 1,526 wagons and 4,602 men. Despite these incomplete figures, it is probable that Council Bluffs did get "four-fifths of all the emigration from Iowa, Illinois, Indiana, and States East and North of

[19]*Ibid.*
[20]Council Bluffs *Weekly Nonpareil,* February 18, 1860.
[21]Council Bluffs *Weekly Bugle,* April 18, May 9, 1860.
[22]*Ibid.,* May 9, 1860.

them . . ." totaling between 10,000 and 15,000. Seventy-nine steamer arrivals prior to August 1 and more than $21,000 collected for freighting services indicated that business was good.[23]

The trend established in 1860 continued the following year. Streets were jammed with wagons. Steamboat arrivals became too regular to be news. Emigrants destined for the Far West, particularly California, appeared to dominate the crowds. Westward movement did not cease during the Civil War years. Economic opportunities continued to beckon emigrants. Not only were gold and silver tantalizing, but also the climate, rich soil, and commercial possibilities drew the restless to western territories. Emigration through Council Bluffs was unabated in 1863, partly due to the disruption caused by the war and fear of the draft.[24] Dexter Bloomer recalled that the emigrants passing through Council Bluffs in 1863 were "almost uniformly opposed to the prosecution of the war and to the policy of the government in putting down the rebellion."[25]

As was customary, the outfitting houses did a heavy business despite rumors circulated by eastern outfitting towns that prices were higher at Council Bluffs. One emigrant buying supplies at Council Bluffs recorded in his diary: "Council Bluffs is not a very large place, but is a very busy one. We are surprised at the amount of business done here and at Omaha."[26] Many newcomers were lured by the prosperity of an emigration market and settled in the Council Bluffs area.

Railroad propaganda intended to attract travelers to Council Bluffs was deflated by the Panic of 1857. An important event, however, helped to rekindle the interest in railroad construction —Abraham Lincoln's visit to Council Bluffs in the summer of 1859. After a campaign trip to Kansas, Lincoln, accompanied by Secretary of State O. M. Hatch of Illinois, visited Council Bluffs. Some have speculated as to why Lincoln came to Council Bluffs.[27] One explanation is that he came to see two families there whom

[23]*Ibid.*, May 9, August 8, 1860.

[24]Robert G. Athearn, "Across the Plains in 1863: The Diary of Peter Winne," *Iowa Journal of History and Politics*, XLIX (July, 1951), 221 ff.

[25]Bloomer, *op. cit.*, XI (April, 1873), 424.

[26]Athearn, *op. cit.*

[27]"The Visit of Abraham Lincoln to Council Bluffs," *Annals of Iowa*, Third Series, IV (July, 1900), 460-461.

he had known in Springfield, Illinois—the Pusseys and the Officers. A second explanation involved real estate. Norman Judd, manager of Lincoln's debates with Douglas, had asked Lincoln for a loan of $3,000, and as security, offered seventeen choice lots in Council Bluffs. It is reasonable to suppose that Lincoln took advantage of the proximity to Iowa during his campaign trip to examine the real estate offered by Judd. Others have pointed to Lincoln's desire to study the railroad question in terms of a proposed transcontinental route—passing across western Iowa and eastern Nebraska.[28] Grenville M. Dodge, a surveryor for the Chicago and Rock Island Railroad, discussed this possible route with Lincoln.

Dodge recommended the 42d-parallel route because it was the most practical and economical. He felt Council Bluffs was the logical starting place because the railroads had been building from Chicago to that point. As far as Lincoln was concerned, the attraction of the 42d-parallel route was perhaps enhanced by the fact that Judd owned Council Bluffs real estate and was asking him for a loan of $3,000 on it.[29]

Speculation aside, the most significant aspect of Lincoln's trip to Council Bluffs was his unofficial designation of that city as the terminus of a future transcontinental railroad. He was escorted to a high bluff on the north edge of town from where he could see ten miles north, ten miles south, and five miles west across the Missouri River. His comment on this occasion will long be remembered in Council Bluffs: "Not one, but many roads will some day center here."[30] A newly generated interest in railroad projects was evidenced by a boundless energy, and a fight by its citizens, during the mid sixties, to make Council Bluffs the leading railroad center of the West.

II

During the 1860s Council Bluffs asserted itself as a leading

[28]Glenn Chesney Quiett, *They Built the West: An Epic of Rails and Cities* (New York: Cooper Square Publishers, 1965), pp. 3-12.

[29]Grenville M. Dodge, *How We Built the Union Pacific Railway* (Washington: Government Printing Office, 1910), pp. 47-49; Quiett, *op. cit.*, p. 6.

[30]J. R. Perkins, *Trails, Rails, and War* (Indianapolis: The Bobbs-Merrill Company, 1929), pp. 46-50; Carl Sandburg, *Abraham Lincoln: The Prairie Years,* II (New York: Harcourt, Brace & World, Inc. 1926), 200.

outfitting town and looked with scorn on its "paper town" rivals. It was no longer necessary to pamper and lure emigrants. The emigrant population was basically comprised of foreigners and farmers with families. Wagons were loaded with household items and furniture, while farm machinery was attached to the back axle. Stovepipes often penetrated the canvas tops, and the smell of beefsteak at mealtime tantalized passers-by. Cows, calves, sheep, and barking dogs followed behind.[31] "Streets full of wagons and emigrants. Never saw the like. Most all headed for Idaho," wrote Bloomer.[32] Emigrants were able to buy supplies ranging from wagons and oxen to frying pans. Once the emigrants left Council Bluffs, they experienced additional difficulty in traveling. One emigrant on his way to Idaho recalled these hardships:

> . . . getting across the Missouri River from Council Bluffs to Omaha, the kinds and degree of discomfort were unspeakable. . . . The ferry boat was flat, rude, unclean, more like a raft than a boat; the approach to it on the Iowa side was a steep band of sticky, slippery, black mud, down which we all walked or slid—as best we could, our baggage and blankets being pushed or hurled after us in indiscriminate confusion.[33]

Newspaper coverage of emigration was overshadowed by news of railroad advancements and celebrations. However, emigrant guidebooks boasted of Council Bluffs' assets:

> Council Bluffs . . . does as much business as any city on the eastern border of the state containing three times the number of inhabitants. . . . And when one of our heavy houses fails to make sales of a thousand or more dollars a day, the proprietors begin to look blue and say, 'Times are dull—nothing doing.'[34]

In addition to outfitting the overland travelers, Council Bluffs residents worked toward more amitious goals: to become the leading railroad center of the Northwest, serving the needs of larger groups of emigrants moved by trains to western lands.[35]

[31]Oscar O. Winther, *The Transportation Frontier, Trans-Mississippi West, 1865-1890* (New York: Holt, Rhinehart and Winston, 1964), p. 17.

[32]Bloomer Diary, April 25, 1864.

[33]Nathan P. Dodge, "Early Emigration Through and to Council Bluffs," *Annals of Iowa,* Third Series, XVIII (January, 1932), 175-6.

[34]W. S. Burke, *An Outline History of Council Bluffs and its Railroads* (Chicago: Horton & Leonard, 1867).

[35]Levi O. Leonard and Jack T. Johnson, *A Railroad to the Sea* (Iowa City: Midland House, 1939), pp. 83-4.

Residents launched an impressive propanganda campaign, hoping to attract various railroads to the city. They campaigned through two newspapers—the *Bugle* and the *Chronotype,* papers of different political persuasions that nevertheless agreed on the benefits of railway expansion to Council Bluffs. The press capitalized on Council Bluffs' reputation as an outfitting center and suggested that railroads be constructed along this "natural highway" to the west.

As inducements to railway companies, local residents voted bonds, donated land and money, and actually provided free labor to speed the completion of the various lines into their city. The effectiveness of the propaganda combined with financial inducements prevented rival towns from gaining importance as railroad centers. Celebrations and land-breaking ceremonies instilled a high level of interest among the townspeople, and served as springboards for promotion of Council Bluffs. Newspapers publicized these events, attempting to attract the attention of eastern financiers, new residents, and employees. Yet, despite all the talk about railroads, Council Bluffs did not obtain railroad connections before 1867. Until that date it relied on the traditional means of communication and transportation—stages, steamboats, and hacks.

Realization of a transcontinental railroad was an important event in Council Bluff's development. When Asa Whitney, New York businessman and Oriental trader, petitioned Congress in 1845 to construct a transcontinental railroad, the idea had provoked much discussion. Whitney proposed that Congress grant a sixty-mile strip between Lake Superior and Oregon to any company willing to build. Expansionist movements of the forties and fifties, the Mexican War, the discovery of gold in California and Colorado, all contributed to a favorable governmental response.

Among the problems faced by promoters of the transcontinental railroad was the method of financing and selecting a route. Some advocated financing the railroad by private means; others recommended a minimum of government assistance. The magnitude of the project finally convinced railroad promoters that government aid would be an absolute necessity.

Sectional considerations influenced the debate over the route.

Southerners clamored for a route favoring their interests—a line either along the Butterfield Overland mail route or one following the Canadian or Red Rivers. Northerners favored a route through South Pass. They pointed to the commercial importance of Chicago, and the influx of population into Minnesota and surrounding territory, making a northern route more desirable. By 1860, public support for a Pacific railroad was sufficient to warrant a railroad plank in the Republican platform declaring "that the Federal Government ought to render immediate and efficient aid in its construction."[36] It was not until the Southerners left Congress that action on a railroad bill was possible. In the face of an actual war situation, the theory arose that a railroad was a military necessity. Besides, considering the large war expenditures, the cost of financing a railroad seemed minimal.

In 1858, Congress had authorized a committee to study the feasibility of a Pacific railroad. Upon the committee's recommendations, the second session of the Thirty-sixth Congress took action on a Pacific Railroad bill. President Lincoln felt that the Union Pacific Railroad was a military necessity and was essential to keep the Pacific Coast (California) in the Union. The measure finally sent to the President was a compromise because it attempted to satisfy the various special-interest groups and the demands of a growing railroad lobby.

The act stipulated that the government form a Union Pacific Corporation which would be given ten sections of land and receive a grant of $16,000 for every mile of track it laid. The subsidy was increased to $32,000 and $48,000 per mile for foothills and mountains, respectively. The Union Pacific Company was required to raise sufficient capital to build the first forty miles, after which the subsidy would start.[37] The specific location of the line was to be decided by competent engineers when the route was defined in general terms. The President was expected to fix the eastern terminus. Lincoln, recalling his interview with Grenville Dodge at Council Bluffs in 1859, summoned him to Washington for a con-

[36]Jack T. Johnson, *Peter Anthony Dey* (Iowa City: State Historical Society, 1939, pp. 83-4.

[37]Jay Monaghan, *The Overland Trail* (Indianapolis: The Bobbs-Merrill Company, 1947), pp. 393-4. James McCague, *Moguls and Iron Men* (New York: Harper & Row, 1964), pp. 31-33.

ference. After carefully studying a report made by engineer Peter
A. Dey of possible routes west of the Missouri River, Lincoln
asked Dodge, who had conducted similar surveys, to help him
make a decision. Dodge said that "after his interview with me, in
which he showed perfect knowledge of the question, and satisfy-
ing himself as to the engineering questions that had been raised, I
was satisfied he would locate the terminus at or near Council
Bluffs."[38]

Other factors influencing Lincoln's decision are obscure. Lincoln may have been influenced by his friend, Norman B. Judd,
who had nominated him for President at the Republican convention in Chicago. Judd owned real estate in Omaha which he
hoped to develop. In addition, Lincoln held a vested interest in
seventeen lots of land at Council Bluffs, and a railroad terminus
in the vicinity would enhance the value of this property. Lincoln
issued his first order on November 17, 1863:

> I, Abraham Lincoln, President of the United States, do hereby fix so
> much of the western boundary of the State of Iowa as lies between the
> north and south boundaries of the United States township within which
> the city of Omaha is situated as the point from which the line of railroad
> and telegraph in that section mentioned shall be constructed.[39]

What point specifically did the President have in mind? In his
proclamation he may have meant to designate Council Bluffs as
the eastern terminus; however, his description does not specifically define that city. The managers of the Union Pacific favored
Omaha and began utilizing that area for supply depots for the
construction crews and the engineering corps. One historian
pointed out that "even in its youth, Omaha was a real center of
transportation."[40]

After the celebration marking the beginning of construction,
Union Pacific officials asked Lincoln for a new order since the
first order was not specific enough. Lincoln complied with the request and issued the second executive order on March 7, 1864.

> I, Abraham Lincoln, President of the United States, do, upon the appli-
> cation of said company, designate and establish such first-named point
> on the eastern boundary of the State of Iowa east of and opposite to the

[38]Dodge, *How We Built the Union Pacific Railway*, p. 57; McCague, *op. cit.*,
pp. 70-71.
[39]*Ibid.*, p. 51.
[40]Johnson, *op. cit.*, p. 107.

east line of section 10, in township 15 south, of range 13 east, of the sixty
principal meridian in the Territory of Nebraska.[41]

Although the second executive order designated the geographic area where the Union Pacific was to originate, it did not
pinpoint the location. Omaha's claim to the eastern terminus was
backed by Union Pacific officials, while Council Bluffs residents
believed that both executive orders referred to their town without
specifically naming it. Besides, wasn't one of their own residents,
Grenville Dodge, waging a battle on their behalf? Council Bluffs
residents were assured that their city would be the "star of the
Northwest," as its leaders had predicted before the Panic of 1857.

In the period before the Civil war, Council Bluffs residents
had become suspicious of railroads. This suspicion stemmed from
a fiasco with the Mississippi and Missouri Railroad. After the
company's organization in December of 1852, promoters had solicited "local aid." Subsequently, $300,000 was pledged in bonds
to the M & M company. In return, the M & M had agreed to
build east from Council Bluffs, but only four miles were graded at
a cost of $4,000. Soon after, the operations were suspended, and
the remaining $296,000 was never accounted for. The panic of
1857 suspended all operations.

A renewed interest in railroads occurred in the mid 1860s
when residents realized the financial and commercial benefits to be
gained from the railway ties between their city and St. Louis and
Chicago. With the assurance that a transcontinental railroad was
to be constructed west from the Missouri River, eastern railroads
competed to link their lines to the trunk line. Council Bluffs, stra-
tegically located in this respect, plunged into an exhausting
campaign to attract eastern railroads. A railroad convention was
held on May 19, 1858 at Council Bluffs to discuss a year-round
supply route between St. Louis and Council Bluffs and to
promote the construction of a railroad from Council Bluffs to St.
Joseph, Missouri. Four Iowa counties, two Nebraska counties,
and three Missouri counties sent delegates to the convention.
Since it was largely a local effort, financing of the project
depended upon local aid. Residents of Council Bluffs and Pottawattamie County voted bonds and gave the title to a right of way

[41]Dodge, *How We Built the Union Pacific Railroad,* p. 51.

through the county. The company's charter, granted by the State of Iowa in July, 1858, authorized the company to build "from Council Bluffs to some point on the Missouri State line to connect with a railroad from St. Joseph to said line." Unfortunately war abruptly halted building plans.[42]

At a railroad convention in St. Louis in 1865, the directors of the Council Bluffs and St. Joseph Railroad engaged a noted engineer, Willis Phelps of Massachusetts, to complete their line. In order to obtain Phelps' services, the directors practically made him the proprietor of the road.

Early in 1866, the directors of the railroad predicted that cars would begin running to Pacific City, Mills County by July, and that the entire road would be completed by September. The early completion depended in part on the availability of the necessary timber for ties and bridges. The editor of the *Bugle* assisted the work of Willis Phelps:

> Citizens along the line should be liberal towards the contractor in furnishing ties and other timber. They should bear in mind that the high price paid for ties by the Pacific Railroad Company, will not and cannot be paid by the contractor on this road. . . .[43]

Residents proved to be generous and accommodating. The money required for construction—the cost of labor, provisions, and materials—had been raised in Pottawattamie County. From fifty to a hundred teams owned by people in and around Council Bluffs assisted in hauling iron to complete the road. Neighboring residents of Mills and Fremont counties, outside the mainstream of overland travel, were not as helpful.

Despite Phelps' predictions, there was still no locomotive in Council Bluffs or western Iowa by August, 1866. A new promise was made stating that Council Bluffs would be linked to St. Louis, via the Hannibal and St. Joseph Railway by the first of June, 1867. The editor of the *Nonpareil* was jubilant.

> Dinna ye hear the whistle blow? The advance guard of the Council Bluffs and St. Joe Road is only three and a half miles from town. . . . Can't we get some kind of demonstration next week to celebrate the coming of the first train into Council Bluffs? . . . This is an event we have labored and waited and prayed for, these many years, and it should not be permitted now to pass by in silence.[44]

[42]Genevieve Powlinson Mauck, "The Council Bluffs Story," *The Palimpsest,* XLII (September, 1961), 416-417.
[43]Council Bluffs *Bugle,* December 7, 1865.
[44]Council Bluffs *Nonpareil,* December 13, 20, 1866.

Floods and high water during the spring and summer, the slow movement of iron and other supplies from the east, and limited capital meant another delay.

Finally, on January 9, 1867, the Council Bluffs and St. Joseph Railroad was formally opened to Bartlett, Fremont County, about twenty-five miles south of Council Bluffs. A number of residents reveled in their first ride. The editor of the *Bugle* encouraged his readers to take a similar adventure. "It will pay any man who is housed up during the busy months, to take a trip over this road, and look at the vast fields o beautiful, luxurient corn, wheat and oats."[45] With the completion of the Council Bluffs and St. Joseph Railway on August 18, 1868, Council Bluffs was finally linked to St. Louis warehouses on a year-round basis.

Merchants in western Iowa hoped to stimulate competition between merchants in St. Louis and Chicago. Farmers in western Iowa anticipated increased prices for their agricultural products through competitive bidding between the two cities.

At the same time, eastern railroad companies were encouraged to build west quickly to cash in on supplying the Council Bluffs merchants. St. Louis firms were advised to support the construction of rails between that city and Council Bluffs in order to maintain its customers there. Encouragement of this nature was hardly needed, since Council Bluffs was the logical place to link to the Union Pacific Railroad.

Out of a series of mergers, recharters, and consolidations, the Chicago and North Western Railway emerged in June, 1864. Its predecessor, the Lyons and Iowa Central Railway, had disappeared after the panic of 1857. In July, 1864, the Chicago and North Western was authorized to build from Boone to the Missouri River. The directors of this line probably planned to build to Council Bluffs from the beginning, but wanted to obtain favors, so they remained coy. The Council Bluffs press urged Pottawattamie County to offer inducements.

> It is yet an unsettled question whether this road will come down the Boyer to Council Bluffs, or cross the river at De Sota, but we have reason to believe, . . . that if a reasonable inducement is offered by Pottawattamie County the road will make its connection with the Union Pacific at Council Bluffs. This road will reach the Missouri several years in ad-

---

[45]*Ibid.,* August 8, 1867; July 6, 13, 1865; October 31, 1867.

vance of any other coming across Iowa, and that town upon the river
which secures the advantage of its terminus, will derive an impetus
therefrom with which rival points will find it difficult to compete.[46]

Grenville Dodge, chief engineer of the Union Pacific, and
Thomas C. Durant, vice president of the Union Pacific, used
their influence to have the Chicago and North Western build to
Council Bluffs. These Union Pacific officials wanted a speedy
completion of a year-round supply route so supplies need not be
shipped to Omaha via the Missouri River, which was frozen four
months of the year. Railroad equipment from Chicago and other
eastern points could be shipped to Omaha any time by rail.[47]

Officials of the Chicago and North Western came to Council
Bluffs on July 9, 1866 to make a definite proposition. If $30,000
and the right of way through the county were donated, the railway
would guarantee to build to Council Bluffs. During a meeting on
July 9, 1866, this proposition was considered. Stages were erected
at either end of Burhop's Hall, one for the band and the other for
the guest speakers. Railroad officials, John I. Blair and W. W.
Walker, and prominent Council Bluffs citizens were featured
speakers. Marshall Turley headed the list of subscriptions with
his donation of eighty acres of land for a depot and other railroad
buildings. Thirty thousand dollars were donated by various firms:
one business firm donated $2,000; eleven subscribers pledged
$1,000 each; other contributors pledged $100. D. C. Bloomer felt
the large donation was unnecessary, since the railroad directors
had always intended to build to Council Bluffs.[48]

Several weeks after the meeting, the Chicago and North Western advertised in the Chicago *Times* for 5,000 workers to help
complete the line to Council Bluffs. The Chicago *Tribune* predicted that by June 1, 1867, the *Tribune* would be placed on every
breakfast table in Council Bluffs and Omaha on the morning
after publication.[49]

In Sepember, 1867, townspeople witnessed ground breaking

[46]Council Bluffs *Nonpareil,* September 21, 28, 1865.
[47]Stanley P. Hirshson, *Grenville M. Dodge* (Bloomington: Indiana University Press, 1967), p. 158.
[48]Council Bluffs *Bugle,* July 19, 1866; Bloomer, *op. cit.,* XI (April, 1873),
441.
[49]Frank P. Donovan, "The North Western in Iowa," *The Palimpsest,* XLIII
(December 1962), 552-5.

ceremonies for the depot. A similar ceremony the previous January had marked the completion of the line to Council Bluffs. Congratulatory telegrams were read and the director and superintendents of the construction company were honored. Grenville M. Dodge used the occasion to spur the citizens to greater achievements:

> If we but will it and use the ability, energy, enterprise, and capital we have among us, we can within the next two years—if financial matters pursued prosperous, and no great revolution overtakes our country—concentrate here five great trunk railroads, that shall bring to and through us the trade and traffic of the North, East, West and South. I therefore appeal to you, today, to awake from this sleep that has possessed us, and each one and all of us determine from this day henceforth to place our shoulders to the wheel, and use all our ability, capital, and enterprise in building up here a city and a railroad centre, that shall be second to none in the State of Iowa, and which shall be the metropolis of the Missouri Valley.[50]

The Chicago & Rock Island Railroad also began building towards Council Bluffs during the 1860s. It was the parent company of The Mississippi & Missouri Railroad which had been built from Davenport to Iowa City and surveyed to Council Bluffs by January 1856. The Civil War halted construction on the road. In December, 1865, the M & M was sold to the Chicago and Rock Island Company. Since Council Bluffs residents had lost their investment in the 1850s in the M & M, the *Nonpareil* exulted, "There will be a rattling among the dry bones on this line of the M & M." The Rock Island, headed for the Missouri, competed with the Chicago and North Western. In the fall of 1867, construction was booming. Some 1,700 men were working, and the company was advertising for 3,000 more. The following elated report was printed in the *Bugle*.

> We have so much whistling now, from the locomotives of the Union Pacific, Chicago and Northwestern, Council Bluffs and Sioux City, and the Council Bluffs and St. Joe Railroads that when the Chicago Rock Island and Pacific Railroad gets here, we will hardly discover the accession to the whistling.[51]

John F. Tracy, president of the Rock Island, used every opportunity to popularize the road. For instance, the editor of the *Bugle* was invited to ride on the Rock Island special between Des

[50]Council Bluffs *Nonpareil*, January 22, 1867.
[51]Council Bluffs *Bugle*, December 19, 1867.

Moines and Marengo, ninety-two miles east. He boasted about the comfort on the Rock Island.

> This road is one of the best, if not the best in Iowa. It is as smooth and level as a 'house floor,' and being constructed with continuous rail . . . that eternal click and jerk heard and felt on roads constructed with rail connected by 'chains,' is not experienced upon it.[52]

On May 12, 1869, the first train of the Chicago, Rock Island and Pacific entered Council Bluffs. Cheering residents were awed when the Rock Island's sensational "silver" (reportedly nickel-plated) locomotive arrived from the East. This locomotive had been purchased in Paris where it had been the toast of the 1867 Exposition. The fire company, ladies' societies, brass band and artillery squad participated in the celebration. Part of the celebration included laying the cornerstone of the Ogden Hotel. This hotel, famous for its luxury, was named for William B. Ogden, a Chicago railroad financier.

During the first few months of through service to Council Bluffs, the Rock Island won the approval of the traveling and shipping public for its fast schedules and comfort. The "fast Pacific express" made the run from Chicago to Council Bluffs in eighteen hours and averaged twenty-seven and one half miles per hour.[53]

The Burlington and Missouri was the third railroad to arrive at Council Bluffs from the East. In the fall of 1868 its president, James F. Joy, proposed Council Bluffs for its terminus, if the citizens would donate twenty acres of ground for a depot. Nearly a year later construction was only seventy-five miles away and many were sure it would soon lead into Council Bluffs. President Joy had become a heavy stockholder in the line between Council Bluffs and St. Joseph; and the Burlington and Missouri formed a juncton with that line at Pacific Junction, running into Council Bluffs upon the same track. On December 4, 1869, the first Burlington train entered the city.[54]

After the Union Pacific had been released from its obligation to construct a branch to Sioux City, the Sioux City and Pacific

[52]*Ibid.*, April 16, 1868.
[53]Council Bluffs *Bugle,* June 17, 1869.
[54]Bloomer, *op. cit.,* XII (January, 1874), 49.

Railroad Company was organized in August 1864 and began laying track at California Junction. In early 1868 the line was completed into Sioux City. From California Junction the Sioux City
and Pacific Railroad was connected to Council Bluffs by the Chicago and North Western, which later acquired a majority of Sioux
City and Pacific Company stock.

Promoters and residents had reason to believe that Council
Bluffs had succeeded as a town by 1869. The population had
swelled from several thousand in 1859 to 10,020 in 1870. Two
large pork packing plants had been built. A new courthouse had
been erected, and business houses "numbered by the hundreds."
The dream of being a major staging area serving the needs of the
westward-bound emigrants was realized:

> There is no necessity now for talking and writing about Council Bluffs as
> we talked and wrote twelve years ago. The clouds that then overhung our
> destiny have been removed, and the sun of the city's glory is shining fully
> upon us and all we have to do is to direct its rays to our advantage and
> future greatness.[55]

After an exhaustive campaign, the citizens of Council Bluffs had
successfully persuaded five railroads to build to their city. The
dream of some ten years earlier of becoming the "Star of the
Northwest" was realized.

[55]Council Bluffs *Bugle,* January, 1870.

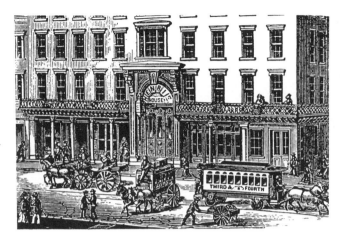

In 1859 the city had several omnibus lines providing a crude form of transportation. Later there were "14 lines of omnibuses and stages running from 5 to 30 miles in the country."

The Miami and Erie Canal was a boom to Cincinnati's commercial development. It carried great quantities of grain, flour, and livestock from inland towns to the city's distilleries, meat packing establishments, and other business concerns. However, the passengers barges operating on the canal had difficulty attracting passengers as they were slow and expensive.

Urban Transport and the Expansion of Cincinnati 1858 to 1920

by Richard Rhoda

The term "urban sprawl" is often used to describe the very rapid spatial expansion of urban areas in the United States, and Cincinnati has been no exception in this process. Even though most urban areas have been expanding tremendously for well over one hundred years, the changes since World War II have received the most attention.

There are several factors responsible for the spread of urban areas two of which are basic and were outlined by John Adams in his model of the residential structure of urban areas.[1] First, the rate of expansion of urban areas is connected to building cycles with the peaks and valleys of these cycles being closely related to economic fluctuations. During economic booms, cities experience rapid expansion as housing is vigorously constructed at the urban periphery. On the other hand, construction activity and the rate of expansion proceed at a much slower pace during economic recessions.

Second, urban spatial expansion is dependent upon the existing methods of intraurban transport. When intraurban travel is slow or difficult, urban activities tend to cluster together, resulting in compact cities with relatively high densities. However, innovations in transportation technology, such as streetcars and automobiles, greatly increase the ease and speed of intraurban travel and contribute to the rapid territorial expansion of urban areas.[2]

Even though it is understood that urban expansion in the Cincinnati area followed economic patterns as mentioned by John Adams, it is especially interesting to review and assess the importance of intraurban transportation on the extension of the Cincinnati urban area during the period from 1858 to 1920.

By modern standards, Cincinnati occupied a very small area in 1858. The city limits were roughly the Ohio River on the east and south, the Mill Creek on the west and McMillan Avenue on the north. With a population of almost 160,000 Cincinnati was the sixth largest city in the United States. Relatively few people lived in the Mount Adams or Mount Auburn sections of the city. The vast majority lived in the crowded basin area where population density approached 30,000 per square mile;[3] for comparison, the 1970 basin density was 7,700 and that of Manhattan in 1860 was 32,000.

The location of employment opportunities and modes of intraurban travel were the major factors responsible for population concentration in this small

131

area. In 1858, almost all commercial and industrial jobs, whether porkpacking, shipbuilding, ironworking, or a variety of other manufacturing activities, were located in the basin area. Most people had to walk to and from work and therefore had to live within a mile or two of their jobs. Consequently, the demand for housing was greatest within walking distance of basin employment opportunities. In response to this demand, builders constructed most housing within the basin area which partially explains why the 1858 population was concentrated there, while the surrounding hills were sparsely populated.

Although walking was the common means of intraurban travel for most Cincinnatians in 1858, a few other travel opportunities were available. The Hamilton and Dayton Railroad enabled a few citizens to commute to Cincinnati from the suburbs of Glendale and Hartwell, but this form of transportation was used only by the wealthy. Various horsedrawn omnibuses which were relatively expensive and not much faster than walking, were in operation. As a consequence they attracted few passengers. For example, the three mile ride from Ludlow and Clifton Streets to Sixth and Main downtown, took an hour and cost twelve cents, an amount equivalent to several dollars today.[4] The passenger barges operating along the Miami and Erie Canal were also slow and had difficulty attracting passengers.[5]

The effect of transportation facilities on the "urban sprawl" of Cincinnati from 1858 to 1920 can be best demonstrated by dividing the period into two basic eras—the period between 1858 and 1888 when the primary modes of intraurban travel were walking and the use of horsecars and inclines, and the period from 1888 to 1920 when electric streetcars were the prime movers of the people. During the former period, the primary mode of travel in Cincinnati was walking even though a number of new means of intraurban transportation became popular. The most important of these was the horsecar which first went into operation along Walnut Street on September 14, 1859.[6] Horsecars were a definite improvement over previous forms of intraurban travel and additional lines were soon constructed. By the end of 1860, five horsecar lines were operating within the basin area and eastward along Front Street (Eastern Avenue) as far as the city limits.

Although horsecars were relatively effective for travel over flat terrain, they had great difficulty surmounting the hills which surrounded the Cincinnati basin area. Extra teams of horses or mules had to be used to pull the cars up the hills and the ascent was very slow. Despite these difficulties, horsecar service was extended to Mount Auburn in 1864 and later lines were extended to Walnut Hills in 1872 and to Avondale in 1874.

The problem of surmounting the hills encircling the basin area was not efficiently overcome until the coming of inclines. A total of five inclines was built in Cincinnati.[7] The first was the Main Street Incline which began to serve Mount Auburn in 1872 followed by the Price Hill Incline two years later. The Elm Street (Bellvue or Clifton) and the Mount Adams Inclines began operating in

132

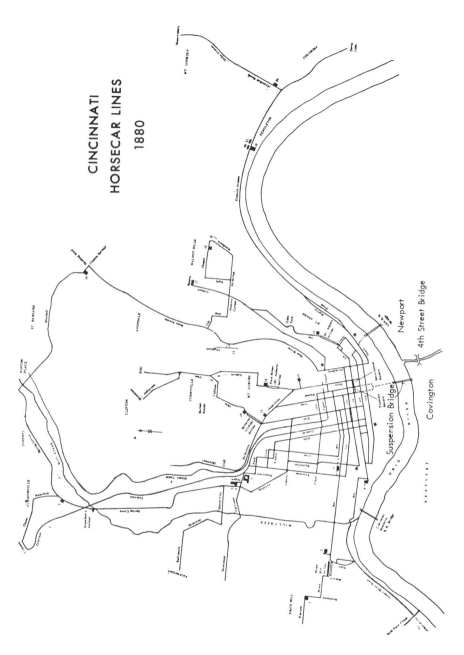

CINCINNATI
HORSECAR LINES
1880

Source: *Roy J. Wright and Richard M.
Wagner*, Cincinnati Streetcars No. 1,
Horsecars and Steam Dummies
(*Cincinnati, 1968*), p. 17.

CINCINNATI
POPULATION GROWTH
1860 — 1890

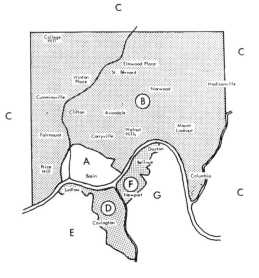

AREA	NAME	POPULATION 1860	POPULATION 1890	CHANGE	% OF TOTAL CHANGE
A	Basin	140,000*	180,000*	40,000*	19.0%
B	Zone Surrounding Basin In Ohio[a]	43,202*	147,496*	104,294*	49.6%
C	Rest Of Hamilton County	33,208	47,058	13,850	6.6%
D	Covington and Ludlow	16,471	39,840	23,369	11.1%
E	Rest Of Kenton County	8,996	14,321	5,325	2.6%
F	Newport, Bellvue, Dayton	10,046	31,860	21,814	10.4%
G	Rest Of Campbell County	10,863	12,348	1,485	0.7%
	TOTAL	262,786	472,923	210,137	100.0%

* Estimate
[a]Nonbasin area of City of Cincinnati; and Storrs, Mill Creek, Spencer, and Columbia (western part) Townships
Source: U.S. Census of Population, 1860 and 1890

1876 and the last incline to be constructed, the Fairview, was completed in 1892.

Horsecar lines were extended from the tops of the inclines to surrounding areas and by 1880 these two modes of transportation served a relatively wide area. Rapid residential development resulted in these areas with the communities of Price Hill, Cumminsville, Clifton, Corryville, Mount Auburn, Avondale, Mount Adams, and Walnut Hills experiencing considerable growth. New residents of these areas could easily travel to and from the employment, business, and cultural opportunities in the basin area.

Besides horsecars and inclines a few other transportation improvements appeared in the Cincinnati area between 1858 and 1888. Construction of the suspension bridge across the Ohio River began in 1857 from the designs of John A. Roebling, famed builder of the Brooklyn Bridge.[8] Completion of the bridge between downtown Cincinnati and downtown Covington in 1867 greatly facilitated travel between the two cities. Another transportation improvement during this era was the construction of the Fourth Street bridge over the Licking River connecting the cities of Covington and Newport.

A steam powered "dummie" railroad, which began operation between Pendleton and Columbia in 1866,[9] was extended to Mount Lookout in 1872, and operated until 1897. However, steam dummies were not an effective means of intraurban travel because they were noisy and relatively dangerous, and consequently had a minor impact on the expansion of Cincinnati.

Cable cars also made a brief appearance in Cincinnati. The first line operated between the basin area and Walnut Hills in 1885. By 1888, two other cable car lines were serving Mount Auburn, Avondale, Corryville, and Clifton.[10] Although these lines were a definite improvement over the earlier forms of intraurban travel, their life was soon terminated by the coming of electric streetcars.

The pattern of population growth in the Cincinnati area between 1860-1890 bears out the fact that improvements in transportation facilities were contributing to "urban sprawl." The Cincinnati urban area included parts of three counties: Hamilton County in Ohio and Kenton and Campbell Counties in Kentucky. The population of these three counties increased by 210,137 between the censuses of 1860 and 1890. Less than ten per cent of this growth was recorded in the predominantly rural portions of the three counties beyond the Cincinnati-Covington-Newport urbanized area. The vast majority of the population growth was absorbed by areas immediately encircling the Cincinnati basin area. Roughly half of the total population growth during the thirty-year period was recorded in the area of Hamilton County which immediately surrounds the basin and includes the nonbasin portion of the city of Cincinnati, as well as communities in Storrs, Mill Creek, and Spencer Townships, and the western portion of Columbia Township. The population of this area more than tripled between 1860 and 1890. The development of horsecar lines and inclines, connecting these areas to downtown Cincinnati, contributed a great deal to this

135

A city ordinance passed on September 3, 1886 granted the cable railway the right to construct and maintain a route beginning at Fourth and Sycamore streets and ending at Rockdale and Main (now Reading Road) avenues in Avondale.

Before the incline planes were erected such suburbs as Mt. Auburn, Mt. Adams, and Price Hill were visited "by few but the residents." The inclines opened these areas to development and many considered them to have "done more real work for the benefit of the city than some half dozen other ambitious schemes. . . . The planes bring the country . . . to the very gates of the city and are the natural supplement of a . . . system of street passenger cars."

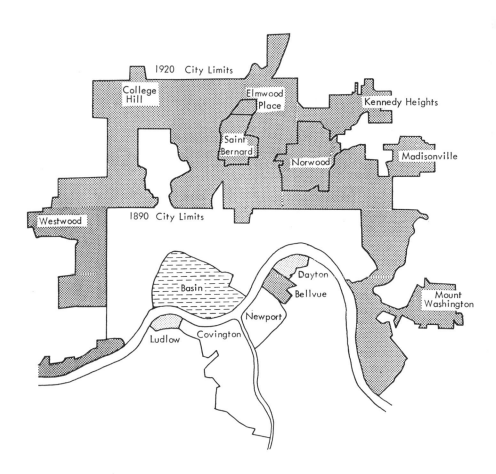

CINCINNATI
POPULATION GROWTH
1890 — 1900

Over 80%

30% to 80%

0% to 30%

Decline

1920 City Limits

College Hill

Elmwood Place

Kennedy Heights

Saint Bernard

Norwood

Madisonville

Westwood

1890 City Limits

Basin

Dayton

Bellvue

Mount Washington

Newport

Ludlow

Covington

rapid growth as did the development of industries in the Mill Creek Valley.[11]

Communities in Kentucky directly across the river from the basin area also experienced rapid growth. These communities absorbed over one-fifth of the total population growth in the three county area. The combined population of Newport, Bellvue, and Dayton more than tripled while that of Covington and Ludlow more than doubled. A substantial portion of this growth can be attributed to the construction of bridges which greatly increased accessibility to downtown Cincinnati.

Although most of the population growth was absorbed by communities surrounding the basin area, this process of decentralization could not keep pace with overall population growth. The basin area was forced to absorb an additional 40,000 residents during this period which increased the population to 180,000 by 1890. The density, which approached 40,000 per square mile, was almost twice as high as the most crowded areas of the city in 1970. Crowding and congestion had noticeably increased which is not surprising because even though transportation improvements were made during the period, walking was still the dominant form of intraurban travel, especially for the working class. Jobs continued to be concentrated in the basin and incoming working class families very often acquired housing within walking distance of their employment. This resulted in increased crowding and congestion, approaching intolerable levels.[12]

The problems of crowding and congestion in the basin area of Cincinnati were not alleviated until the coming of electric streetcars which characterized the second transportation era from 1888 to 1920. They were so much faster, quieter, safer, and cheaper than earlier forms of transit that they were an immediate success. The first electric streetcar operated along McMillan Street in Walnut Hills in 1888 and by 1894 they served all areas of Cincinnati as well as many surrounding communities.

Streetcars greatly increased the accessibility of areas on the urban periphery and contributed to the rapid growth of these regions. Developers subdivided and built voraciously during the 1890's. Of the 343 subdivision plots filed at the Hamilton County Registry of Deeds between 1890 and 1922, fully one third were filed during the 1890-1893 building boom.[13] Most of these new subdivisions were located along streetcar lines.

The impact of the electric streetcars on the spatial expansion of Cincinnati can be assessed by investigating the pattern of population growth during the streetcar era. Between 1890 and 1900 many basin area residents moved to surrounding communities resulting in a population decline of almost 10,000 during the decade. Nonbasin areas within the 1890 city limits registered a small population increase of about six per cent between 1890 and 1900. This slow rate of growth can be attributed to the fact that most suitable residential land in this area had already been developed by 1890. A lack of available land also limited the population growth during the decade within the city limits of

139

Covington and Newport.

On the other hand, areas immediately beyond the 1890 city limits of Cincinnati, Covington, and Newport experienced relatively rapid growth during the 1890's. The populations of Ludlow, Dayton, and Bellvue, Kentucky increased by thirty-five per cent, forty-three per cent, and one hundred per cent respectively. The most rapid growth during this decade was recorded in the region just beyond the 1890 Cincinnati city limits where the population increased by over 110 per cent.[14] Streetcars are largely responsible for this growth because they expanded the accessibility of this area to other parts of Cincinnati. Employees could easily travel from their homes beyond the 1890 city limits to their jobs located elsewhere in the urban area.

The most spectacular growth during the 1890's was registered in Norwood. Its population jumped from less than 1,000 in 1890 to 6,480 in 1900. Although streetcars contributed to this very rapid growth, the main impetus was industrial development. The completion of the Cincinnati, Lebanon and Northern Railroad through Norwood in conjunction with low tax rates and adequate water supply brought numerous industries and factory workers to the small city and caused the population increase.[15]

It has been argued that streetcars stimulated the growth of certain areas by increasing their accessibility. If this argument is valid, growth during the streetcar era should be concentrated along the streetcar lines. To test this assertion, the pattern of population growth for relatively small spatial units needs to be analyzed. Unfortunately, the lack of appropriate data for the early years of the streetcar era makes this type of analysis very difficult. On the other hand, census tract data compiled by Quinn, Eubank, and Elliot make a relatively crude analysis possible for the last twenty years of the streetcar period.[16] Although these census tract data lack a desired degree of spatial detail, they can be used to make a rough assessment of the impact of streetcars on the distribution of population growth.

Study of streetcar lines (circa 1911) and rapid growth areas between 1900 and 1920 indicates all rapid growth areas were well served by streetcars except for Kennedy Heights and Mount Washington which were connected to downtown Cincinnati by interurban railways. This provides further support for the argument that streetcars stimulated the growth of certain regions by expanding their accessibility. Even though rapid industrialization in areas such as Norwood and Oakley was probably the major reason for their rapid growth between 1900 and 1920, it should also be pointed out that the development of streetcars also contributed to these communities' growth.

The territorial expansion of the Cincinnati urban area between 1858 and 1920 was closely associated with intraurban transportation improvements. Horsecars and inclines facilitated urban expansion from 1858 to 1888. Electric streetcars were largely responsible for the territorial growth of the region between 1888 and 1920. The continued expansion of the Cincinnati urban area

Even though horsecar lines extended to Mt. Auburn in 1864 and later to Corryville and Avondale, residential development of these areas remained sparse for the horsecars had great difficulty surmounting the hills which surrounded the Cincinnati basin area.

Electric streetcars expanded the accessibility of the basin area to other parts of the city. Employees could now live beyond the city limits and commute to their jobs located downtown or elsewhere in the metropolitan region.

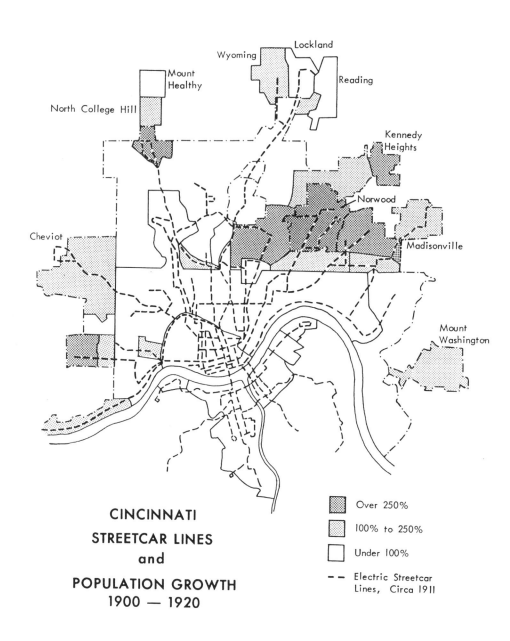

Wyoming
Lockland
Reading
Mount Healthy
North College Hill
Kennedy Heights
Norwood
Cheviot
Madisonville
Mount Washington

CINCINNATI
STREETCAR LINES
and

POPULATION GROWTH
1900 — 1920

Over 250%

100% to 250%

Under 100%

Electric Streetcar
Lines, Circa 1911

in the years since 1920 is also closely related to innovations in intraurban transport. The "urban sprawl" experienced during this period can largely be attributed to the increased use of private automobiles in conjunction with road improvements and the development of the freeway system in the Cincinnati metropolitan area.

RICHARD RHODA, a Ph.D. candidate in Geography at the University of Iowa, has worked for the federal Environmental Protection Agency and the Agency for International Development. He recently returned from a four month assignment in West Africa.

(1) John S. Adams, "The Residential Structure of Midwestern Cities," Annals, Association of American Geographers, LX (March, 1970), 37-62.

(2) For a very interesting study of the role of streetcars in suburban development see Sam B. Warner, Jr., Streetcar Suburbs— The Process of Growth in Boston (Cambridge, Mass., 1962).

(3) For the purpose of this study, the basin area is defined as the following 1940 census tracts: 1, 2, 3, 4, 5, 6, 7, 8, 9, 10, 11, 14, 15, 16, 17, and 91. This area is roughly bounded on the south by the Ohio River; on the west by State Avenue; on the north by Harrison Avenue to McMicken Avenue, along McMicken to Liberty Street, and along Liberty to Gilbert Avenue; on the east the area is bounded by Gilbert Avenue to Eggleston Street and along Eggleston to the Ohio River just east of the L & N bridge.

(4) Arthur G. King, "Historic Clifton," The Cincinnati Historical Society (Cincinnati, n.d.) mimeo.

(5) Writers' Program, Cincinnati, 393.

(6) Roy J. Wright and Richard M. Wagner, Cincinnati Streetcars No. 1, Horsecars and Steam Dummies (Cincinnati, 1968).

(7) Roy J. Wright and Richard M. Wagner, Cincinnati Streetcars No. 2, The Inclines (Cincinnati, 1968).

(8) Joseph S. Stern, Jr., "The Suspension Bridge: They Said It Couldn't Be Built," Bulletin, Cincinnati Historical Society XXIII (October, 1965), 211-228.

(9) John H. White, Jr., "By Steam Car to Mt. Lookout: The Columbia and Cincinnati Street Railroad," Bulletin, Cincinnati Historical Society, XXV (April, 1967), 93-107.

(10) Roy J. Wright and Richard M. Wagner, Cincinnati Streetcars No. 3, Cable Cars and the Earliest Electrics (Cincinnati, 1969).

(11) Zane Miller, Boss Cox's Cincinnati (New York, 1968), 25-27.

(12) Ibid., Chapter One.

(13) City Planning Commission, The Official City Plan of Cincinnati, Ohio (Cincinnati, 1925).

(14) This area includes the parts of Cincinnati between the 1890 and 1940 city limits as well as Norwood, Saint Bernard, and Elmwood Place.

(15) Writers' Program, Cincinnati: A Guide to the Queen City and Its Neighbors (Cincinnati, 1943), 326-328.

(16) James A. Quinn, Earle Eubank and Lois E. Elliot, Population Changes— Cincinnati, Ohio, and Adjacent Areas 1900-1940 (Columbus, Ohio, 1947).

143

When Milwaukee Streetcars Were Horse-Drawn

By Herbert W. Kuhm

WHILE Milwaukee was relatively small and compact, transportation posed no serious problem. Those not affluent enough to own or hire a horse and carriage used "shanks' mares," which is to say, they simply walked. But as the city grew, its burgeoning populace of almost 45,000 in 1860 reached for more *lebensraum*, and with this came the need for some form of urban transportation. As if in answer to this, horse-drawn streetcars, running on iron rails, came into being. One east coast mayor, in an euphoric overstatement called such transportation "a phenomenon that will go down into the history of our country as the greatest achievement of man."

And so it may have seemed to the eager crowds that lined the streets of downtown Milwaukee on May 30, 1860 to witness the first of many horsecar trips, which eventually would develop into an extensive rapid transit network. An enthusiastic response from the public prompted the laying of additional rails and the purchase of more cars within the year.

Actually, Milwaukee was slow to adopt the then new horsecars, the first of which had appeared as early as April, 1831 in New York City. Two years later New Orleans became the second city to have a street railway, followed then by Boston, Philadelphia, Baltimore, Pittsburgh, Cincinnati and Chicago.

Despite the regular use of such streetcars in other cities, it was not until the 1860s that Milwaukee established its own horse railroads. Interest in this modern type of city transportation became a reality when, in the summer of 1859, George H. Walker, Dr. Lemuel W. Weeks, Walter S. Johnson and F. A. Blodgett received city council permission to lay tracks. Thus was the beginning of the River and Lake Shore City Railway Company. This company's rolling stock was franchised to operate on any thoroughfares in the area bounded on the west by the Milwaukee River, on the south by the Menomonee River, the lake to the east and city limits (East Kenwood Boulevard) to the north.

By May, 1860 most of the Lake Shore railway lines were laid and ready. Two cars had been ordered from the Kimball & Gordon Works in Philadelphia. These buff cars, with light blue panels, arrived in Milwaukee on May 27 on the brig *D. Ferguson.* Three

30

118

This photo of East Wisconsin Avenue, taken about 1885, illustrates that horse-drawn vehicles, including streetcars, were a common means of transportation for many residents.

days later on May 30, four horses pulled each car, made their way along East Water Street (North Water) from the Walker's Point bridge on the Menomonee River to Division Street (East Juneau Avenue), and then proceeded east to Prospect Avenue. In order to serve the Civil War volunteer camp at Lafayette Place and Prospect, the railway was extended up Prospect. With the end of the war in 1865, horsecar service terminated at Albion Street.

The Lake Shore Railway also had a branch that extended from the corner of East Water (North Water) and Wisconsin east to Jefferson. Cars then proceeded north to Biddle (Kilbourn), east to Van Buren, and again north to Division (Juneau). The last leg of the route ran from Division east to its turntable at Prospect and Albion.

The immediate success of the River and Lake Shore line caused a wave of petitions for additional charters to deluge the common council. The West Side Passenger Railway Company was incorporated in November, 1859 and expected to serve the area bounded by Burleigh Avenue on the north, North Thirty-fifth Street at its most western edge, West Burnham to the south and to the east the Milwaukee River and the Lake. However, its actual development collapsed in ward politics. Although the West Water Street Horse Railway Company and the Cold Spring City Railway Company also sought charters, the outbreak of the Civil War postponed their construction.

Residents of the west side continued to clamor for street railway facilities. Finally in March, 1865, the Milwaukee City Railway

31

Company was incorporated by John Plankinton, Frederick Layton, Samuel Marshall, Charles F. Ilsley and Walter S. Johnson. They immediately took over the Lake Shore Railway which was then experiencing financial problems. Within a short time, the west and south sides were being served by a network of street rails. In May, 1866 tracks over the Oneida (Wells) Street bridge connected the old Lake Shore lines on the east side and the Milwaukee City's new rails on the west.

The late 1860s saw a downturn in profits for the Milwaukee City. In 1872, the entire operation was sold to Isaac Ellsworth, a director and superintendent of Milwaukee City, who reorganized the company to a once again profit-making venture.

Ellsworth concentrated the streetcar system on the west side of the Milwaukee River. He removed the rails from East Water (North Water) Street and installed them on West Water (Plankinton). From there the network spread to the south and west sides of the city, cutting across Chestnut Street (West Juneau Avenue), State, Seventh and Eighth, Washington, Walnut, West Water, Third, Reed (South Second), National and Fourth Avenue (South Ninth). By 1880 the company consisted of thirty-two cars, 250 horses and eighty men, and owned fourteen and one-half miles of rail.

Unable to expand the system due to a lack of funds, Ellsworth sold his interests in 1881 to Peter McGeoch. He in turn built and operated the streetcar lines on Eleventh (South Sixteenth) and Muskego Avenues, and another to Forest Home Cemetery. McGeoch extended the Third Street line north to the city limits at Burleigh, and the Chestnut (West Juneau) route was advanced to Twenty-seventh Street. The Walnut Street branch was extended, and a new line was laid across the Martin (State) Street bridge, running from Broadway to Michigan Street and on to the Chicago and Northwestern Railway depot at the lakefront. In December, 1888 the railway was purchased by a Wall Street syndicate and reorganized as the Milwaukee City Railroad Company.

On June 1, 1874 the Milwaukee Common Council granted a franchise to John Plankinton, Sherburne S. Merrill, Samuel M. Green, Stephen A. Harrison and John H. Tesch under the name of West Side Street Railway Company. The following year the railway was incorporated and took over the city's original grant for operation from the original owners. Five horsecars were put into operation on the company's opening, Thanksgiving Day, 1875. The route consisted of double tracks between West Water (Plankinton) and Eleventh Streets on Grand (Wisconsin) Avenue, on Eleventh to Wells, and from Wells to Twenty-second Street. A single track had been laid on Wells between Twenty-second and the city limits at Thirty-fourth Street. In 1876 Washington Becker took control of the company and expanded its service lines. By 1878 another track was laid along the already existing rails on Wells Street from Twenty-second to Thirty-fourth Streets, and two cars were added to the stock. Another double track was laid in 1879

32

on Twelfth from Wells to Walnut Street, and an eastern branch stretched from West Water (Plankinton) on Wisconsin to the lake. The company then owned twelve cars. In 1880 the Walnut Street line was extended to Twentieth and the Twelfth Street line was advanced from its Walnut Street terminus to the intersection of Beaubian Street (Garfield) and Teutonia Avenue. By 1881 the company operated five miles of double track, employed forty-five men and owned 160 horses.

Speculating on the future potential of electric traction, Becker reorganized his company as the West Side Railway Company in 1888 and applied to the common council for a franchise amendment to use such power.

After Isaac Ellsworth had removed the Lake Shore rails from the east side streets in order to concentrate on more profitable west side routes, efforts were begun to establish another railway system for residents of the old Juneautown area. Winfield Smith, Christian Preusser and Ferdinand Kuehn were actively involved in such promotion. In the summer of 1874 the Cream City Railroad Company was incorporated by Frank B. Van Valkenburgh, Christian Preusser, Daniel Schultz, James B. Turek and Lewis Duerr. The first branch of this horsecar system ran from the foot of Water and Mason Streets to Farwell Avenue and Brady Street, the site of its car barns. By December, 1876 an extension was made from Division (East Juneau) to Kinnickinnic, following East

Isaac Ellsworth bought the Milwaukee City Railway Company in 1872 and made it a profit-making enterprise.

33

This streetcar firm was reorganized as the Milwaukee City Railroad Company in 1888. Today the streets are paved and mass transportation now has motorized its "horsepower."

Water (North Water) and Clinton Street (South First). A car barn was established at the end of this line.

Routes were further expanded when, in 1879, Cream City purchased the lines of the Forest Home Railroad Company. The latter operated steam-driven cars from Clinton (South First) Street to Forest Home Cemetery during the summer months. The locomotive was disguised with a decoy stuffed mule so as not to frighten horses. With this purchase, Cream City doubled its track and substituted horse power for the steam engines. The company's rolling stock then consisted of fifty cars; its barns held 200 horses, and it employed fifty to seventy-five men. Cream City was an innovator among its peer railways because it was the first to use girder rails, automatic switches and heat its cars in the winter.

In October, 1889 the company amended its franchise in order to use electric power, but it was sold in April, 1890 before any work could be started. The company was reorganized as the Cream City Railway Company. After five months of construction problems, it was sold to the North American Company, which transferred the entire operation to the Milwaukee Street Railway Company.

As the city grew, so did the horsecar companies. By 1889 a total of 41.42 miles of rail had been approved by the city for the three major companies. Milwaukee City was the largest and oldest of these rapid transit companies and owned 17.86 miles of track. Second in rank was the Cream City with 14.56 miles. And, trailing behind was the West Side Company with only nine miles of rails. However, this latter company's horsecars traveled in a densely populated area of the city and its routes were the most profitable of the three.

34

The earliest horsecars consisted of a closed vehicle with a single passenger compartment which provided longitudinal seats for about twenty to thirty passengers. The seats were usually of molded three-ply wood veneer, covered with rattan.

One horse was able to draw a small single-ended car. There was a platform for the driver at the front, but passengers entered from the rear. Some cars had a full platform at the rear, but more often only a small step which led to the rear entrance door was provided. Because these smaller cars were usually operated by a driver only, passengers walked from the rear entrance to the front to pay their fares. Later, cars came equipped with a device that allowed coins to be deposited at the side of each seat. The change would then roll to the front to be received by the driver. More heavily traveled passenger routes usually used larger cars which were pulled by two horses. These vehicles had full platforms at the front and rear, and were manned by a driver and conductor.

Car ventilation, especially in the cold weather, was poor, and the air malodorant. During the 1860s car manufacturers began to adopt the monitor roof which had a raised deck running the length of the car body and low windows in the sides of the deck for better light and air. Horsecars had a distinctive odor, a blend of horse sweat, floor straw, and smoky coal oil lamps. Rear platforms held those riders who wanted to smoke in transit.

Streetcars were in great demand by workers going to or returning from work, and the passengers' comfort or the pulling power of the horses was often disregarded. It was not unusual for horsecars during the rush periods to be overloaded with passengers packed in the car's interior and clinging to the front and rear platforms. In inclement weather, the cars were jammed with riders unmindful whether the horses were equal to the load.

Horses were expected to work ten to fifteen miles a day, and they managed to pull the cars at an average speed of five to six miles per hour except on lines that were hilly. These lines were often called "horse killers." For example the Third Street line had a steep incline from Cherry to Galena at the Schlitz brewery. Others, equally steep, were Winnebago Street from Seventh to Eleventh; State from Sixth to Eighth Streets; Wisconsin Avenue from East Water (North Water) to Broadway, and Chestnut (West Juneau) at the Pabst brewery from Seventh to Ninth Streets. In icy or snowy weather, extra horses were stationed at the bottom of these hills to aid in the car's ascent.

The ratio of horses to cars hauled was often eight to one. Horses were stabled for over nineteen hours each day and usually rested one day in seven. Their food consumption was about thirty pounds of grain and hay each day. These animals were expensive as a source of power in that they worked only a limited number of hours a day. Street railway work was hard on horses, and most were good for only three to five years of service before they were sold for less arduous work. Many ended up in a tannery or glue factory.

35

Rapid transit companies had a tremendous investment tied up in their animals and the cost of maintaining their extensive stables was high. The Cream City had its barns at Brady and Farwell as well as at Mitchell and Kinnickinnic. The West Side line operated two stables on Wells Street. One was at Twenty-second Street and another between Eleventh and Twelfth Streets. The latter was made of brick, 80 by 125 feet, and three stories high. The structure was equipped with water and gas, and had enough space for ninety horses and twenty cars. On West Water (Plankinton) the Milwaukee City railway operation had its car barns.

Extreme weather conditions affected all the horsecar lines. Horses tired easily during hot weather and more frequent relays were required. But in winter, snow and ice imposed even greater difficulties. Such conditions were conducive to car derailment due to ice on the track, and slippery roadways were dangerous in that horses might fall and break a leg. When the tracks were covered with ice, it was not unusual for the male passengers to alight and push the car, or merely lighten the load for the horse.

In pre-horsecar days, snow removal was virtually unheard of. It was just packed down by the traffic and ignored. Sleigh drivers viewed horsecars on the streets as an invasion of their rights. The car rails were thin bars of iron laid on top of crossbars of wood. This made sleighing hazardous because the runners could get caught in the rails and tip the vehicle. However, snow packed streets derailed the streetcars. "Snow removal" in the 1870s and 1880s grossly overstated the operation. The snow was merely shifted by cumbersome eight-horse plows and banked a few yards from the tracks.

The Milwaukee City Railway's main barn and office was on Plankinton between Wisconsin and Wells, while these large barns were on Second Street just north of Wisconsin. This scene dates back to about 1880.

36

Washington Becker successfully initiated the first electric streetcar in Milwaukee in 1890.

Riding the horsecars in winter held few charms, for the heating of the cars was a problem. Prior to kerosene stoves, derisively called "stinkpots" by the riders, the floors of the early cars were simply strewn with hay to about midcalf, and it was not unusual for wisps of hay to cling to trouser legs or sweeping skirts as passengers left the cars.

Summer was a particularly good time for "Sunday riders" who took the horsecar to some attractive picnic spot at the end of the line. The people aboard reflected a mood wholly different from the workday riders. Everyone was decked in their "Sunday best," with the ladies in their peekaboo shirt-waists and flower-rimmed hats, the men in striped seersucker suits and straw hats with gayly colored bands. Each group had its wicker basket or shoebox of sandwiches, along with Brownie camera, baseball bats and mitts, and the ubiquitous mandolin or accordion. Other Sunday riders favored the large cemeteries such as Forest Home, Calvary or Union, which were located at the then outskirts of the city. Horsecars took the tediousness out of the journey to the family plot which, in those days, was usually planted with an abundance of flowers needing frequent trimming and watering.

The growth of the city demanded more and faster modes of transportation than the horsecar could provide. In April, 1890, Washington Becker of the West Side Railway Company proved the merits of electric traction in successful experiments on his Wells Street line. The time was ripe for the electric trolley and the handwriting was on the wall for the horsecar.

Although the horsecar was not as first predicted, "the greatest achievement of man," it was an important factor in the community's growth. The horse railroad systems allowed workers to commute several miles to work, and thus helped to change the containment of cities and the habits of workers.

37

STREET RAILROADS IN COLUMBUS, OHIO,
1862-1920

LARRY J. SAYLOR

The street railroad became the dominant mode of transportation in virtually all American cities of more than 10,000 population over a fifty-year period beginning around 1860. During this era, one of rapid urban growth, the physical and economic conditions of street railroads were closely involved with those of the cities. Such was the case in the history of Columbus, Ohio. In the time between the start of its first line in 1862 and the decline of the subsequent system in the 1920's, Columbus grew from a population of 18,629 to one of nearly a quarter of a million and extended its boundaries from ones about ten city blocks beyond those of the 1834 charter to ones approximating the built-up area of the city in the 1970's. From the original horse-drawn line of about one and a half miles, the city's street railroad network grew to include 133 miles of track traversing seventy-one miles of routes in 1917. In 1920, 140 interurban trains each day departed or arrived over 767 miles of electric interurban lines radiating in all directions from the city and in part operating over its streets.[1] This study will trace the development of the street railroad system in Columbus, including the effects of governmental control on its form, operation, and economy, its relation to the growth of the city, and its response to technological and cultural change.

I. The Beginning

Prior to the advent of the street railway, transportation within Columbus was largely provided by animals and private wagons, and no local public transportation system appears to have existed. Interurban stages, however, would stop at any point on their routes through the city to pick up and discharge passengers. In 1832, the first steam railroad had entered the city, and by 1850 several more had joined it. The arrival of the Columbus and Xenia Railroad and the construction of the city's first passenger station in 1850 created demand for some type of local transportation to this connection. In 1853, the first horse-drawn omnibus service to and from the station was initiated. Routes to Worthington and Canal Winchester were begun soon thereafter. Hackney coaches, omnibuses, and pigmy omnibuses served the still relatively small-scale transportation needs of the city until 1863.[2]

The first experimental horse-drawn street railway line had been installed in New York City in 1831, and the first permanent line in that city in 1852. The idea spread rapidly to other cities in the remainder of the 1850's.[3] In 1854, the first ordinance authorizing construction of a street railway in Columbus was passed by the city council, but no action was taken toward construction and the ordinance lapsed. Then on 11 November 1862, an ordinance was passed chartering the Columbus Street Railroad Company and empowering it to build tracks and operate in the public right of way on High Street, State Avenue, and Town Street. The tracks on High were to extend from North Public Lane (Naughten Street) to South Public Lane (Livingston).[4] The ordinance specified the route and established a time limit within which the line must be in operation or the franchise forfeited, but little else was included. It did not seek to regulate fares or schedules, require the company to contribute to street maintenance, or hold the city immune from suit in connection with the operation of the line; but all of this and more was to come later. The initial years of the Columbus Street Railroad were prosperous ones because of the Civil War and its free-spending soldiers, and the company initially did well. Tracks were extended south and north on High Street, and cars were leaving each end of the line every six minutes. The original car barn and stable were erected on High Street north of Goodale, which was then far out in the country.[5]

As early as 1863 the council began to regulate the operation of the lines. The ordinance of 29 June 1863, authorizing several track extensions, included a time limit for construction and made this extension conditional on the acceptance of various regulations by the company, to apply to both the existing and the proposed lines. Plans for all construction were to be submitted to the city civil engineer and reviewed by him; ac-

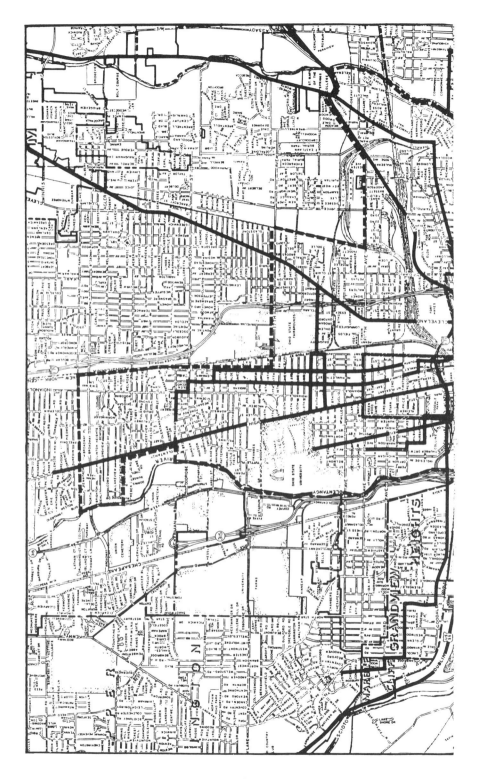

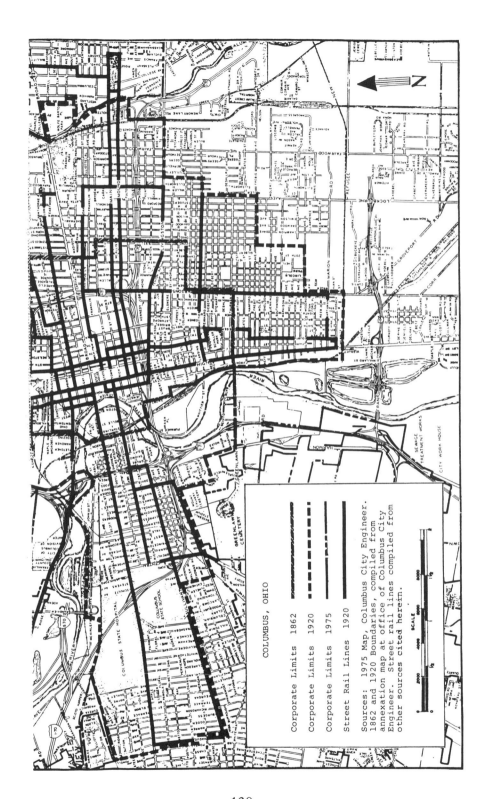

COLUMBUS, OHIO

Corporate Limits 1862	———
Corporate Limits 1920	—·—·—·
Corporate Limits 1975	▬ ▬ ▬
Street Rail Lines 1920	▬▬▬

Sources: 1975 Map, Columbus City Engineer.
1862 and 1920 Boundaries, compiled from
annexation map at office of Columbus City
Engineer. Street rail lines compiled from
other sources cited herein.

tual work was then to be supervised by the engineer to insure proper grades, drainage, and free traffic movement. A fine of fifty dollars per day was to be imposed for any impediment to free use of the streets or obstruction to water flow caused by the railroad. The city also reserved the right to stop the operation of any line until such problems were corrected. This ordinance was accepted by the company, and its provisions were included in all subsequent legislation affecting street railroads.[6]

In April 1867, the company sought further extensions and was granted them, along with additional regulations. The city reserved the right to remove track if necessary to effect street or utility repairs; the proviso added that "if possible" this would be accomplished without interfering with the schedule of the lines. The city was also to be held not liable for any damage to the tracks due to broken water, gas, or sewage pipes, for changes in grade of the street, or for other improvements or maintenance by the city. The company was to be held liable for any damages caused by its operation or maintenance of its lines under this franchise. The company was also bound to obey all current and future regulations passed by the council, which at that time prohibited stopping in a cross-walk except briefly to load or discharge passengers, allowing women and children to enter or leave a car while it was in motion, and standing and blocking traffic. Other regulations required the company to maintain 100 feet between cars traveling in the same direction and to limit motive power to horses and mules. Street cars were granted right of way over other vehicles on their tracks, but only after first giving ample warning. A twenty dollar fine was to be imposed for any violation of these regulations, and the party in violation was not immune to civil suit. The company was also ordered to pay a yearly franchise fee of five dollars on every car operated. If the company acquired any other street railroad company, all of these provisions were to apply to the operation of both. The fare was fixed at seven cents or five tickets for twenty-five cents; and cargo other than passengers was limited to baggage, for which proportionally extra fare could be charged. The franchise was granted for a twenty-year term. This ordinance of 1867 set the precedent for and tone of all future street railway franchises in the city of Columbus. Later in 1867, when the city determined to repave High Street, the company was required to pave the street between its tracks to match the pavement of the remainder of the street, and was required to repave that part of any street from which it removed its tracks.[7]

Following the Civil War business had fallen off, and the $5,700 assessment resulting from the High Street improvements nearly forced the company to give up the business. But substitution of lighter rails and one-horse cars for the heavier tracks and two-horse cars formerly in use enabled the company to make up the loss within three years. By 1874, however, public agitation against the company's poor maintenance of streets and its irregular service on South High nearly resulted in the repeal of its charter. Sale and reorganization of the company mollified public opinion for a time, but the company's difficulties were to reappear in later years. The Columbus Street Railroad Company was soon joined by the North Columbus Railroad Company, which constructed a line from the northern terminus of the existing High Street line to north Columbus. The Friend Street Railroad Company was created in 1868, followed by the Oak Street Railway in 1872. Both of these lines extended to the east from downtown, the former as far as Franklin Park, then the fairground. The Greenwood and Green Lawn Railroad Company, chartered in 1872, built a line across the new Broad Street bridge and on Broad to the west corporation line at Souder Avenue, then south to the Green Lawn Cemetery outside the city. The company was also required to comply with State Board of Public Works requirements on the National Road. Thus by 1875 lines radiated from the Statehouse in the four compass directions to the edges of the existing city. The franchises granted these new companies incorporated all of the regulations previously imposed on the Columbus Street Railroad Company and added several provisions not yet affecting the older company. It was specifically stipulated that the city bore no responsibility to maintain the street railroad lines, and the speed limit for all cars was fixed at six miles per hour. In addition, a new state law required the written approval of the owners of the greater part of the street frontage along proposed lines before such lines could be approved by the city council; these approvals were solicited by the companies and filed with the city clerk.[8] While opposition to the extension of street railroads had sometimes been generated in the more conservative Northeastern cities, there is no record of similar feeling at this time in the growing Western city of Columbus. Most owners were eager to realize the increase in land value brought about by proximity to a street railway, and the companies experienced little or no difficulty in obtaining the required signatures. What adverse public sentiment did develop appears to have been limited to objection to the sometimes slipshod maintenance and operation of existing lines, and not to their initial extension.

In November 1874, despite some lingering dissatisfaction with the company, the Columbus Street Railroad Company was authorized to extend a line up Neil Avenue to what was then the agricultural college. At this time all the recent restrictions were imposed on the new line, and the cars were additionally required to have signal lights at night. Fares were continued at five tickets for twenty-five cents, and the company was allowed to charge no additional fare on the extension. For the first time the city also regulated the schedule. Cars were required to operate at half hour intervals on the Neil Avenue line by day, and by night as directed by the president of the college. In June 1875, as part of an ordinance authorizing a track extension, the State and Oak Street Railway was for the first time required not only to pay for street paving between its tracks but also to reimburse property owners along the line for part of the assessment previously paid by the property owners for street paving, the amount paid to be proportional to the amount of the street occupied by the tracks.[9] The fact that this stipulation in effect required the company to pay twice for street pavement seems not to have deterred the city council and certainly did not displease the property owners.

In July 1875, the council authorized the East Park Place Company to lay tracks parallel to those of the Columbus Street Railroad on High Street. The latter company threatened an injunction, and to counter this threat the Park Place Company marshaled a large labor force and began construction in the middle of the night. As a part of this ordinance the company was specifically required to keep its pavement in good repair and was not allowed to charge a fare on the line greater than that charged on any part. The company at that time charged a double fare to the fairgrounds, a practice which was stopped in October 1875.[10] This and subsequent legislation enforced the council's strict opposition to a zone system of fares.

A tunnel under the tangle of railroad tracks on High Street at Union Station was completed in December 1875, and the Columbus Street Railroad Company was authorized to lay tracks through it with the provision that any other company would be allowed to use those tracks upon payment of just compensation to the company. The company was also required to remove all surface tracks over the tunnel and was limited to a five cent fare over its entire line.[11] This ordinance was the first requiring one company to allow another to use its tracks; court cases at that time had determined that a company could not demand compensation based on loss in revenue due to competition caused by joint use of lines.[12] Together with the authorization of parallel competing tracks on High Street, this ordinance signified the arrival of the period of greatest competition in Columbus street transportation.

An 1876 ordinance allowing the North High Street Railroad to extend its tracks from the old north corporation line at Fifth Avenue to the new one at Arcadia Street included new language dealing with health and safety, requiring that "no car or cars shall be used on said railroad if so worn out, broken, or so constructed as to imperil the lives, limbs, or health of passengers; but such cars must always be subject to the satisfaction and approval of the sanitary inspector, or health officer of said city."[13] All subsequent ordinances treated the matter in similar language. This ordinance also permitted the North High Company to use the North High Street tunnel jointly with the Columbus Street Railroad Company.[14]

In August 1878, a petition was submitted by "fifty prominent citizens declaring the High Street Railway a public nuisance" and asking the council to take action to repair the street. The City Solicitor then served notice on the company that if it did not repair the street within twenty days application would be made for an injunction to bar further operation of the line. The company complied and no further action was taken, but the seriousness of the city was noted by all concerned.

The East Park Place, Friend Street, and Columbus Street Railroad Companies merged in December 1879 to form the Columbus Consolidated Street Railroad Company, then acquired the North Columbus Street Railway and Chariot line in 1880. The consolidated company doubled the tracks in High Street and began to form what was to become a virtual monopoly of Columbus street transport.[15] In 1885 the Columbus Board of Trade reported that "The Consolidated Street Railway Company operating all the lines of street railroad in the city, excepting a single line west of the river, is one of the most enterprising corporations in Columbus . . . . With an energy characteristic of the gentlemen composing this company, the works of extending and improving the usefulness of the street railway traffic was pushed with vigor, until to-day we have service and accommodations unsurpassed by any city of this size in the country."[16] The city council concurred in these mergers by ordinance, and made no attempt to restrain the formation of a monopoly. However, the council continued to increase its regulation of the street railroads. In the initial 1879 ordinance dealing with the newly consolidated company, the council required that the company pave between and within one foot *outside* of its rails, pave between double tracks, and permit the still independent North High company to use portions of its lines. Even more significantly this ordinance initiated the universal transfer. For a single fare of five cents a passenger was to be issued a transfer good for travel on any connecting line owned or operated by the Consolidated Company; and upon application by the North High and State Street Companies, the Consolidated Company would be required to issue transfers to those independent lines, with the fare to be divided equally among the companies. From this time all franchises included a mandatory universal transfer clause, including the franchises of the interurban lines which operated in part on the streets in later years. In 1881 the Consolidated Company acquired the North High Company, eliminating part of this bookkeeping problem. The ordinance authorizing the linkage of these lines further required that cars be run over the entire North High line at ten minute intervals from 6 A. M. to 11 P. M.[17]

The State and Oak Street Railroad Company had been formed in 1872 to construct a line from the State Street bridge, over several streets downtown, to connect with the East Park Place line at Monroe Avenue. Financial difficulties halted construction of the line at Seventh Street, and by 1882 the line was in such bad repair that the public breathed a collective sigh of relief when it was acquired by the Consolidated Company that year. Necessary changes were immediately authorized to allow this line to connect with those of the Consolidated Company, and it was extended to Franklin Park.[18]

The Consolidated Company continued to expand for the remainder of the decade. In 1883 the company acquired a parcel of land near Franklin Park and built a large stable and carhouse to augment its older one at High and Chittenden, which had in turn replaced the original barn at High and Goodale. Several extensions of existing lines were granted; the company also acquired the South High Street Railway and Chariot Company in 1886 and extended its tracks on South High Street to the south corporation line at Innis Avenue in 1888. Tracks on several existing lines were doubled during this period. Ordinances in 1886 and 1888 stipulated that if the company failed to make repairs the city could cause such repairs to be made and place a lien against the company; it was also required that a motorman or conductor be on the front platform of each car at all times. Service on High Street was established by ordinance at ten minute intervals, and at five minute intervals during the rush hours of 6:30-8:30 A. M., 11:30-1:30 P. M., and 5:30-7:30 P. M. from the Court House to Eighth Avenue. The 1886 South High franchise was granted for twenty-five years, the period typical of all franchises to this time; but the city reserved the right to revoke the franchise if the company failed to meet its terms. In 1889 several extensions and double trackings were authorized, the Schiller Street line was built, and the company was required to deposit snow removed from its tracks evenly over the remainder of the street, and was relieved of its responsibility to run cars over the apparently unprofitable South High Street line south of the courthouse, while being allowed to maintain the right to use the right of way. It was also required to "maintain first-class service and replace one-horse cars with two-horse cars on all lines as soon as the necessities of travel require."[19]

II. Electrification

Only one experiment with motive power other than horses and mules was attempted in Columbus before 1882. In 1871 a company had been incorporated and authorized to build a line to North Columbus, to be powered by a steam dummy engine. The line was built for this company under contract by Samuel Doyle, a land developer, and in October 1873 a dummy engine pulled three carloads of prospective customers to Doyle's Summit Street Addition. The road was operated at a loss for a short time into 1874, then abandoned and its tracks removed. The East Park Place company was authorized to use steam power on its Long and Broad Street line in November 1872, but apparently never did so. The dummies which were authorized were limited to five horsepower and six miles per hour and were required to burn either coke or anthracite to minimize smoke.[20] Although cable traction was used on a large scale by a number of cities from about 1873 onward, a cable line was never installed in Columbus. Experimentation with electrically powered railways had been carried on since the 1860's by Thomas Edison and others, and in 1882 Charles J. Van de Poele built a short experimental line in Chicago. Other experimental lines in South Bend, Kansas City, New Orleans, and several other cities were installed in succeeding years; however these were all relatively short lines not comprising complete urban systems. But the technology developed rapidly; and in 1888 a city-wide system of forty cars began operation in Richmond, Virginia, and the race to electrify was on.[21]

The Chittenden Avenue line of the Columbus Consolidated Company attained the distinction of being the first electric line in the city. Sidney Short, an early experimenter, built the line from High to the new fairgrounds in time to amaze crowds on their way to the state fair of 1887. The system was not entirely successful, but as the technology was refined electricity was soon adopted for general use.[22] In August 1890, the Glenwood and Green Lawn Company obtained authorization and introduced electric power on West Broad Street and other lines. The Consolidated Company then secured permission to use "electric motors, gas engines, or cable traction" on all existing and future lines. These ordinances also embodied the city's first regulation of electric street railways. The companies were required to use straight wooden or neatly painted iron poles and to submit a plan of all improvements to the city engineer for approval, and were relieved of their previous responsibility for street cleaning when electric power was introduced. The destruction by fire, in February 1891, of the Consolidated Company's Chittenden Avenue car barn and stable, together with twenty-five cars, may have speeded the transition to electric power. In its place the company built a power house "sufficient to furnish twenty thousand horsepower" which began supplying power to the High Street line, already electrified in January of that year. The Long Street line was electrified in September 1891, and Main Street and Mount Vernon Avenue in November. As early as January 1891, ordinances granting line extensions limited the companies to "electric or other improved motive power equivalent thereto"; and the horse-drawn car rapidly disappeared from the streets of Columbus.[23]

William R. Pomerene, writing in Columbus in 1917, was to attribute in part the poor financial condition of the street railways in America to this early rush to electrify. In adopting a young technology, the companies had purchased systems that would be obsolete within five years and would, due to popular demand, be replaced three times within fifteen years. This replacement resulted in a considerable "lost capital account" which was to haunt the lines well into the next century. By comparison, European systems waited until after 1900 to electrify and avoided this experimental period.[24]

In 1891 the city constructed a bridge over the railroad tracks at Fourth Street and authorized the Consolidated Company to lay tracks over it for a fee of $250 annually. Construction of tracks on the new street surface was to keep pace with street pavement by the city. As in the case of the earlier High Street tunnel, any other company was to be allowed to use these tracks upon equitable compensation to the Consolidated Company. In March 1892, the speed limit on High Street was raised to fourteen miles per hour, including stops; and later in the year various streets were double tracked and service on High was required at fifteen minute intervals.[25]

In June 1892, the Consolidated Company was sold to the newly incorporated Columbus Street Railway Company. The same entrepreneurs had also acquired the Glenwood and Green Lawn Street Railroad Company and so finally combined all existing street railroad lines in Columbus under a single ownership if not a single company.[26]

In August 1892, the Leonard Street Railway Company, which in 1893 combined with the Glenwood and Green Lawn as the Crosstown Street Railway Company, was authorized to build an electric street railway system on Neil, Cleveland, and numerous other streets. Part of this line was allowed an eight mile per hour speed limit, the remainder fourteen miles per hour. Extensions of the Columbus Street Railway system on Goodale Avenue and Naughten Street were authorized, the latter establishing fifteen minute service intervals from 6:00 A. M. to midnight, and further stipulating that if the company abandoned the line it would remove all poles, wires, and track and repave the street within thirty days. The company was also required to reimburse property owners for its share of the street paving assessment within ten days of construction. In May 1892, the first combined street and interurban line in Columbus was founded. The Columbus and Westerville Railway Company was enabled by ordinance to build a street railroad in Cleveland Avenue from the north corporation line to Mount Vernon Avenue and thence to High Street, and the line was to extend north to the Village of Westerville under authority granted by the county commissioners. The time limit for completion of this line was extended several times by the council, judging that "It is in the best interests of said city that said railroad be completed," after an injunction to prevent its construction was brought by two landowners north of the city. The litigation took two years to resolve, but the line was finally operating in October 1895. This was apparently the only time in the history of Columbus street railways that construction was delayed by citizen opposition, and it is significant that even in this instance the opposition did not come from residents of the city. The Columbus and Westerville line was permitted to charge twenty-five cents for a round trip between its termini, or fifteen cents one way, but was required to adhere to the standard fare structure of six tickets for twenty-five cents or a single fare of five cents for trips entirely within the corporation limits. In addition the company was to issue transfers to intersecting lines of other companies, dividing the fare. Adherence to the intra-city fare schedule and the requirement to issue transfers was imposed on all street and interurban lines operating on the streets of Columbus subsequent to this time.[27]

Virtually all legislation by the city council was designed to regulate the operation of the street railways, not to determine their location. Locational decisions were left to the initiative of businessmen. Minor departure from this pattern had occurred in the construction of the Fourth Street bridge and High Street tunnel, but these projects had served primarily to facilitate existing patterns rather than to create new ones. The city's single venture into locational planning met with interesting results. In July 1893, the city council determined that there should be a street car line from High Street west on Mound Street to the corporate limits; however, no company had proposed to build such a line. The council then took the unprecedented step of establishing the desired route by ordinance, then advertising for bids from companies to construct, own, and operate such a line. Bids came in, and the low bidder was selected on the basis of proposed fares. The Columbus and Harrisburg Street Railway Company, the winner, had been organized specifically for this project. A one year deadline was established by council for construction and operation of the line, a generous length of time in comparison with the limits set by other contemporary ordinances. However, the fledgling company ran into financial difficulties and repeatedly requested delays in the deadline. In June 1884, a frustrated council repealed the franchise fee in a final effort to enable the company to remain solvent. The line was never constructed, and thus ended the city's attempts at influencing locational decisions.[28]

In 1894 construction was completed on a new $400,000 viaduct over the railroad tracks on High Street at Union Station, burying the older tunnel two stories underground. The viaduct, lined with structures designed by Daniel Burnham, pro-

vided a monumental interchange between steam rail and
street rail transportation and replaced Columbus's most
prominent eyesore with one of its most prominent landmarks.
In 1899 the Columbus Street Railway Company boasted that
its cars "Pass directly in front of these entrances for all parts
of the city," and that the company provided a waiting room
and a representative in Union Station.[29]

The year 1895 saw two extensions to the independent Crosstown Street Railway system in the western part of the city,
both ordinances reflecting changing conditions and thought at
that time. The burgeoning tangle of wires over the streets of
Columbus led to the requirement that the company's wires and
poles be placed so as not to interfere with any existing street
railroad, telephone, or electric transmission line. The cars
were required to conform to any future speed limit set by the
council; such a limit was not built into this ordinance. Service
was required over these lines on twenty-four minute intervals for three years, and thereafter at whatever schedule the
council should direct. In accordance with an 1890 ordinance
directing that "no franchise of any street railroad company
shall be extended in time or renewed until after the time prescribed in the original grant, and whenever a line is extended
on a street or avenue that extension shall expire at the same
time as the original grant," these extensions were granted for
periods to conform with the earlier Glenwood and Green Lawn
franchises under which Crosstown was operating.[30]

Olentangy Park was built by the Columbus Street Railway
Company in 1896 along the river just outside the north corporation line, near where the company had previously located its
end of the line car barn. The park was the Columbus manifestation of the "trolley parks" which were being constructed
throughout the country by street railway companies. These
parks were intended not only to turn a profit in their own
right, but perhaps more importantly to reverse the flow of
traffic over the street railway lines during otherwise slack
periods. The *Illustrated Guide to Columbus*, published by the
street railway company in 1899, devotes half of its forty-eight
pages to praise of Olentangy Park. Here could be enjoyed un-
spoiled nature, where "every young man and maiden and
every young child may find exquisite enjoyment in strolling

midst these bluffs and glens." For those seeking more intense activity the company had provided a boat house, four bowling alleys, a dance pavilion, and a 2,000 seat theater featuring such attractions as " . . . Diana, the marvelous Mirror Dancer; Geo. H. Fielding, European Juggler; the Goldsmith Sisters, European Artists; Witney Brothers, Musical Wonders; Zeno, Karl & Zeno, Contortionists; . . . Ollie Young, Club Swinger; Professor Howard and His Trained Ponies" and many others. Intended to be entertainment for the whole family with no alcohol or profane conduct condoned, the park was "Only twenty-five minutes ride from the heart of the city," and carfare included the price of admission.[31]

III. The New Century

By the turn of the century Columbus had grown to have 125,560 people, over seven times its 1860 population. The city was served by more than eighty-eight miles of street railways and had 103 miles of improved streets; and 126 steam trains entered or left the vast railyards at the city's center each day.[32]

The only interurban electric line yet operating in the city was the eight year old Columbus and Westerville line, which was eventually acquired by the Columbus Street Railway Company and became part of the city system. In 1900, there began, however, a spate of interurban building which by 1908 had produced over 2,798 miles of interurban electric road in Ohio, over a thousand miles more than were built in any other state in the nation.[33] Most of the interurban lines operating through Columbus did so in part on city streets, and in so doing provided competition for the Columbus Street Railway system. As has been noted, these lines were required to provide intra-city service at the standard five cent fare, and to issue transfers to any other intersecting line within the city.

The Columbus, New Albany, and Johnstown road, built in 1901, served Gahanna, while the Columbus, Urbana and Western extended a line along the river to the Scioto storage dam. The former arranged with the Columbus Street Railway Company to use its tracks and power within the city. The Columbus, Buckeye and Newark Company, the Columbus, London and Springfield line, and the Columbus, Grove City &

Southwestern, which served Orient, were founded around 1900 and were among the smaller companies agglomerated by the giant Ohio Electric Railway Company in 1907.[34] Ohio Electric, while never profitable, acquired lines throughout the state and provided an integrated system serving nearly every city of over 10,000 population in the state. In Columbus, as in other major cities, Ohio Electric built a downtown interurban terminal, first on Rich Street and later on Third. To connect with this terminal, interurban cars operated on the "interurban loop," made up of lines in Rich, Third, Gay, and Water Streets encircling the downtown area. Several other companies, including the Columbus, Delaware and Marion, the Ohio & Southern Traction Company, and the Scioto Valley Traction Company remained independent, in part due to the city council's requirement that they be permitted to use the tracks and terminal facilities of any other company, providing that at least ten miles of their total route not be parallel to the lines of the other company, and that they pay compensation mutually agreed upon, or if necessary, determined by arbitration. The Columbus, Newark and Zanesville line, incorporated in 1902, eventually was acquired by Ohio Electric, while both the OE and other large and small lines throughout Ohio, Illinois, and Michigan formed the Central Electric Railway Association to provide coordinated service throughout the North Central United States.[35]

Developments in the control of street railways by the City of Columbus continued during this period. In 1901 some of the charters under which the Columbus Street Railway Company had been operating came due for renewal, and the council took this opportunity to consolidate the "various and inconsistent conditions" found in many franchises. In the spirit of the 1890 ordinance requiring all franchises of a company to expire concurrently, all of those affecting the company were pulled together into a single "blanket charter," which also made uniform on all lines the burden of street repair, maintenance, operating conditions, fares, and transfers, and established operating schedules for most lines. The ordinance instructed the company to make connections with the lines of the Columbus Central at several points, allowed no route changes without prior approval of the city, and required the company to deposit $1,000 with the city Board of Public Works as an emergency fund for repairs. As before, any other company could use the lines of the company and could make such changes as installing a third rail if necessary. Transfers were to made available to any intersecting line of any company, continuing the universal transfer provision, and fares to any part of the city were not to exceed the single fare of five cents or seven for twenty-five cents. Children under six were to be allowed to ride free, and children between six and twelve were to pay five cents. When the company's gross receipts reached $1,750,000 the fare was to be reduced to eight for twenty-five cents. The company was also ordered to conform to new technical criteria, including use of concrete roadbed on new or rebuilt lines and protection of the city's utility pipes from electrolysis. All construction was to be under the design-review and supervision of the city Board of Public Works. Among other provisions, the company was permitted to carry mail and packages and to operate cars exclusively for such purpose if similar in appearance to passenger cars, but was strictly prohibited from hauling freight cars within the city. Work cars used exclusively in track and pavement maintenance, however, were permitted. A franchise fee of two percent of gross receipts from fares within the city was levied. Provisions similar to these appeared in all interurban and city franchises granted in Columbus after this time, and future extensions and improvements which were authorized conformed to both these provisions and the time period of the blanket franchise granted each company.[36]

The year 1901 also saw the first requirement that cars be heated when the temperature dropped below thirty-five degrees.[37] Despite this concession to the passengers' comfort, the motormen's vestibules remained open and unheated for many more years. One Columbus motorman, Frank Nixon, suffered frostbitten feet in his open cab, and as a result departed for the warmth of southern California. His second son, Richard, was born there in 1913.[38]

The year 1910 brought a strike against the Columbus Street Railway system that was probably the most serious labor disturbance in the city's history. The period was one of labor unrest throughout the country and of the growth of unionization in American industry, and this pattern prevailed in Columbus. In April 1910, most of the six hundred employees of the line demanded recognition of their union, collective bargaining, an increase in pay, time and a half beyond the nine and a half hour work day, and reinstatement of several men discharged for their involvement in union organization. The company refused to deal with the union, and although the company promised to make several of the demanded changes the union later claimed that it had not lived up to its promises. A strike resulted on 14 July in which track and cars were dynamited, rioting developed, fights occurred between union men and "scabs" and the National Guard was called to restore order. A business boycott of all who rode the cars was called by the union, and a number of policemen were discharged by the city for refusing to ride the cars. Finally, on 18 October the broken union called off the strike, and the company graciously agreed to rehire any men "who had not been identified with the violence, if their places had not already been filled."[39]

The decade 1900-1910, which saw the massive construction of interurbans, produced few track extensions of the two city street railway companies in Columbus. However, ridership continued to increase until the middle of the decade and remained constant thereafter, and most of the remaining single lines in the city were double tracked and rebuilt to improved specifications during this period. One independent line, the Central Market Street Railway, was chartered in 1901 and built lines to the north and south corporation lines, but was acquired by the Columbus Street Railway Company in 1907.

A history of consolidation of small companies similar to that of the street railways had occurred in the electric power and heating industries in Columbus. Two companies providing electricity were combined to form the Columbus Edison Company in 1903, while four companies marketing electricity and hot water for heating were consolidated as the Columbus Public Service Company in 1904. Both of these combined companies were acquired by the Columbus Street Railway Company, which became the Columbus Railway, Power, and Light Company in 1904. The Company also acquired the lines of the Columbus Central Street Railway, finally uniting all Columbus street railways and electric power companies under a single ownership, and making Columbus the only major Ohio city with such combined ownership. The interurbans all contined to operate under a separate ownership from the C. R. P. & L., although the company provided several interurban lines with power.[40]

IV. The End of the Beginning

By the end of the first decade of the century, both the interurban and city systems were virtually complete. The "scare of 1907," a financial recession, had hurt the profitability of both systems briefly, and a state law limiting interurban fares to two cents per mile had affected the profitability of that industry in general.[41] However, the passenger electric rail lines as a whole continued to return profits, to make improvements in lines and service, and to attract their share of passengers from the growing urban population well into the second decade. Several extensions of interurban lines took place in Columbus during this decade, and the downtown interurban station was completed in 1913. Another, improved station was constructed in 1918.

For several years beginning about 1912 the Columbus Railway, Power and Light Company operated double deck cars on the congested High Street line. These cars, seating eighty-three and capable of carrying up to 171 sitting and standing passengers, were larger than any then in use outside New York and Philadelphia.[42] In 1915 the company purchased the latest crane car, two snow handling cars, and a street flusher car, the latter operated in part under contract for the city. The city, too, made improvements during this period, completing an underpass at the railroad tracks on Eleventh Avenue and requiring the street railway company to pay $5,000 for its use. A new Mound Street bridge, built in 1918, caused the temporary removal of the Indiana, Columbus, and Eastern's interurban line at that point.[43]

During the "World War," however, the street railroads faced mounting problems arising from financial conditions, rigid governmental control, and changing technology. While the Civil War and its free-spending soldiers had aided the growth of the infant street railway industry in Columbus, World War I had the opposite effect. The war produced a general inflation, which resulted in higher maintenance and construction expenses as well as demands for higher wages. The long-term franchises of the period firmly fixed routes, schedules, and fares, and did not permit the companies to adjust to shrinking revenues on some lines or to increases in operating costs. In 1912 the fare on Columbus lines had been reduced to eight tickets for twenty-five cents, or three and one-eighth cents, when the company's receipts had reached the level stipulated in the 1901 franchise. This left the company with a very narrow profit margin and the advent of the war threatened to push it below the break-even point. In 1918 the company requested a higher fare schedule from the council, but no action was taken. Then in July of that year its employees struck for higher pay; the company asserted that it was unable to pay more, and the dispute was referred to the National War Labor Board. The board mandated a wage increase adding $560,000 per year to the company's payroll, and in response the company notified the council that it was immediately raising its fare to five cents with an additional cent charged for a transfer. The company also claimed a perpetual franchise on its six most profitable lines on Main, High, Long, Broad, and several other streets, declared itself a tenant at sufferance on all other lines and surrendered those franchises. The company also brought suit in the Federal district court to prevent the city from enforcing the terms of the seventeen year old franchise. The court refused the injunction, and the National War Labor Board subsequently refused to reconsider its award to the employees. Many passengers, meanwhile, either offered to buy tickets at the old rate or refused payment altogether.

In January 1919, the management of the company was reorganized and the old rate restored. The council then authorized a change to six tickets for a quarter, but a referendum was called and the increase defeated by the electorate. After further unsatisfactory service, the council passed an increase to six cents or five for a quarter; and an effort to call another referendum was unsuccessful.[44]

This increase was still not sufficient, by the company's estimation, to cover its bonded indebtedness, pay dividends, and maintain its system. Because of its inability to pay interest on its outstanding bonds, the company had nearly been placed in receivership before the reorganization and fare increase. Putting the Columbus experience in a national perspective adds some weight to the company's claims. The three and one eighth cent Columbus fare was the lowest charged by sixty-two companies surveyed in a 1917-1919 study of U. S. and Canadian street railways. While the fare increase was welcome, it was one of the smallest increases granted during the inflationary period. On the other hand the strike, which was one of nineteen in the country that year, appears to have been based on a legitimate grievance. Of the U. S. companies surveyed, forty-three paid higher starting wages than did the Columbus firm, while those of only six were lower and six the same. Top wages showed a similar pattern: forty-two were higher, five lower, and six equal.[45] Thus it appears that the low fares in Columbus were due in part to the regulated monopoly of the company, but also in part due to the low wages paid company employees. In effect, the employees were subsidizing the riding public.

While a large number of street railroad companies throughout the country were placed in receiverships during these difficult years, the Columbus company escaped this peril and by 1923, with no further fare increases, could report that "It should . . . be gratifying to the public, as well as to the stockholders, to realize that this Company, operating under rates of car fare about forty per cent lower than is the average fare in all other cities of this country, has been able to attain the earnings necessary to the strengthening and continued upbuilding of the transportation service of Columbus."[46] The company stated, too, that it needed fifty new cars and one and a third miles of new track, neither of which it could afford.

The company also survived two other threats symptomatic of the times. In 1920 the company treasurer and others were indicted for embezzling $323,000 of company funds, an amount exceeding one quarter of the street railway division's gross annual receipts. A judgment against the treasurer, which the U. S. Supreme Court refused to reconsider, was made in 1924 and the company received a settlement of $856,000.[47]

In addition to the hazard of private corruption, the company faced the danger of public ownership. Public ownership of the street railway system had been debated in Columbus as early as 1906[48] and popular call for public ownership reached a peak during and immediately after the war years. San Francisco had purchased its street railway system in 1912, Seattle in 1919, and Detroit in 1922.[49] Various politicians were elected in Columbus during the 1920's who, in the estimation of the Columbus company, were antagonistic toward utilities and favored public ownership.[50] Nevertheless, no strong movement toward public ownership ever developed in Columbus during the era of the street railroads, probably due to general public satisfaction with the service and low fares offered by this closely regulated monopoly. In recognition of this possibility, however, the Columbus, Delaware, and Marion franchise of 1921 included a clause reserving to the city the right to terminate the grant and purchase or lease all of the property of the company, with the price to be fixed by a jury if the city and company were unable to agree. Similar provisions have appeared in all franchises since this time, although the option has never been exercised.[51]

The war years were also hard ones for the interurban roads. By 1921 the Ohio Electric Railway Company was in shambles; its pieces were reassembled by smaller companies operating on a more local scale, and the lines continued to operate through the decade.[52]

However, the 1920's brought an onslaught that the electric railroads were ultimately to prove incapable of dealing with. The gasoline engine had made possible both the automobile and the motor bus, which together were the undoing of the electric systems.

The "jitney" bus, a privately operated, sometimes homebuilt vehicle deriving its name from the popular slang for a five-cent piece, sprang up in most cities in response to the rising street car fares after the war. The buses were able to hold their fare down by operating only on the most profitable lines, and often only during rush hours, as they were initially free from the regulation of routes and schedules imposed on the street railroads. Initially claiming a considerable portion of the street cars' traffic, in several cities the jitneys were credited with driving the street railroads out of business. The street railroads fought desperately for municipal regulation of their new competition and eventually were successful.[53] In 1933, the Columbus Railway, Power and Light Company voiced its opinion on buses: "Buses may find a permanent place as feeders to electric lines but they will never take the place of them. They are performing a service auxiliary to the cars operating in sections of the city where traffic is not dense enought to warrant the running of an electric car line. Once traffic becomes dense enough to make an electric line pay, the bus cannot render adequate service."[54] Elsewhere this same brochure revealed that the system had carried 39 million passengers in 1932, down from 81 million as recently as 1925.[55] Although in 1932 The Columbus Railway, Power and Light Company was the largest employer and tax payer in Columbus, only one quarter of its revenues at that time were coming from its street railroad operations, and by this date over one half of the interurban mileage in Ohio had already been abandoned.[56]

Lewis M. Schneider, writing in 1965, made a more accurate summation of the situation: "The abandonment of street cars prevented financial collapse of the industry. Almost overnight, an industry with high fixed costs of maintenance of way, generation of power, and, in some cases, engineering and construction of rolling stock found itself buying standardized products from a limited group of manufacturers .... In many cities, two-man streetcars were replaced with one-man buses. Routes were no longer tied to the inflexibility of the steel rail, and could follow the population growth into suburbia."[57] The real villain of this piece was the automobile. When suburban growth was freed from dependence on street railroad transportation, the city was freed to sprawl across the countryside, in turn depriving the street cars of the concentrated point-to-point traffic in newly developed areas which they needed to grow. The proliferation of automobile ownership even among residents of the built-up area of the city took away a large part of the lines' existing business. This development, coupled with the fact that the motor bus could do what was left of the street railroads' job more economically and flexibly, spelled the epitaph of the street railway. The automobile and the motor bus in combination also killed off the interurban lines. By the end of the 1930's the interurbans were gone and the street railroads were dying. By 1956, every street car line in Columbus had been either abandoned or replaced by motor and trolley buses,[58] and in another ten years even the trolley buses were gone.

V. Epilog

Even in 1885 the Columbus Board of Trade had reported that "New lines are being pushed forward as the growth of the city and the demands of the people warrant the same. One only needs to look out over the streets laid with railroad track to become satisfied that this enterprise has done wonders to build up the city in the suburbs."[59] The street railroads thus reacted to, facilitated, and shaped the growth of the city. The early true suburbs of Gahanna, Bexley, Westerville, Worthington, and Upper Arlington owed their growth to the extended street railways and interurbans that came later, but even the "suburbs" of which the Board of Trade was speaking, now considered to be near the center of the city, were made possible by the relatively inexpensive, convenient transportation offered by the street railways. Conversely, the growth rate of the downtown business and commercial district was even faster than that of the city because of the radial street railway transportation available from the dispersed residential areas. The interurbans even made it possible for the farmer, the farmer's wife, or the formerly remote rural villager to go downtown for a day of business or shopping and get home in time for supper. William Pomerene, in 1917, attributed the peripheral growth of American cities not only to the existence of street railroads but to the absence of zone fare systems and to the availability of the universal transfer. In European cities, where zone systems were common and transfers were not, the older densely populated city center was maintained.[60]

That the street railroads should today be considered as constraints to development would seem ironic to these early writers, but with the advent of the automobile these systems proved to be too rigid to long compete. Long-term, inflexible franchises and unfavorable economic conditions hastened their downfall, but the real cause of the demise of the street railroads was America's unrestrained acceptance of the automobile and of the changed order that it created. Street rail transportation exists today only where, for physical, social, or economic reasons, that older order has not changed.

NOTES

[1]Osman Castle Hooper, *History of the City of Columbus, Ohio* (Columbus: Memorial Publishing Company, *ca.* 1920), I, 53, 236; Columbus, Ohio, *Annexation Map* (1972).

[2]John J. Janney, "Street Transportation," in *History of the City of Columbus,* ed. Alfred E. Lee (New York: Munsell and Co., 1892), p. 304.

[3]Camille J. Botte, "The Development of the Street Railway," M. A. Thesis Ohio State University 1928, p. 6.

[4]Columbus, Ohio, *General Ordinances* (1882), p. 337.

[5]Hooper, *History of Columbus,* I, 233.

[6]Columbus *Ordinances* (1882), pp. 339-48.

[7]Janney, "Street Transportation," p. 308.

[8]Columbus, *Ordinances* (1882), pp. 348-53.

[9]Columbus, *Ordinances* (1882), pp. 354-61.

[10]Janney, "Street Transportation," p. 310.

[11]Columbus, *Ordinances* (1882), pp. 363-65.

[12]Henry J. Booth, *A Treatise on the Law of Street Railways* (Philadelphia: T. and J. W. Johnson and Company, 1892), p. 172.

[13]Columbus, *Ordinances* (1882), p. 368.

[14]Columbus, *Ordinances* (1882), pp. 367-70.

[15]Janney, "Street Transportation," p. 309.

[16]The Columbus Board of Trade, *The City of Columbus* (Columbus, Ohio: Press of G. L. Manchester, 1885), p. 65.

[17]Columbus, *Ordinances* (1882), pp. 374-77.

[18]Janney, "Street Transportation," p. 312.

[19]Columbus, Ohio, *General Ordinances* (1896), sec. 2155, 2920, 4155, 4620, 4755, 4936.

[20]Janney, "Street Transportation," pp. 310-13.

[21]Botte, "Development," p. 30.

[22]Hooper, *History of Columbus,* I, 232.

[23]Janney, "Street Transportation," p. 309.

[24]William R. Pomerene, *Trams and Trolleys* (Columbus, Ohio: Champion Press, 1917), p. 7.

[25]Columbus, *Ordinances* (1896), sec. 6704.

[26]Janney, "Street Transportation," p. 309.

[27]Columbus, *Ordinances* (1896), sec. 6652, 6804, 6954, 7480, 7632, 7633.

[28]Columbus, *Ordinances* (1896), sec. 7594, 8105, 8356, 8576, 9808.

[29]The Columbus Street Railway Company, *Illustrated Guide To Columbus* (Columbus, Ohio: The Columbus Street Railway Company, 1899), p. 22.

[30]Columbus, *Ordinances* (1896), sec. 10050, 10053.

[31]Columbus Street Railway, *Columbus*, pp. 28-48.

[32]Hooper, *History of Columbus*, I, 53.

[33]George W. Hilton and John F. Due, *The Electric Interurban Railway in America* (Stanford: Stanford University Press, 1960), p. 255.

[34]Hooper, *History of Columbus*, I, 235-36.

[35]Hilton, *Interurban Railway*, p. 265; Hooper, *History of Columbus*, I, 235-36.

[36]Columbus, Ohio, *The Columbus Code* (1919), sec. 17801.

[37]Columbus, Ohio, *The Columbus Code* (1919), sec. 17801.

[38]Theodore H. White, *Breach of Faith: The Fall of Richard Nixon* (New York: Atheneum Publishers, Readers Digest Press, 1975). pp. 59-60.

[39]Hooper, *History of Columbus*, I, 260-62.

[40]Hooper, *History of Columbus*, I, 233.

[41]Hilton, *Interurban Railway*, p. 266.

[42]U. S. Department of Commerce, Bureau of the Census, *Central Electric Light and Power Stations and Street and Electric Railways, 1912*, p. 359.

[43]Columbus, *Code* (1919), sec. 30584.

[44]Hooper, *History of Columbus*, I, 233-34.

[45]Delos F. Wilcox, *Analysis of the Electric Railway Problem* (New York: Afferton Press, 1921), pp. 191, 533; Columbus Railway, Power, and Light Company, *Annual Report* (1918), p. 15.

[46]C. R. P. & L., *Annual Report* (1923), p. 12.

[47]C. R. P. & L., *Annual Report* (1920), p. 14; C. R. P. & L. *Annual Report* (1924), p. 13.

[48]Conrad Wilson, "The Best Street Railway System," *The Ohio Magazine*, November 1906, pp. 450-56.

[49]Tax Foundation, Inc., *Urban Mass Transportation in Perspective* (New York: Tax Foundation, Inc., 1968), p. 13.

[50]The Columbus Railway, Power and Light Company, *Questions and Answers* (Columbus, Ohio: The Columbus Railway, Power and Light Company, 1933), p. 32.

[51]Columbus, Ohio, *The Columbus Code* (1930), sec. 32, 680.

[52]Hilton, *Interurban Railway*, p. 255.

[53]Botte, *Development*, p. 65.

[54]C. R. P. & L., *Questions and Answers*, pp. 16-17.

[55]C. R. P. & L., *Questions and Answers*, p. 25.

[56]Hilton, *Interurban Railway*, p. 255.

[57]Lewis M. Schneider, *Marketing Urban Mass Transit*, quoted in Tax Foundation, *Urban Mass Transit*, p. 14.

[58]Columbus, Ohio, *Transit Franchise: Columbus Transit Company* (1956), p. 2.

[59]Columbus Board of Trade, *Columbus*, pp. 65-70.

[60]Pomerene, *Trams and Trolleys*, pp. 25-29.

THE HORSE-DRAWN STREET RAILWAY:
The Beginning of Public Transportation in Shreveport, 1870-1872

By Debbie Wikstrom

The street railway, which Burton J. Hendrick, a prominent American economic historian, called a "peculiarly American contrivance," was the first successful form of urban public transportation.[1] The first horse-drawn streetcar line in the world began operation in New York City in 1832. These first cars were primitive, similar to stagecoaches, and mules, usually past retirement age, provided the only reliable power. Streetcar companies in the major Northern cities made improvements in the facilities, and before 1860, thirty new lines opened in cities throughout the United States.[2] According to Hendrick, "practically every city of any importance" had a street car line by 1870.[3]

Shreveport in 1870 could not be considered an important American city. Editors of Shreveport's newspapers defended its right to even be called a "city" at all.[4] A small number of Shreveport citizens were anxious for their community to merit that name, and from 1870 to 1872, different groups formed three companies to build street railway lines through the downtown district and to outlying areas. Of these companies, two succeeded in building and operating their lines, and only one proved to be a profitable enterprise.

Shreveport's potential growth was the basis for most claims to cityhood. Civil War refugees continued to move into East Texas and the Shreveport area, which was relatively untouched by the war. The city expanded so quickly that the gas company could not meet the increased demand. At this time, natural gas was the principal fuel used for lighting. For seven weeks in the first three months of 1870, the company left the streets in darkness so that it could supply gas to private consumers.[5] Shreveport became a popular cotton market and shipping center. The new farmers in the area brought their produce to Shreveport for shipment, but they did not patronize retail firms. Instead rural stores supplied these people with staple goods they needed, and they purchased larger items from New Orleans or New York by mail. The *South-Western*, a local weekly newspaper, advised that citizens of Shreveport should make the town more attractive to "our country friend" in order to become "a city in every sense of the word."[6]

One group of men took a step to modernize Shreveport. On June 11, 1870, stockholders of the Shreveport City Railroad Company met in Brewer's Hall for an organizational meeting. The men accepted the company charter, which the city council had granted on June 8, by a unanimous vote. The charter gave the company a twenty-five year right-of-way for a streetcar line in Shreveport, with a tax exemption for the first five years. Stockholders also elected company officers for the year: Judge A. B. Levisee, president; Edward Jacobs, vice-president; L. L. Tompkies, treasurer; and Thomas Phillips, secretary.[7] The company did not decide on the route for the line until it had received recommendations from a surveying team a month later. The decision was that the line would begin at the intersection of Spring and Texas Streets, follow Texas Street to Texas Avenue, and end at the corporation line, then at Jordan Street.[8]

83

The stockholders chose Captain James F. Utz, a local dealer of heavy machinery and a director of the company, to act as purchasing agent. For two months, he traveled to New Orleans, St. Louis, Chicago, Cincinnatti, and New York.[9] He bought iron railing for the tracks in New Orleans. This arrived in Shreveport in early July aboard the steamship *Rudolph*.[10] When Captain Utz returned home on the *Tidal Wave* in mid-August, he informed the *South-Western* "that the cars, three in number, are perfect beauties--- light enough for one mule to draw with twenty passengers up a considerable grade with perfect ease. They will seat comfortably fourteen passengers, and twenty-five can find room by a little 'scrounging.'"[11] The cars arrived in early November by way of New Orleans on the *Silver Bow*.[12]

The contract for laying the track was awarded to John Rooney. His crews began grading Texas Street and Texas Avenue in early October, 1870, and by October 26, they began to lay the crossties and iron rails. Rooney stated at this time that he and his men would have the line completed in forty days.[13] By November 2, the track had been completed as far as McNeil Street, about one-fourth of the entire route.[14] In late December, Rooney finished construction at a final cost of about $35,000.[15]

Rooney put the first car on the track about eight o'clock the night of Saturday, December 31, 1870, twenty-five days later than he had forecast. A crowd gathered along Texas Street, and greeted the car with cheers. As the car passed a theater, a band, "accidentally or by design," began to play "Dixie." The *South-Western* commented that this was "an important event to our city, (and may we not without any twinge of conscience call Shreveport a city now?)."[16] Two of the cars were on the track that night, each making three circuits of the track with a full load of selected prominent citizens. Rooney asked that the newspaper "return his thanks to the citizens for their enthusiastic demonstration on the first appearance of the cars on the track," and promised "to do all in his power for the accomidation [sic] of the public."[17]

Sunday, January 1, 1871, Rooney put all three cars into operation making hourly runs. The cars earned thirty dollars each in five-cent fares on the first day.[18] Judge Henry G. Hall, a Shreveport lawyer recorded his reaction to the new street railway in his diary entry for January 3: "attended to some business but concluded to go home at 2 on the streetcar! This institution was put into operation Saturday night----carries me 5 quarters and I walk the 6th home."[19] The *South-Western* felt that "the novelty of the thing" was the reason for the first success, but throughout the next weeks, Judge Hall mentioned having to walk because the streetcars were so crowded.[20]

As far as appearance was concerned, the streetcars were open on the sides with five or six seats each. The seats were hard, straight-backed, and faced forward in a single line with no aisle. Poles suspended the flat roof over the seating area. The driver was the only attendant, and was also responsible for collecting fares. The company allowed male passengers to sit at front with the driver if they wished. Bells attached to the one mule announced the approaching car to the people. The metal wheels turned on the single track, which was four feet four inches wide and lay in the middle of the street. The cars changed direction at each terminus on the turntables there. They could pass each other at a switching area half way between the Spring Street and Jordan Street ends of the track. The barns, offices, and depot building were located at the city limits at the Jordan Street terminus.[21]

84

The Shreveport City Railroad Company did not manage the line itself those first few years. Rooney, the contractor, leased it in 1871 at a cost of $2,500 per year.[22] For the year 1872, William Heffner, who had just moved to Shreveport from Kentucky, leased the road for $5,850. He renewed his lease in 1873 for $6,010.[23] In November and December, 1873, the City Railroad advertised for sealed bids for the leasing of the road for the coming year. Only one bid was submitted and was so low that the company decided to run the line itself in 1874.[24]

In the fall of 1871, another group organized a company to provide the Fairfield suburb of Shreveport with streetcar service. This was a residential section which was rapidly growing. By September 9, $20,000 worth of stock had been subscribed. Judge Henry G. Hall, who lived in Fairfield, entered the company and was in charge of the surveying party. Judge Hall was familiar with the organization and construction of the City Railroad through his friendship with Judge Levisee, president of that company. The new company decided on the terms it wanted to incorporate into its charter at a meeting on October 18.[25] These terms were similar to the ones included in the charter of the Shreveport City Railroad Company. On October 24, the city council granted the Fairfield Streetcar Company a twenty-year right-of-way for a line along the Norris Ferry Road from its junction with Texas Avenue, with a five year tax exemption. The city council also changed the name of Norris Ferry Road to Fairfield Avenue at this time.[26]

John G. Gragard purchased the iron, cars, and equipment for the Fairfield line in New Orleans. Crews began laying the track by February 1, 1872. The cars, manufactured in New York by the Stephenson Company, arrived in Shreveport on April 9 aboard the *Lady Lee*.[27] Service began on Tuesday, June 18, 1872. The cars made hourly runs to coincide with the schedule of the City Railroad, with an additional fare of five cents. W. Dyer leased the line from the company and acted as manager. In March, 1875, the company established a market in the depot building stocked with groceries and other supplies to save suburbanites the added trouble and expense of making frequent trips into town.[28]

In January, 1872, a third group formed a company to continue the street railway from the Texas Avenue terminus of the other two companies for another three-quarters of a mile down Texas Avenue. By February 25, the company had sold subscriptions amounting to $11,750, but needed $5,000 more to begin seriously considering the construction of the line. On June 11, the company had not been able to collect the necessary funds, and stockholders met to discuss the situation. The group elected L. L. Tompkies to preside over the meeting. Tompkies chose T. F. Bell to act as secretary, and appointed a committee of three to raise pledges for the amount needed, and collect ten percent of the subscriptions which they sold. Unable to get the necessary support, the Texas Avenue Railway Company failed before it really began.[29]

There was not much faith in the success of the streetcar when the Shreveport City Railroad began operation. Newspapers recorded the development as being in spite of "the prognostications of the skeptical."[30] Stockholders of the company, however, considered their investments sound, and supported the other ventures. Of the Shreveport City Railroad Company stockholders, twenty-five percent held stock in one of the other two companies. The Fairfield line had twenty-eight percent of its stockholders investing in

85

157

another line. Over half of the backing of the proposed Texas Avenue
Railway Company came from the stockholders of one of the other lines.
Oddly enough, no one person invested in all three lines.[31] The success of
the first street railway company forced early doubters to reverse their
positions. The *South-Western* mentioned that the success of the Shreveport
City Railroad Company was "quite a reflection on the 'power of the press'"
since none of the city's newspapers had supported it.[32] The *South-Western*
complained earlier of the inconvenience that the construction of the line
brought to travelers on Texas Street, but after the opening of the City
Railroad Company, the newspapers began to support streetcar endeavors.[33]
The *Times* tried to help the proposed Texas Avenue Railway raise the money
it needed by encouraging both stockholders in the other two companies and
landowners along the proposed route to invest.[34]

The *Times* included landowners in its plea because property values
increased wherever the streetcar went. Judge Henry Hall was very satisfied
with the price he received when he sold a lot west of Fairfield Avenue eight
days before the opening of the Fairfield Street Railway.[35] Advertisements
for selling land began to state the convenience of the street railway in
getting to town from the distant lots. Even advertisements for the sale of
lots along Texas Avenue outside the city limits during the time that the
Texas Avenue Railway was being considered included mention of the advantages
of the streetcar.[36]

The Shreveport *Times* on December 7, 1872, reprinted a story from the
New York *Herald* dealing with the testing of a new steam-powered streetcar.
At the same time, an epidemic of "epizooty", so-called by the *Times,* infected
the horse and mule population of Shreveport, threatening the continued
service of the street railway companies.[37] Epizoon is a contagious respira-
tory infection, and is often referred to as a cold or influenza in equine
animals. The disease lasts from one to four weeks, and in the most severe
cases, requires a three week convalescence period.[38] At first, the streetcar
lines gave limited service. On December 16, 1872, the Fairfield Avenue
streetcars were forced to shut down. Two days later the City Railroad also
discontinued service, with seven of its eight mules ill. By December 21,
the companies were able to resume running on schedule, making "hourly trips
during the day, weather permitting."[39]

The weather also hindered service of the street railway lines, especially
during the winter months. If any ice or snow was on the track, cars failed
to run. Even when not operating, the street railway was an asset to the
people since walking on the tracks was easier than walking through the snow.
As Judge Hall recorded in his diary on January 14, 1871, "Snow on the ground
and hard freeze. One car being stopped for a time, Col. Tompkies walked
down on the track."[40] In rainy weather, the mules had trouble pulling the
cars down muddy streets. The animals would often stumble, and Judge Hall
recorded an incident when, "A mule fell down while pulling the car."[41]

The streetcar companies had a leisurely attitude and gave each customer
individual service. Cars would make unscheduled runs upon request, as they
did for a theater performance after regular hours on January 1, 1876.[42]
Special attention was given to the ladies. Smoking was forbidden if a single
female were present.[43] Maude Hearne O'Pry, a Shreveport journalist,
emphasized the chivalry of the early streetcar companies, and at the same
time, criticized the young ladies of the 1920's saying:

86

158

The gallant conductor of the car would patiently wait for
some lady fair to return to her home to get a letter she wished
to mail or to seek a dainty kerchief. It was not lipstick or
rouge then. Mercy me, what a disgrace that would have been![44]

The poor condition of roads in Shreveport hurt the street railway
companies. No streets were paved, and frequent rains kept most of them
muddy. It was hard to keep the tracks from sinking out of sight.[45] A
South-Western reporter gave this description of the streets: "We floundered
in the darkness from one mud-hole into another in such rapid succession
that we began to entertain serious fears that we would be swallowed up and
heard of no more forever."[46] The newspapers attacked the city for not
keeping the roads in better shape throughout the early 1870's. Finally
the city paved the streets with wooden planks, but this was not satisfactory.
The planking was uneven and easily broken. At last beginning in June, 1877,
Texas Avenue was paved with rock.[47]

The downtown streets were kept in good enough condition to save the
city railroad. In May, 1873, business was good enough to justify replacing
the one old switch with two new ones. This allowed the company to double
its service, running four cars at the same time rather than two.[48]

The Fairfield Railway Company did not see such success. Louis Hennick,
in his book, *Louisiana: Its Streets and Interurban Railways*, attributed
this failure to the yellow fever epidemic in 1873 in Shreveport, and a
business depression which occured the same year.[49] In truth, there were
still other reasons for failure. The main problems of the Fairfield line
seem to have been poor management and maintenance of the tracks, partially
due to the condition of the streets. The business depression of these
years did not greatly affect Shreveport, according to the Shreveport *Times*.
Though Shreveport lost some of its newfound Texas trade, it was able to
hold up under the economic pressure.[50] The yellow fever epidemic cut
drastically into the white population of Shreveport, but a special census
taken in 1875 showed Shreveport's population as 7,066, a fifty-five percent
increase over the U. S. Census of 1870.[51] The epidemic did cause the deaths
of some of the leading stockholders in the company, Judge Henry Hall among
them, but since the lessee ran the company, this should not have contributed
to its decline. The growth of Fairfield was slowed for a time, but there
were still enough people to support the line. On April 1, 1875, the
residents of Fairfield held a meeting at the depot building to discuss the
future of the streetcar company, and "to prevent the contemplated inter-
ference."[52] The condition of Fairfield Avenue was worse than most of the
other streets in Shreveport. As the Shreveport *Times* reported, there was
one hole "in which a wagon almost disappears from view," and already in the
hole there were "two or three dead horses and several 'bogged' wagons."[53]
Cars frequently ran off the track. The *Times* recorded such an accident
which took place on July 19, 1875. The car was derailed and overturned.
The one passenger and driver received only minor injuries, but the *Times*
states that "If the car had been crowded at the time of the accident serious
injuries might have resulted."[54] The Fairfield Railroad Company declared
bankruptcy on December 28, 1876, and the sheriff seized it. Captain Peter
Youree purchased the line at auction on February 3, 1877, for $500 and
made no attempt to restore it.[55]

87

159

An authority on Louisiana street railways has noted that the business depression and Yellow Fever Epidemic of 1873 were reasons for the failure of the Texas Avenue Railway Company.[56] This company, however, was formed in January, 1872, and was unable to raise the necessary funds long before these two things began in 1873. The citizens of Shreveport were skeptical of the success of one streetcar line in their small city. Doubting even more the chances of a third line, the people were not willing to give the idea the financial support it needed to develop.

The Shreveport City Railroad Company was the only surviving horse-car line in Shreveport. This small line, though perhaps insignificant in itself, began the public transportation system of Shreveport. The City Railroad did not remain the only streetcar in Shreveport. In 1890, the electric cars of the Shreveport Railway and Land Improvement Company began operation. The two companies had a strong feeling of rivalry, and in 1892, the Shreveport Railroad installed electric cars, and put their mules out to pasture. In 1902, the Shreveport City Railroad Company and the Shreveport Railway and Land Improvement Company merged to form the Shreveport Traction Company. The Shreveport Railways Company took over the Shreveport Traction Company in 1914. In 1937, streetcar service in Shreveport stopped, and trackless trolleys and buses began to do its work.

The streetcar companies laid the foundation for the buses of the Shreveport Transit Company of today, with the Shreveport City Railroad Company as the cornerstone.[57] On the Fifty-Seventh Anniversary of the street railway in Shreveport, Captain H. B. Hearne, president of the Shreveport Railways Company, stated, "Were it not for the Street Railway our city could never have expanded to its present size and importance."[58] This was the basic contribution of the streetcar, allowing Shreveport to expand and spread out as it grew.

NOTES

1. Burton J. Hendrick, The Age of Big Business, New Haven, 1919, 119.
2. Alvin F. Harlow, "Street Railways," The Dictionary of American History. ed. James Truslow Adams, New York, 188-189.
3. Burton J. Hendrick, The Age of Big Business, New Haven, 1919, 119.
4. South-Western, October 12, 1870, and Shreveport Times, December 23, 1871.
5. South-Western, January 12, 1870, and March 2, 1870. Also, Louis C. Hennick and E. Harper Charlton, Louisiana: Its Streets and Interurban Railways, Shreveport, 1962, 63; hereafter cited as Hennick, Interurban Railways.
6. South-Western, January 18, 1870, and Shreveport Times, February 8, 1872.
7. South-Western, June 8, 1870, and June 15, 1870. Also, Hennick, Interurban Railways, 64.
8. South-Western, June 8, 1870, and Maude Hearne O'Pry, Chronicles of Shreveport, Shreveport, 1928, 134; hereafter cited as O'Pry, Shreveport. Also, Hennick, Interurban Railways, 64.
9. South-Western, June 22, 1870, and August 17, 1870. Also, O'Pry, Shreveport, 134.
10. South-Western, July 6, 1870. Also, O'Pry, Shreveport, 134.

88

160

11. South-Western, August 17, 1870. Also, O'Pry, Shreveport, 134.
12. South-Western, November 9, 1870.
13. Diary of Henry Gerard Hall, MS, Xerox Copy, Centenary College, and October 26, 1870; hereafter cited as Hall, Diary. Also, South-Western, October 26, 1870.
14. South-Western, November 2, 1870.
15. Shreveport Times, December 4, 1872.
16. South-Western, January 4, 1871.
17. Ibid.
18. Ibid.
19. Hall, Diary, January 3, 1871.
20. Hall, Diary, January 8, 1871, January 10, 1871, February 5, 1871, and February 2, 1871.
21. South-Western, October 26, 1870; Shreveport Times, May 2, 1873; Maude Hearne O'Pry, "Highlights of the Shreveport of Yesterday," Shreveport Magazine, VII (March, 1926), 13; and Captain H. B. Hearne, "Fifty-seven Years of Street Railway Service," Shreveport Magazine, VIII (June, 1927), 14.
22. South-Western, January 4, 1871.
23. Shreveport Times, December 4, 1872.
24. Shreveport Times, November 13, 1873, and December 16, 1873.
25. Hall, Diary, November 23, 1870, September 2, 1871, September 9, 1871, October 11, 1871, October 14, 1871, October 16, 1871, October 17, 1871, October 18, 1871. Also, Hennick, Interurban Railways, 64, and O'Pry, Shreveport, 134.
26. Ordinances of the City of Shreveport, Shreveport, 1902, 50.
27. Shreveport Times, February 18, 1872, February 1, 1872, and April 11, 1872. Also, O'Pry, Shreveport, 134, and Hennick, Interurban Railways, 64.
28. Shreveport Times, June 16, 1872, and June 18, 1872, and March 26, 1875.
29. Shreveport Times, January 18, 1872, and February 5, 1872, March 26, 1872, and June 11, 1872.
30. South-Western, October 26, 1870, and Shreveport Times, May 3, 1873. Also, Captain H. B. Hearne, "Fifty-seven Years of Street Railway Service," Shreveport Magazine, VIII (June, 1927), 14.
31. Lilla McLure and J. Edward Howe, History of Shreveport and Shreveport Builders, Shreveport, 1937, 124-125; and Shreveport Times, February 25, 1872.
32. South-Western, November 2, 1870.
33. Ibid., October 26, 1870.
34. Shreveport Times, February 22, 1872.
35. Hall, Diary, June 10, 1872.
36. Shreveport Times, June 19, 1872 and March 6, 1873.
37. Ibid., December 7, 1872.
38. U. S. Department of Agriculture, The Yearbook of Agrilculture--Animal Diseases, Washington, D. C., 1956, 534-536.
39. Shreveport Times, December 17, 1872, December 19, 1872, and December 21, 1872.
40. Hall, Diary, January 14, 1872, January 28, 1873, and January 29, 1873.
41. Ibid., January 13, 1871.
42. Shreveport Times, January 1, 1876.
43. Ibid., September 9, 1875.

89

44. Maude Hearne O'Pry, "Highlights of the Shreveport of Yesterday," Shreveport Magazine, VII (March, 1926), 13.
45. Hennick, Interurban Railways, 64.
46. South-Western, January 26, 1870.
47. Shreveport Times, June 1, 1877, March 6, 1873, January 4, 1871; and South-Western, February 12, 1870.
48. Shreveport Times, May 3, 1873.
49. Hennick, Interurban Railways, 64.
50. Shreveport Times, March 31, 1875.
51. Ibid., August 24, 1875.
52. Ibid., April 1, 1875.
53. Ibid., March 6, 1873.
54. Ibid., July 20, 1873. Also Hennick, Interurban Railways, 64.
55. Hennick, Interurban Railways, 64.
56. Ibid.
57. Ibid., 65-85.
58. Captain H. B. Hearne, "Fifty-seven Years of Street Railway Service," Shreveport Magazine, VIII (June, 1927), 14.

90

(Eight Minutes
to New York)

Boring ever deeper
into the soft, sticky
mud beneath the
riverbed, the men who
built the tunnels

All pictures accompanying this article are courtesy of The Port Authority of New York and New Jersey.

The Story of the Hudson & Manhattan Tubes

*daily risked death
from drowning and
"caisson disease" in
an all-out effort to
conquer the Hudson.*

By FRANK E. JOHNSON

AL VIEW of NEW TUBE.

OO TUNNEL.

Photo Only COPYRIGHT 1900.
By The H. HAGEMEISTER CO. N.Y.

When they were first built, America's railroads originated in and served the East Coast. Their inexorable route westward was one of the nation's great epics, but when a transcontinental rail network finally became a reality, the route still extended eastward. New York City was then, as always, the gateway.

Most cities could be linked with the rest of the nation fairly easily, but New York was more challenging. The Hudson River allowed it to become one of the world's major deep-water ports, but the river remained a formidable obstacle to the railroads. During the first half of the 19th century, railroad chiefs, engineers, and city government proposed various ways to link the city with New Jersey, and none of them amounted to much. The river had been plied for many years by ubiquitous ferries, which seemed adequate to handle the passengers that swelled the railroad terminals on the west side of the river.

But the ferry was conditionally adequate. In a short time it proved to be less and less capable of meeting the demands of increased railroad passenger haulage. Sometimes the journey across the river was longer than a passenger's previous train trip, though the Hudson's breadth was only little over a mile. During the winter, when snow and fog lessened visibility, the ferries were hazardous. In a cold spell ice floes coursed down the river, clogging traffic and creating even greater congestion. And ferries were dangerous, too. John Augustus Roebling, the designer of the Brooklyn Bridge, would lose his life to one. While supervising the bridge's construction—which he would never live to see completed—he suffered a crushed foot as a ferry docked, and he subsequently died miserably of lockjaw.

Though the Brooklyn Bridge had been started in 1870, its construction had only then been possible as a result of concurrent engineering breakthroughs. One of these was the caisson, a pressurized tank that allowed digging to go on in mud or sandy soil. Compressed air would be sent down from above-ground apparatus, entering the caisson underground and keeping it pressurized. Workmen would enter the caisson

13

165

With this article we welcome Frank E. Johnson, assistant editor of Reader's Digest General Books Division, to the pages of AHI. His sources on this topic include William G. McAdoo's Crowded Years and Joseph Gies's Adventure Underground: The Story of the World's Great Tunnels.

through an airlock, permitting the inside pressure to be maintained at all times. On some caissons, the compressed air could be used to expel the diggings, thus turning the caisson into a gigantic sluice.

James Buchanan Eads, engineer and Civil War hero, was one of the first bridge designers to use caisson construction effectively. His life's triumph was the magnificent St. Louis bridge over the Mississippi River, one of the largest in the world at the time. Work on the bridge began in 1867. Eads adopted the technique of first lowering caissons to the river bottom from barges, then excavating and securing a firm foundation upon which three great arches rested. The bridge was finally completed in 1874 and was named the Eads Bridge in his honor; for this, Eads won international acclaim.

Among the observers of this triumph was San Francisco-born DeWitt Clinton Haskins, then on a transcontinental tour of the United States and on a brief stop-over in St. Louis. Haskins had been successful as a financier in California, and had accumulated a fortune. With a flair for promotion, he carefully studied Eads's caissons as they were slowly lowered into the river. Haskins reasoned: Why not use the caissons as a means of driving a tunnel through the river bottom instead of using them merely as a foundation?

Ironically, tunnel construction was a far older—and technically more advanced—science than bridge building in 1870. By that time tunnel engineers all over the world were using the theodolite, a device that measures the exact location of underground diggings by comparing reference angles to those obtained with sighting instruments. As a contrast, steel girders used in bridgework represented a newer and considerably more exact science.

But tunnel designers encountered a snag when they tried to work in soft river bed mud. It was easy to penetrate the yielding silt—all too easy. A tunnel dug by conventional means would quickly collapse in this sort of material. In addition, treacherous water pockets within the riverbed could give the mud an almost fluid consistency, threatening the lives of anyone working in such volatile material.

It was precisely in the soft riverbed silt that Haskins wanted to test his theory. He would use a caisson to dig a tunnel; not vertically, so as to support a bridge,

14

but horizontally, creating a newly formed path as it bored through the mud. Haskins would pressurize the caisson to "hydrostatic" levels by making the internal pressure of the caisson equal to the outside water pressure, and then increase it slightly. Careful maintainance of the pressure could keep water out of the diggings. After a length of cut was made, the resulting tunnel would be temporarily supported by timber, and finally by giant iron rings on the sides of the tunnel, replacing the timber as the tunnel advanced.

This approach was completely revolutionary. Skeptics were quick to point out the difficulty of regulating hydrostatic pressure, to say nothing of the logistics involved in getting the iron rings in place and lining the tunnel with them. More conservative thinkers believed the ferries to be perfectly adequate, all the more so because many of the railroads owned them and did not want to see their investment threatened by any newfangled tunnel.

Though he was aware of the obstacles confronting him, Haskins rushed excitedly to New York. There he met Trevor W. Park, another financier, and the two men raised over $10 million worth of capital for a tunnel company, to be called the Hudson Tunnel Railroad Company. The corporation was founded in 1871, although the first caisson was not sunk until November 1874. Haskins had sold a lot of people on his idea; that is, with the notable exception of the railroad bosses. At the Hoboken, New Jersey terminal, where the caisson was slowly descending, the Lackawanna Railroad put a temporary halt to the diggings by obtaining an injunction. The reason? The route for the tunnel paralleled that of the railroad's Christopher Street ferry.

The injunction enabled the railroad to hold up construction of the tunnel for five years. It is ironic to observe that, though the tunnel theoretically would compete with the railroad-owned ferry, it could only be beneficial to the Lackawanna by allowing its traffic to move more quickly and attract more passengers. But the Lackawanna was preoccupied with a tunnel of its own through Bergen Hill in 1874. Much blasting had been required to provide ventilation shafts and there had been some cost overruns; the Lackawanna wanted to be sure of all the money they could get.

Finally, work on Haskins' tunnel began once more in September 1879. During this time of inactivity a bit of encouraging news had come from abroad. In Belgium a small tunnel had just been completed under the Scheldt estuary in Antwerp using similar compressed air excavation techniques. This information was heartening to Haskins because the clay-silt composition of the Scheldt riverbed was similar to that of the Hudson's, and this tunnel's success implied that others were sure to follow.

President William G. McAdoo (left) and Chief Engineer Charles M. Jacobs of the Hudson & Manhattan Railroad Company.

Haskins' tunnel was started at both sides of the Hudson; the two bores would meet halfway under the river near the state line. The Hoboken caisson reached a working level of about sixty feet; but its progress was often impeded by water leaks, which necessitated giant pumps to keep the workings dry. At sixty feet the caisson was turned horizontally and the tunnel digging commenced. The caisson itself was a cylinder of thick boilerplate iron, fifteen feet long by six feet in diameter. A door located at each end of the caisson opened outward on the tunnel side and inward on the shaft side. Workmen entered through an air lock, waited until at least twelve pounds of air pressure developed inside, and then opened the tunnel door to commence digging. Above-ground compressors worked constantly to keep up pressure, which could be brought to forty pounds per square inch or more.

There was one serious drawback to compressed air when used in caisson excavation. Its effect on the human body, if exposed to high pressures for an extended time, could be fatal. Under high compression nitrogen in the air would fill the blood vessels, and during decompression it would then form bubbles all over the body by seeping through the arteries, producing excruciating pain and frequently death. Workers on

the tunnel soon began to double up with pain, and progress slowed.

The Hudson River was not the only place where this condition—nowadays called nitrogen narcosis but known appropriately in Haskins' time as "caisson disease"—occurred. It had struck workmen on Eads's bridge in St. Louis, and would plague the sandhogs under the Brooklyn Bridge; it could happen anywhere that caissons were used. During construction of the London subway, or "underground," as the British call it, its effects were properly diagnosed by one Dr. Foley —who holds an obscure but nevertheless highly important place in medical history—in 1863. Foley reasoned that the effects of caisson disease were proportional to the speed of decompression. The symptoms could thus be relieved by recompression for the victims and prevented by slow decompression for all workers in the caissons. Used properly, Foley's treatment would insure the safety of many tunnel and bridge workers.

However, Foley's method was unpopular with hard-nosed construction bosses, and many of them actually refused to employ it. The disadvantages were that slow decompression after working in the caisson took an hour or more, lost time from work on the tunnel. Furthermore, many bosses maintained that tunnel

15

167

work and similar construction was obviously risky: Didn't the workmen realize this when they signed up for jobs in the first place? Caisson disease and other appalling accidents first occurred long before the days of powerful unions and minimum wages. The tunnel pushed on, though hundreds of workers were to lose their lives to compressed air until Foley's method of decompression became widespread (and, more important, mandatory) in construction sites.

But other events in the tunnel would finally put a temporary halt to the misery in the caisson. In their eagerness to push through the soft clay as quickly as possible, engineers on Haskins' project neglected to keep pressure in the caisson close to hydrostatic. This not only worsened the incidence of caisson disease, it threatened the whole project, because each time the door to the "heading"—or face of the caisson—was opened, air could leak out from the diggings through the silt. So long as the silt was thick enough, and of the proper consistency, there was no immediate danger. But what if the diggings reached a soft spot with too high a working pressure?

Disaster occurred in July 1880, less than a year after construction had been resumed. Leaking air managed to percolate through the mud and cause a "blow-out" of loosened clay, under pressure from the river above. One sharp-eyed workman, Pete Woodland, noticed its beginning as a tiny leak. He dropped his tools, sounded

the alarm, and raced towards the air lock, but the leak suddenly turned into an exploding torrent. Pete instantly made a decision that would turn him into a hero. To protect the other men in the air lock, yet facing certain death himself, he shut the airlock door in front of him and sealed it only seconds before the muddy water came crashing down on him and nineteen other workmen. Eight men in the airlock heard the awful roar and shivered to imagine what their comrades' final seconds must have been like.

To Haskins, the blow-out was more than an impediment; it was the *coup de grace* for his ailing company. Only about 1,800 feet had been completed, and his financial backers pulled in the reins. Two years later the company went bankrupt, and the order was given to seal up the unfinished tunnel. Ruined by the disaster, Haskins faded away into obscurity.

Six years later a British concern, S. Pearson and Son, took over the rights of the bankrupt Hudson Tunnel Railroad Company and tried to push the diggings through under the Hudson. Pearson moved forward only a few feet before it likewise went bankrupt. Meanwhile several railroads were convinced that a Hudson River tunnel was impractical, and President George B. Roberts of the Pennsylvania, with the aid of prominent engineer Gustave Lindenthal, came up with a plan for a high-level railroad bridge over the Hudson. His initial proposal was rejected by the U.S. Army as an obstacle to shipping. A second plan which he presented to the other railroads in 1890 called for an enormous, three-level bridge with a single stone arch extending for over 3,000 feet. Fourteen tracks

The entrance to the 19th Street station on the uptown branch in 1908. This station was closed in 1954.

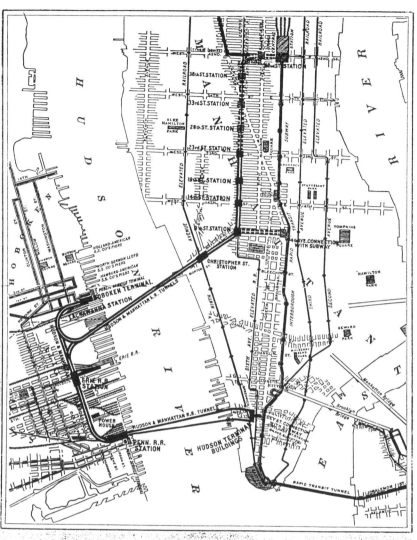

Map of the Hudson & Manhattan Railroad Company's tunnels.

17

169

One tube is completed (right) and work begun on another.

would have used the structure, and it could be had, said Roberts, for the stiff sum of $100 million.

It should be kept in mind that $10 million had been almost enough to get Haskins across the Hudson under water. No other railroad magnates shared Roberts' enthusiasm. "For that kind of money," they replied, "let 'em ride ferries." Besides, they said, the bridge would have crossed the river at 59th Street, and most city planners believed that New York would never extend that far uptown!

In the interim an event took place on the other side of Manhattan that was to prove extremely conducive

One of the shields used in burrowing under the river.

Sand hogs at work tunneling under the Hudson River.

to the eventual success of the Hudson tunnel. Charles M. Jacobs, a leading construction engineer, had been delegated to build a gas tunnel underneath the East River, which would pipe gas manufactured from a plant on Long Island to Manhattan. Compared to the Hudson River tunnel, building it was even more difficult because the East River bed was a mass of crumpled rock, and the tunnel frequently met with solid walls of bedrock that had to be blasted out with dynamite (this was largely unnecessary under the Hudson). The tunnel was begun in June 1892, and it took two years to complete, during which a major depres-

The first train through the tunnel under the Hudson River.

19

sion on Wall Street in 1893 almost halted all work. But one of the triumphs on completion was that, though two separate tunnels had begun from either side of the river and during construction their course had been impeded by blow-outs, solid rock, and widely-varying strata, they met under the river scarcely an inch out of alignment.

Other successes abroad also revived interest in the Hudson tunnel project, in particular the London tube tunnels under the Thames, which were finished in 1897. But the greatest incentive of all came from Georgia-born (1863) William Gibbs McAdoo, who would later become Woodrow Wilson's Secretary of the Treasury. McAdoo combined legal acumen with some experience in running railroads, for he had taken part in a venture to electrify the streetcars of Knoxville, Tennessee. He moved to New York in 1892 and there met Jacobs shortly after the East River Gas Tunnel had been completed. Both men were optimistic about finishing up the work already begun by Haskins. Via a bond sales scheme, McAdoo purchased the plagued Hudson River Railroad Tunnel Company and re-named it the New York & New Jersey Railroad Company, making Jacobs the chief engineer for the project. Later the company consolidated with several others, and it became known as the Hudson & Manhattan Railroad, or the "H & M" for short. One important discrepancy from the start was that the tunnel would operate independently from the railroads, although naturally it would be so placed that it served the terminals on the western shore. This posed several advantages for the company. First, the tunnel clearance could be drastically reduced since conventional-sized cars and locomotives would not be required; instead, much smaller subway cars would be utilized in "multiple units": Each would be able to travel in either direction with a cab at both ends. Secondly, the smaller tunnel clearances would allow the speed of construction to increase because the size would not have to be as large. Finally, by operating independently from the railroads, they ended the threat that by further consolidation or buying up interests, the railroads would eventually monopolize the company.

McAdoo and Jacobs had learned enough from Haskins' failure not to use compressed air and caissons, and instead employed a "shield"—used with great success on the London subway tubes—which consisted of an inclined ram pushed by hydraulic jacks. Work started in 1902 from the Jersey side, and because of the new process, progress was quite rapid; the average rate was about three feet a day. In just under two years the tunnelers found what they were looking for: the bulkhead that marked the end of Haskins' tunnel, extending from Morton Street in Manhattan. The work-

men promptly notified McAdoo at his desk, and after he quickly changed into oil-skins to accompany a number of other dignitaries, William G. McAdoo proudly walked under the Hudson from New Jersey to New York, the first man ever to do so.

After this breakthrough a second, companion tunnel was built to parallel the first one uptown to Sixth Avenue. In addition, two other downtown tunnels were constructed from the Pennsylvania Railroad terminal in Jersey City to Church Street in Manhattan. At Jersey City the new station was called "Exchange Place," so that the Pennsylvania's New York-bound passengers could continue on to Manhattan. The usefulness of Exchange Place lessened in 1910 when the Pennsylvania's new terminal in New York was completed. Nevertheless, Exchange Place continued as a link between Hoboken and Wall Street; it still exists today, whereas the former Pennsylvania terminal building in Manhattan does not.

No major difficulties were encountered with the downtown tunnels, but a blow-out did occur during the construction of the second uptown tube. A workman on the night shift had evidently been dissatisfied with the progress of the tunnel and had opened up a door on the shield to speed its advance. As the shield was under high pressure at the time, a wall of mud shot through the open door, and the diggings collapsed. One workman perished as the tunnel was overwhelmed. Jacobs devised a series of canvas sails to be lowered from a river barge over the leak. These, when weighted down and secured, effectively plugged the leak and allowed digging to resume.

With the aid of expert engineering from Jacobs, solid financial backing from McAdoo, and modern equipment and techniques borrowed from other successes, the Hudson tunnels were soon to be finished. Newspapers ceaselessly carried stories of the street on Sixth Avenue being torn up in preparation for the tracks beneath; after the first tunnel was completed, excitement mounted as the system moved closer to being finished. The two uptown tunnels reached 19th Street and Sixth Avenue by 1908, and on February 25 of that year the first subway train left from this location for Hoboken. A number of dignitaries had been invited for the occasion; the station and the eight-car train were cloaked in darkness. Seated at his desk in the White House, President Theodore Roosevelt was given the honor of signalling the electrification of the third rails. At exactly 3:30 p.m., Roosevelt telegraphed the order, and the station was a blaze of light.

21

As the train sped swiftly through the tunnel, with McAdoo and many of New York's leading personalities aboard, several elegantly clad passengers were apprehensive as to the security of the iron supports on the side of the tunnel. But the ride did not last long, for a circle of red, white, and blue lights marked the exact point of the New York-New Jersey boundary. The train paused at this decorative location; New York Governor Charles Evans Hughes jubilantly shook hands with New Jersey Governor Franklin Fort. The two states were "formally married."

After a running time of only eight minutes, the train pulled into Hoboken, and a crowd of 20,000 people filled the square outside the station. The phrase "Eight minutes to New York!" was on everyone's lips, and it seemed to be an apt slogan for the company. Later, however, the running time took considerably longer than eight minutes when the system was completed, and McAdoo's own motto, "The public be pleased!" more appropriately symbolized the dawn of a new customer-oriented era in railroading.

A year later the downtown tunnels were completed to the Church Street station, which later came to be known as "Hudson Terminal." In 1910 the Sixth Avenue uptown line reached the 33d Street terminal. By 1911 McAdoo extended the system to Newark,

ABOVE: Air-conditioned PA-3 rapid transit cars of the PATH system. OPPOSITE: One of the first trains carrying passengers under the river.

bringing the total cost of the Hudson tubes upon completion to nearly $70 million. Meanwhile the Pennsylvania Railroad, influenced by McAdoo's success, dug a mainline railroad tunnel under the Hudson, ending at their new 34th Street terminal in Manhattan. At the same time the Pennsylvania's subsidiary, the Long Island Railroad, burrowed under the East River, and in 1917 a great arch span over the East River's Hell Gate connected the system with upstate railroads. This allowed the Pennsylvania, via the New Haven Railroad, a through route from New England south to Philadelphia and Washington, D.C.

These glittering accomplishments quickly made McAdoo's innovations pale by comparison, and as railroad passengers quietly got down to the routine business of commuting, the various new routes soon got taken very much for granted. But not McAdoo himself. Quitting his job as president of the Hudson & Manhattan Railroad in 1913 to serve in the Wilson

Plaque commemorating the opening by President Theodore Roosevelt of the Hudson & Manhattan Railroad Company's tunnel under the Hudson River in 1908.

22

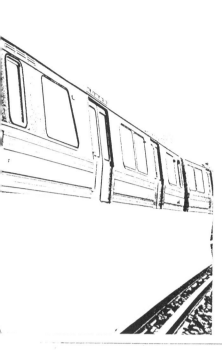

Administration, he became Wilson's son-in-law the next year when he married Eleanor Wilson. McAdoo's skills as a railroad administrator were realized during World War I when he was appointed director-general of the United States Railway Administration (USRA). He missed the Democratic presidential nomination in 1924 only by a hair's breadth and served as senator from California until 1932. At the time of his death in 1941 he was quite a wealthy man as a result of his profitable Hudson & Manhattan Railroad. Though in an effort to show how purportedly small his holdings were McAdoo pointed out that he received "only $50,000 a year," in truth his stock ownership amounted to much more than that.

Shortly after the Hudson tube tunnels were built, they were unofficially nicknamed "the McAdoo tunnels," a name which William G. McAdoo, a hater of notoriety, personally disliked. Yet a tunnel, bridge, or highway named after a man signifies his triumph over nature, and in this sense perhaps McAdoo should have felt honored. But success had been made possible by the unsung sufferings of men like Pete Woodland, and of others who perished from caisson disease; it would be difficult to name the tunnel for all of them.

Forty-six years after the first subway run under the Hudson, the 19th Street station was closed; one can no longer visit the historic site. It is ironic that, practically on the centennial anniversary of Haskins' original Hudson River tunnel project, the Hudson tubes were paralyzed by a sixty-three-day-long strike, a grim reminder that the glorious days of railroading are now over.

23

175

The Horse Distemper of 1872
and its
Effect on Urban Transportation

by

Sister Denise Granger

Through most of the nineteenth century, urban transportion
depended upon the horse. In October of 1872 a raging epidemic of
"horse distemper" paralyzed the east coast of the United States,
demonstrating that the horse was not nearly as dependable as it
was generally believed to be. Entire cities came to a standstill;
food could not be delivered; fires could not be extinguished; it
was virtually impossible to travel within the city except on foot.
The distemper first appeared in Canada and spread through New
York State, across New England, and as far south as Pennsylvania.
By October 23rd, the disease was reported in Syracuse, Boston,
Springfield and Philadelphia. The Secretary of the Treasury,
George S. Boutwell, reacted to the situation by prohibiting the
importation of Canadian horses into the United States.

The symptoms of the "horse distemper" were a sore throat,
slight swelling of the glands, a loss of appetite, a severe hacking cough, fast pulse, quickened respiration, great feebleness on
the part of the animal, and a yellowness of the eyes and mucus
membrane. The illness, lasting a period of days, was rarely

43

177

Lith. & Print of A. Brett, 39 Walnut St. Phil.

fatal if properly attended to. It did, however, totally incapacitate the animal.

New York City already was a sprawling urban complex crisscrossed by tracks of the so-called "street railroads," which were horse-drawn streetcars. These railroads were used by businessmen to haul merchandise, by workers going to and from work and by the public which relied on the street railroads for transportation. When the distemper hit, these vital links in urban transportation were broken.

There was such concern over the "Horse Plague" that the New York Herald appointed a reporter to be the "Herald Horse Plague Commissioner," whose assignment was to investigate the stables in New York City. In an article entitled "The Poor Beast," dated October 23, 1872, the "Horse Plague Commissioner" described a ride on the Bleeker Street Lines and an interview with a driver. The reporter asked: "How do your horses feel today, my friend?" The driver answered: "on this line they feel pooty bad, I can tell you, they always feel bad on this line, they aren't treated right and they know that they're only worth $130.00 a piece."[1] The Bleeker Street Line seems to have received bad press all through the reports of the distemper. The route was a notoriously hilly one with a number of corners making it rough on horses and drivers alike.

The reporters assigned to cover the situation found it difficult to obtain information. The people connected with the stables were afraid that if it became public knowledge that a large number of their horses were unable to work, the company's stock would depreciate.

The "Horse Plague Commissioner" of the _Herald_ was kept busy during these days. On October 25th, he reported that sixteen thousand horses in the city were too sick to work, and that many stages and street railways had been halted; the inevitable result was increased irritation to all segments of the population.

The veterinarians enjoyed their finest hour, being in constant demand. In an editorial, the _New York Times_ praised the "Horse Doctors." Although the _Times_ noted that veterinary medicine was not a respected profession, it went on to indicate that "the epidemic among horses has given in the eyes of the public an unusual value to the service of persons skilled in treating the disease of the lower animals...." The editorial continued by stating that although the work of veterinarians was "of infinitely less consequence than treating the malaise of human beings...we do not see why it should be regarded as a matter of inferior dignity...."[2]

However respected the veterinarians were, there was some confusion among them in diagnosing the illness. Some labelled it

pneumonia or diphtheria, while others thought it was either a
liver disorder or blood poisoning. There was general agreement,
though, that the most likely cure was to treat the animal as if
it were a human being with the same symptoms. The Springfield
Daily Republican advised horse owners that "affected animals
should be kept in a well-ventilated stable and fed upon hot,
soft food easily digested such as linseed tea, oatmeal gruel and
boiled oats or barley. The appetite of invalids should be tempted
by carrots, apples or any other delicacy our sick, quadruped
friend is known to have a failing for. In all stables a carbolic
disinfectant should be liberally used."[3]

New York City officials stated on October 23rd that sanitation
was vital to the containment of the disease and that strict measures would be taken to force companies to clean up their stables.
The Herald reported that the city government would quarantine the
worst offenders unless considerable improvement could be seen within twenty-four hours.

On October 26, 1872, the New York Times reported that the
disease was spreading, with "car companies compelled to reduce
their forces; partial suspension of business in the livery stables"
and "travel and business impeded." The Times noted that the
"number of affected horses increased nearly 60% yesterday with
comparatively few horses being left to do the work." For instance, 74 of the fire department's 141 horses were "laid up."[4]

During the last few days of October, the ravages of the horse distemper were most conspicuous. In New York City traffic came to a stop, and it was reported that business was at a standstill on the 29th. In an article entitled "Streets Without Traffic" the New York World reported "a common feeling of common inconvenience and common losses for everyone..."[5] In some sections of the city, notably at the great market centers like Jefferson and Washington businessmen turned to oxen for hauling their vegetables and provisions. On October 30 a prominent citizen, George Templeton Strong noted in his diary that the "horse distemper raged. Saw an ox team on Broadway." He predicted: "They will have to utilize the elephants and camels of Central Park..."[6] On the same day the New York Sun reported that the fruit dealers were devastated due to their inability to get their produce to market; the cotton market too was suffering heavy losses with thousands of bales piled high on the city's wharves.

Most of the truckmen, grocers, and hackmen were hurt financially, but those whose horses somehow avoided the distemper could name their price. A gentleman offered a hackman $12.00 to take him from the Courtland Street Ferry to the 42nd Street Depot, but the hackman refused the offer and waited for a better one. Another journeyman demanded and obtained $42.00 for transporting one load of cotton.

The distemper was a disaster not only to New York, but it was to disrupt business elsewhere. The <u>Springfield Daily Republican</u> reported on Monday, October 28, that the "one absorbing topic of conversation and general inquiry yesterday was the horse disease. Politics were forgotten and, we fear, the thoughts of many of the worshippers were divided between worldly and spiritual matters...."[7] Nine of the thirteen largest stables in town had been hit by the distemper, and the remaining four were expecting its arrival momentarily. A reporter observed that even in stables where every precaution was taken, "the disease seemed to rage quite as severely as those in sections more exposed, and the greatest care provies entirely unavailing."[8] An example was made of the stable of E. W. Burr on Liberty Street, located in the lowest dampest section of the city. Burr's horses were healthy while those on the Hill just up the road had been severely affected.

Later in the week the <u>Republican</u> reported that the ailment had crossed the Connecticut River into West Springfield. Palmer, Pittsfield and Deerfield were also reporting cases, with a rare fatal case having occured in Deerfield.

New York City, Syracuse, Boston, Springfield, Hartford, Newport and Philadelphia were all attacked regardless of the precautions taken or the care given. Cornelius Vanderbilt found

that his vast resources were not enough to stop the disease. His
trotter "Mountain Boy" died on November 15 after a three week
bout with the distemper. When interviewed by a reporter, Vanderbilt sadl- said that "Mountain Boy" "was the kindest, and best of
horses.... I never had a horse I loved so well."[9] Vanderbilt had
the horse buried in the same building that contained his carriage
house and stable.

By early November the situation was critical. All along
the eastern seaboard, businessmen were losing vast sums of money, and
that proved to be a great incentive to seek a solution to the
problem.

The New York City Board of Aldermen took the matter in hand
on November 20, 1872. Representatives from two steamcar companies
met with the Railroad Committee of the Board and a resolution was
passed permitting the use of "dummy engines" on the city's streets
for three months. F. M. Peck, representing the Remington Steam
Car Company, proposed that his car could be stopped safely within
six feet when moving at four miles per hour. He claimed that
the cost of running a Remington Steam Car was less than the
expenditure for a team of horses. The Remington car was tested
on Elm Street, on the Bleeker Street route, and it was able to
overcome the steep grade which was one of the most difficult
in the city.

Not to be outdone was U. M. Camp of the Langdon Steam Car
Company. He claimed that his car provided greater utility in that

it could run either way and could pull three loaded cars at once. The Langdon car had been tested on the Brooklyn and Flushing Road, where it reached a speed of thirty miles per hour. Camp and Peck both stated that their machines were noiseless and as safe as the streetcars then in use.

The Railroad Committee decided in favor of the Langdon car, which was brought into New York City on an experimental basis. On November 22 the Herald reported the trial run. A committee of the Board of Assistant Aldermen boarded the car at city hall, and they headed for the uptown depot of the Bleeker Street Line. Unfortunately, when the car reached Howard Street it hit a horse that was standing on the track. Luckily there were no injuries and the Herald attributed the mishap to the fact that the sand boxes on the steam car were not in working order. Cheering children lined the streets to cheer on the car and it attracted the stares of many passers-by, who all seemed enthralled by the novelty of a horseless street car.

This experiment and others like it all across the country marked the end of the horse-drawn streetcar and the dawning of a new era in city transportation. The steam car, with all its assets—speed, safety, security and economy—was here to stay; it was never to be stopped by a hacking cough, swollen glands, and the other symptoms of the "great horse distemper of 1872." The great calamity

186

of the horse epidemic drew on the energies of man's inventiveness and budding technological know-how to produce a step toward transportation as we know it today.

References

1. New York Herald, October 23, 1872.

2. New York Times, October 29, 1872.

3. Springfield Daily Republican, October 21, 1872.

4. New York Times, October 26, 1872.

5. New York World, October 30, 1872.

6. The Diary of George Templeton Strong: Post War Years, 1865-1875, edited by Allan Nevins and Milton Halsey Thomse (New York, 1952), p. 448.

7. Springfield Daily Republican, October 28, 1872.

8. Ibid.

9. New York Sun, November 16, 1872.

Settlement Across Northern Arkansas as Influenced by the Missouri & North Arkansas Railroad

By LAWRENCE R. HANDLEY*

University of California

THE MISSOURI & NORTH ARKANSAS RAILROAD (M&NA) ran from Seligman, Missouri, to Helena, Arkansas, 303 miles within the state of Arkansas (Fig. 1). The railway was conceived during the era of prosperity and optimism that encompassed the United States in the 1880s. It first flourished as a small branch line, but like many railroads caught in the capitalistic expansion of the times, the M&NA began to extend its main line, an extension from eighteen to 364 miles that meant foreclosure, receivership, and sale eight times in sixty years.[1]

*Mr. Handley received his B.A. and M.A. degrees from the University of Arkansas and is presently working on his doctorate in geography at the University of California at Berkeley. This article was taken from his thesis, "Geography of the Missouri & North Arkansas Railroad."

[1] The Eureka Springs Railway Company was chartered June 26, 1880, and, along with the Missouri & Arkansas Railroad Company, chartered September 21, 1880, was to construct the railroad from Seligman, Missouri, to Eureka Springs, Arkansas. They were combined February 27, 1882, to form the Eureka Springs Railway, which was foreclosed and sold on May 17, 1899, to the St. Louis & North Arkansas Railroad Company which was foreclosed and sold on May 26, 1906, to the Missouri & North Arkansas Railroad Company. On April 10, 1922, it became the Missouri & North Arkansas Railway Company. Foreclosed and sold on April 16, 1935, the railroad became the Missouri & Arkansas Railway Company. On November 15, 1946, it was sold to Mayer P. Gross & Associates who operated sixty-five miles of the railroad from Seligman, Missouri, to Harrison, Arkansas, under the Arkansas & Ozarks Railway. They sold fifty-four miles to the Helena & Northwestern Railway Company and abandoned, in February 1949, 176 miles of track. The Arkansas & Ozarks Railway

The railroad was plagued with problems. The topography of the Ozark Plateau presented many obstacles to construction; the flat Mississippi Valley was poorly drained and suffered periodic flooding. The population was sparse throughout the Ozark Plateau and there was a dearth of natural resources, accounting for the region's limited agricultural and industrial productivity. The railroad was poorly constructed at the outset and maintenance was never adequate. Intermittent mismanagement made success of the railroad an impossibility. The management failed at various times to maintain the railroad properly, to provide improvements, and to deal tactfully with management-employee problems. High wages cut enormous funds out of the revenues each year. It was high wages and mismanagement that led to the longest strike in United States railroad history levied against the M&NA in 1921-1923; another strike, in 1946, led to the abandonment of the railroad. Misfortune always seemed to appear when the railroad was meeting with some financial success. A train accident in 1914 that killed 42 passengers, floods in 1927 and 1945 that washed out miles of track, the deaths of important individuals at crucial times in the railroad's history, and several of the railroad's facilities destroyed by fire, are examples of hard luck that plagued the railroad at inopportune times. Thus, with all these problems the railroad was a business failure from its earliest years.

However, the railroad was a success to the people it served. It became an institution to the people of the Ozark Plateau who centered their entire lives around it. The railroad had a profound effect upon settlement along its line throughout northern Arkansas. It spawned no less than thirty-three settlements, caused the demise of at least five communities, and fostered the rapid growth of several existing towns.

There were some communities, such as Carrollton, Duff,

was abandoned April 7, 1961. The Helena & Northwestern Railway abandoned forty-eight miles of track November 5, 1951, and sold six miles to the Cotton Plant-Fargo Railway in January 1952, which is still operating today.

Settlement, Salt Springs Barrens, and Old Mt. Pisgah, that actually disappeared as a result of the railroad missing them. There were the communities that saw their inception as a result of the railroad, such as Alpena, Everton, Pindall, Gilbert, Baker, Elba, Arlberg, Shirley, Edgemont, Higden, Letona, Armstrong Springs, McClelland, Fargo, Aubrey, Rondo, and West Helena. There were several existing communities along the route of the railroad that showed phenomenal growth as a result of the railroad, such as Harrison, Leslie, Pangburn, Cotton Plant, and Heber Springs.

In 1924 there were twenty-four agency stations along the M&NA with a total population of 76,633, which included Joplin, and Neosho, Missouri.[2] Fourteen of these agency stations were not on some other railroad as well, but their population was only 15,374. The M&NA also operated sixty-six non-agency stations, of which twenty-two had no population or post office, and the other forty-four had populations ranging from six to 304.

It was the phenomenal growth of Eureka Springs, the Arkansas resort town, that led to the inception of the Eureka Springs Railroad in 1880. On February 3, 1883, the Eureka Springs Railway was opened to service from Seligman, Missouri to Eureka Springs, the terminus of the line. Eureka had a turntable on which to turn the locomotives around, it had repair facilities, including an engine shed and a backshop, and it had a two-story wooden depot that also housed the general offices of the railroad on the second floor.

The St. Louis & San Francisco Railway (Frisco) provided at its Seligman station the origination-termination link for the Eureka Springs Railway. The Frisco passenger department spent several thousand dollars a year advertising the advantages of Eureka Springs to health and pleasure seekers, and in conjunction with the Eureka Springs Railway operated several excursions each week during the months of June through September. These brought thousands of

---

[2]Donald Kennedy Campbell, II, "A Study of Some Factors Contributing to the Petition for Abandonment by the Missouri & Arkansas Railroad in September, 1946," *Arkansas Historical Quarterly*, VIII (Winter 1949), 288.

visitors to the resort town which remained a popular watering place from the 1880s until after the turn of the century.[3] The Eureka Springs Railway, it was estimated, transported over 20,000 persons a year to the spa during these years, as the sick flocked to its health-giving waters, which had a reputation of being of especial benefit to those suffering from Bright's Disease.[4]

The railroad made Eureka Springs a center for stagecoaches and freight wagons operating from Harrison and Huntsville and from several towns in southern Missouri. With the extension of the line to Harrison at the turn of the century Eureka enjoyed a short-lived boom as a center for the railroad builders. The resort town also benefited from the fact that the shops and general offices were located there. When the shops were moved to Harrison in 1911, Eureka suffered a terrific economic blow.[5]

Harrison's growth and prosperity owed much to the coming of the St. Louis & North Arkansas Railroad (St. L&NA), which was the successor to the Eureka Springs Railway. Founded in 1836 as the village of Crooked Creek, Harrison in 1871 became the seat of government of Boone County (created in 1869) after a spirited contest with the town of Bellefonte. Its isolation ended on April 15, 1901, when the St.I.&NA opened regular service from Seligman on the Frisco through Eureka Springs to Harrison, a distance of sixty-five miles. To obtain its prized rail connection, Harrison had paid a bonus of $40,000 and had secured the desired right-of-way from the Carroll County-Boone County line to Harrison.[6]

The railroad was, of course, the making of Harrison. The city became a wholesale and retail trading center for a multi-county territory rich in agriculture, horticulture, live-

---

[3] *Poor's Manual of Railroads, 1884-1924* (New York: H. V. and H. W. Poor, 1884-1924); Cora Pinkley-Call, *Eureka Springs—Stairstep Town* (Eureka Springs, Arkansas, 1952), 37.

[4] *Ibid.*

[5] Otto Ernest Rayburn, *The Eureka Springs Story* (Eureka Springs, 1954), 58; Frances R. Donovan, "I Have Found It" (Unpublished manuscript, Eureka Springs, Arkansas, n.d.), 6.

[6] *Harrison* (Arkansas) *Times*, Dec. 23, 1899.

stock, timber, and mining. Adjacent to the tracks in Harrison there developed large warehouses, storage facilities for lead and zinc ore awaiting shipment, a canning plant, lumber mills, and woodworking plants. During World War I zinc mining boomed in the hinterland about Harrison, much of it, brought in by ore wagons, was shipped from there.[7]

In 1911 E. M. Wise, the general manager of the Missouri & North Arkansas Railroad, successor to the St.L&NA, proposed consolidating the shops at Eureka Springs and Leslie and allowed Leslie and Harrison to bid for their location. The people of Harrison raised $26,000, and provided a site on the town square for the railroad's general office building. This secured for the town the railroad's shops and office, which brought numerous employees and a payroll that amounted to as much as $80,000 monthly at times. The Kirby Building on the town square, built at a cost to the M&NA of $15,000, and five miles of new track in the railroad yard on Crooked Creek, a roundhouse, turntable, and new passenger depot, all costing an additional $175,000, also came as a result of the successful bid. It was no surprise that Harrison's population jumped to 3,477 by 1920, a figure that was 117 per cent higher than that of 1910.[8]

Marshall, seat of government for Searcy County, in 1900 was a mountain village of 260 people. The St. Louis & North Arkansas Railroad, reaching it in 1902, sparked a tremendous growth in commerce and population. Indeed, the town's population tripled between 1900 and 1920, by which time it was an important center for wholesale and retail trade, agricultural marketing, and timber products. Its banks and stores served a large mining area in Marion, Boone, Newton, and Searcy counties, although Marshall

---

[7]*Ibid.*, Dec. 28, 1901; George West, "Railway Penetration into Northwest Arkansas," *Eureka Springs Daily Times,* April 24, 1905, p. 28; Dallas T. Herndon, ed., *Centennial History of Arkansas* (Little Rock, 1922), 897.

[8]Ralph R. Rea, *Boone County and Its People* (Van Buren, 1955), 168; James R. Fair, Jr., *The North Arkansas Line* (Berkeley, 1969), 108; *Harrison Times,* Sept. 22, 1900.

itself was never as large an ore-shipping center as Harrison. But, as at Harrison, the railroad gave impetus to new industries at Marshall—a canning factory, a number of sawmills, two stave mills, and a large stockyard. The Flint Rock Berry Growers Association, an organization made up of strawberry growers, greatly benefited from the new means of transportation, shipping in various years close to a million dollars worth of the fruit.[9]

In 1902 Leslie was a Searcy County hamlet of some fifty souls.[10] Lying in a fine agricultural and fruit-growing district and surrounded by abundant timber, Leslie outdistanced even Marshall once the St.L&NA reached it in 1903. For four years, from 1903 to 1907, Leslie was the terminus of the railroad, which located in the little town its repairs facilities, an engine shed, and a wye for turning trains around. By 1910 her population stood at 1,898, the largest town in the county. She suffered a blow in 1911 when the railroad shops were lost to Harrison, but was partly reimbursed for the loss when the St.L&NA opened a large limestone shipping yard there. Leslie also prospered as headquarters for the construction crews that extended the railroad beyond the town to the southeast in 1907-1909. In 1920 the population stood at 1,472, the loss being largely the result of the removal of the railway shops to Harrison.[11]

Leslie became the largest shipper of timber products on the St.L&NA. Sawmills, lumber mills, stave mills, hub plants, and handle factories were established there to exploit the stands of virgin forests all around the town. The first sawmills and lumber mills in the Leslie area had come in 1901 with the arrival of the Great Western Mill Company of Leavenworth, Kansas.[12] In 1904, the year after the railhead reached Leslie, Geyhauser & Galhausen built the

[9]West, "Railway Penetration into Northwest Arkansas, *Eureka Springs Daily Times,* Apr. 24, 1905, p. 28; Orville J. McInturff, private interview, May 27, 1972.

[10]Clifton E. Hull, *Shortline Railroads of Arkansas* (Norman, Oklahoma, 1969), 66.

[11]"Centennial Edition: 1819-1919," Little Rock *Arkansas Gazette,* Nov. 20, 1919, p. 238.

[12]*Marshall* (Arkansas) *Mountain Wave,* Aug. 17, 1901.

first stave mill there.[13] By 1910 the Pekin Stave Manufacturing Company, the Mays Manufacturing Company, and the H. D. Williams Cooperage Company had located at Leslie.[14] During the next decade some of these plants left the area to be replaced by still others so that by 1920 several new companies were operating at Leslie, such as, the Export Cooperage Company, the Leslie Handle Works, Curtis Mills, and A. L. Barnett.[15]

The largest and most interesting of the Leslie plants was the H. D. Williams Cooperage Company, which arrived in 1907. "This industry has been the making of Leslie," one writer said in 1911.[16] The company moved from Popular Bluff, Missouri, and purchased large tracts of timber in the vicinity of Leslie.[17] It brought in many of its former workers, who were mostly Negroes, built about sixty houses for them and a large hotel in a section of Leslie that became known as "Dink Town."[18] The main factory site at Leslie covered sixty-eight acres and could turn out 5,000 barrels a day,[19] which made it the largest of its kind in the world.[20] The Williams Company operated thirteen portable sawmills in its timber tracts, and built a seventeen-and-a-half-mile standard gauge "Dinky" railroad into the mountains west of Leslie to bring out the timber.[21] This shortline railroad, which used shay-type locomotives to pull the trains of log cars, had extremely steep grades and sharp curves.

Employing 1,200 men and 300 teams of horses and mules[22] and turning out thousands of barrels per day, the Williams Company supplied the M&NA with a large percentage of its annual tonnage and was, of course, directly

---

[13]Orville J. McInturff, *Searcy County, My Dear: A History of Searcy County, Arkansas* (Marshall, Arkansas, 1963), 98.

[14]*Ibid.*, 107.

[15]*Marshall Mountain Wave*, Apr. 13, 1923.

[16]Faye Hempstead, *A Historical Review of Arkansas*, 3 vols. (Chicago, 1911), III, 1623.

[17]*Ibid.*, 1622.

[18]McInturff, *Searcy County, My Dear*, 107.

[19]Hull, *Shortline Railroads of Arkansas*, 66.

[20]Hempstead, *Historical Review of Arkansas*, 1623.

[21]McInturff, *Searcy County, My Dear*, 108.

[22]Hempstead, *Historical Review of Arkansas*, 1623.

dependent upon the railroad for movement of its products. During the M&NA strike in 1921-1923, when the railroad was shut down, the cooperage company was also forced to close down. As a result Leslie lost a payroll of over $7,000 a month.

There was no doubt but that the H. D. Williams Co-operage Company had a "remarkable effect upon the settlement" at Leslie. At its peak, between 1905 and 1920, the little railroad town had four newspapers, four banks (of which only the Leslie State Bank is left today), three hotels in addition to the H. D. Williams Hotel, and one of the finest hospitals of its time. Although the town's economy depended primarily on its timber industry, it was also a large shipping point for farm products, especially fruit. Two canneries, two flour mills, and a creamery were located there.[23] The M&NA railroad had also made these possible. Indeed, as one writer has said, Leslie "owes its birth to that road and has had ups and downs directly in proportion as the railroad has been active or dormant."[24]

As the Missouri & North Arkansas Railroad extended southeast into Cleburne County, Heber Springs, the seat of government for the county, was brought into the orbit of the railway. By rail Heber Springs was 182 miles from Seligman and 117 miles from Harrison. The town received its original impetus as a health spa in the early 1880s, but when the M&NA built into it, it had a population of only 552. Thanks to the railroad, it became a lumbering center and its population doubled by 1910 when it stood at 1,125. Lumbering continued as the little town's main industry through the years of World War II. In 1940, for example, five of the ten industries in Heber Springs were related to the timber industry[25]—fashioning rough and planed lumber, handles of various kinds, railway ties, pilings, poles, and creosoted materials.

---

[23]*Ibid.;* Herndon, *Centennial History of Arkansas,* 900.

[24]David Y. Thomas, ed., *Arkansas and Its People,* a History, 1541-1930, 4 vols. (New York, 1930), I, 772.

[25]Henry A. Thane, "Directory of Industries of Arkansas in 1940" (Unpublished senior thesis, University of Arkansas, 1940), 55.

By 1911 Heber Springs had become an operating division point on the M&NA with engine house, repair and service facilities, a turntable, a small storage yard, one of the largest depots on the railroad, and a substantial community of railroad people. Most of the population increase over that of 1910 was directly related to the railroad. The 1920 population was 1,650.

Cotton Plant, in Woodruff County, was the town that benefited the greatest of all the communities of the Mississippi Valley from the M&NA. The town showed population increases of:

Decade	Population Increase	Percent Increase
1890-1900	29	67
1900-1910	523	114
1910-1920	581	58

In 1908 the M&NA Railroad completed the section of track from Helena to Cotton Plant. This section had been finished early, before the track was finished from Searcy and the northwest, mainly to move the cotton harvested that fall. As the name Cotton Plant might imply, it was located in a very rich cotton producing area of Woodruff, St. Francis, Monroe, and Prairie counties. In 1920 the town had four cotton gins, a cotton compress, and several large cotton warehouses.[26] However, with the coming of the M&NA the focus of Cotton Plant's industry became wood products. In 1909 the Standard Stave and Hoop Mill located there.[27] By 1920 there were five large sawmills and seven woodworking factories producing handles, spokes, staves, packing boxes, shingles, cooperage, and the Southwest Veneer Company which was the largest veneer plant in the state.[28]

There were several existing towns that awaited with rampant anticipation the construction of the railroad toward them. Towns, such as Berryville, Green Forest, and St. Joe, were sure that the railroad would pass through

---

[26]Herndon, *Centennial History of Arkansas*, 900.

[27]"State Centennial Edition: 1836-1936," *Arkansas Gazette*, June 15, 1936, p. 164.

[28]"Centennial Edition: 1819-1919," *Arkansas Gazette*, Nov. 20, 1919, p. 180.

them, since they were major political and/or economic centers for their surrounding areas. However, due to difficult topography the railroad management found it unfeasible to pass directly through these towns. The urban dynamics stemming from their economic or political impetus, from their industrious townfolk, and from their longevity allowed these towns to compensate for the railroad passing the main town by.

In 1899 the citizens of Berryville were very enthusiastic about the extension of the Eureka Springs Railway to Harrison because they thought it would have to pass through Berryville. Berryville, Green Forest, and Harrison were to share the bonus they offered to the railroad for building the extension. Berryville was to secure the right-of-way from Eureka Springs to the east line of Hickory Township and to put up $500.[29] However, the surveyors found that to build the line through Berryville would entail too much expense because of rough terrain. The railroad would pass three miles north of Berryville. The town was very shaken by this news, and pleaded with the railroad promoters to build through their town. In 1900 the promoters finally offered to build the railroad into Berryville providing the townsmen would raise and give $50,000 to the railroad.[30] The community said that they could pledge only $40,000 and the railroad would not accept their offer.

The railroad built a depot at the closest point to Berryville, and named it Freeman Switch for J. W. Freeman, a Berryville merchant, whom the railroad offered to install in business there.[31] Upon this proposal J. W. Freeman and Dr. W. P. George took options on much of the land in the Freeman Switch area. The citizens of Berryville were concerned about this proposed community for they saw it might divide Berryville and become a rival.

The townspeople decided to build a "turnpike" or

---

[29]*Harrison Times,* Dec. 23, 1899.
[30]O. Klute Braswell, ed., "Freeman-Hailey Family Mementos Given to Carroll County Museum," *Carroll County Historical Quarterly,* XIII-XIV (December 1968-March 1969), 2.
[31]*Ibid.*

motor road from Berryville to Freeman Switch, but finally made another proposal to the railroad: they would give the $40,000 originally pledged to the railroad, furnish the right-of-way, make the grade, and furnish the materials if the railroad would build a spur and loop into Berryville (Fig. 2).[32] The railroad accepted this offer and in early 1901 started laying track on the Berryville spur. The spur cost the town $12,000 in materials and labor[33] besides the $40,000 raised by the community, quite a substantial sum of money for a town of only 600 people. The community rejoiced that it had a railroad, and 2,000 people attended the celebration of the formal opening of the Berryville spur on June 15, 1901.[34]

Between 1910 and 1920, Berryville's population increased by ninety per cent, as it developed into a prominent political and agricultural center. By 1930 Berryville had two banks, two newspapers, three hotels, a brick factory, a large planing mill, two canneries, and three flour mills.[35] In Figure 2, it can be seen that the railroad spur did not reach the center of town, but only the north edge of the community. As a result Berryville's industries located in this area to be on the loop, and Berryville's major industries are still located there. A residential area also built up in the same area as workers desired to live close to their work, and near the depot several commercial businesses developed to provide the services necessary for this north end of town.

Eight miles east of Berryville and fifty miles from Seligman by rail was Green Forest. This little Carroll County town, incorporated February 22, 1895, raised $4,350 to contribute to the railroad,[36] and secured a right-of-way across Hickory Township to the Boone County line.[37] Her people, with the coming of the rails in 1901, entertained "high

[32]*Ibid.*, 3.

[33]*Green Forest* (Arkansas) *Tribune,* June 22, 1901.

[34]Braswell, "Freeman-Hailey Family Mementos Given to Carroll County Museum," 2.

[35]Thomas, *Arkansas and Its People,* 674.

[36]Braswell, "Freeman-Hailey Family Mementoes Given to Carroll County Museum," 3.

[37]*Harrison Times,* Dec. 23, 1899.

hopes of a boom."[38] But the boom failed to materialize after all because the depot was located a mile southwest of the central business district, and the threat of rival business in the vicinity of the tracks and station dampened the spirit of "old town" merchants and residents. Looking at Figure 3, it is obvious that the threat was very real. Both residences and businesses stretched out toward the train depot, and most of the new industries located near the tracks, between 1901 and 1905.

St. Joe, the origin of whose name has seemingly escaped students, was located in northwest Searcy County, nineteen miles southeast of Harrison by rail. St. Joe "old town" had been a very active mining center since before 1900. The coming in 1902 of the railroad, which passed a mile north of "old town," found "new town" developing two miles east and on the north side of the tracks, where a wagon road to the northeast from St. Joe "old town" crossed the tracks (Fig. 4). St. Joe was incorporated on July 7, 1904, and by 1910 had a population of 159. Although the little town had several sawmills and limekilns, the mainstay of its economy was the shipping of zinc ore from a dozen mines lying within five miles of the townsite (Fig. 5).

Some communities, anxiously anticipating the railway, were not as fortunate as Berryville, Green Forest, and St. Joe. Carrollton in Carroll County, Duff in Searcy County, Settlement in Van Buren County, and Mt. Pisgah in White County are examples of the fate that awaited once prosperous rural trade centers that got bypassed by the Missouri & North Arkansas Railroad.

Carrollton had long been an important north Arkansas town. It became the seat of government for Carroll County (created in 1823) in 1834,[39] a post office being established there the same year. When a large slice was taken off the east end of Carroll County in 1869 for the creation of Boone County, Carrollton was no longer the geographical and political center of the county and lost out to Berryville in

[38]Jesse Lewis Russell, *Behind These Ozark Hills* (New York, 1947), 27.
[39]Rea, *Boone County and Its People*, 21.

1871 when a contest to relocate the county seat took place. Carrollton declined somewhat as a result of these events, but its mortal blow come in 1901 when the railroad passed three miles northeast of the community. A depot was established here at Alpena Pass, which was nothing but a construction camp at railhead just east of the Boone County line (Fig. 6). But railroad promoters soon laid out the town of Alpena and began pushing the sale of lots.[40] Most of Carrollton's businessmen who were still around in 1900 moved to Alpena, and Carrollton declined so rapidly that even its post office was discontinued.

Alpena, however, thrived as a small railroad town, becoming a timber and lumber shipping point. It had in 1920 a bank, a cannery, a sawmill, a hub and spoke factory, besides several woodworking mills.[41] Farm products from eastern Carroll County and western Boone County were shipped from there, and it became a trade center for a large farming and timbering area.

Duff in Searcy County was a small settlement in 1906 with a post office, a store, a blacksmith shop, a flour mill, a sawmill-cotton gin, and a cluster of houses. The railroad in 1902 missed Duff by half a mile, the result being that the village moved about two miles southeast to the construction camp at railhead on the north bank of Buffalo River. This camp was named Gilbert, after Charles W. Gilbert, secretary-treasurer of Allegheny Improvement Company, the construction company then building the railroad, who was afterward president of the M&NA.

Because of the necessity of bridging the Buffalo at this point, work on the railroad centered for some nine months at Camp Gilbert. William Mays moved his store, which also housed the post office which moved too, from Duff to what now became the new town of Gilbert. A hotel was also built there. As a temporary terminus on the railroad, Gilbert became a departure point for prospectors and in-

[40]*Harrison Times,* Nov. 24, 1900.

[41]Herndon, *Centennial History of Arkansas,* 736; Thomas, *Arkansas and Its People,* 668.

vestors to the rich Rush Creek mining district down the Buffalo River into southern Marion County. A Gilbert resident, Mrs. Jesse Moore, says she can remember when ore wagons from Maumee, Tomahawk, and other mines lined up two abreast for three-fourths of a mile waiting to unload their ore into railroad cars at Gilbert (Fig. 5).[42] The new river-railroad town, which had caused the village of Duff to disappear, prospered too as a lumbering and milling center, many of the logs being floated down the Buffalo for shipment by rail.

Settlement lay on the outside, south bank of a giant horseshoe bend in Little Red River in northeastern Van Buren County. In 1907, when the M&NA built through the area, it had a bank, post office, drugstore, five general stores, and a cotton gin.[43] With the railroad coming the community naturally was very optimistic about its future. Gil Cotrell, one of Settlement's leading merchants, offered to donate land for a station and shipping yards. However, the terrain was exceedingly rough along the south side of Little Red River at Settlement, and to have brought the railroad through would have required tunneling through a mountain upstream from the town. Engineers decided it would be cheaper to build bridges across the river above and below Settlement, that is, run the railroad across the horseshoe bend opposite the town, thus isolating it from the railroad (Fig. 7).

A railway construction camp was established across the Little Red from Settlement, and remained about a year as the two bridges were built and the rails were laid through toward Heber Springs in neighboring Cleburne County. In September 1908 a station house opened at the site of the construction camp. It was named Shirley after a Cotton Belt agent who had an office in the new building; his mission was to buy ties for his company and to urge M&NA officials to send all possible business through the Cotton

---

[42]Mrs. Jesse Moore, private interview, May 28, 1972.
[43]Glen Hackett, "Early History of Shirley, Arkansas" (Unpublished graduate paper, Arkansas State Teachers College, 1957), 10.

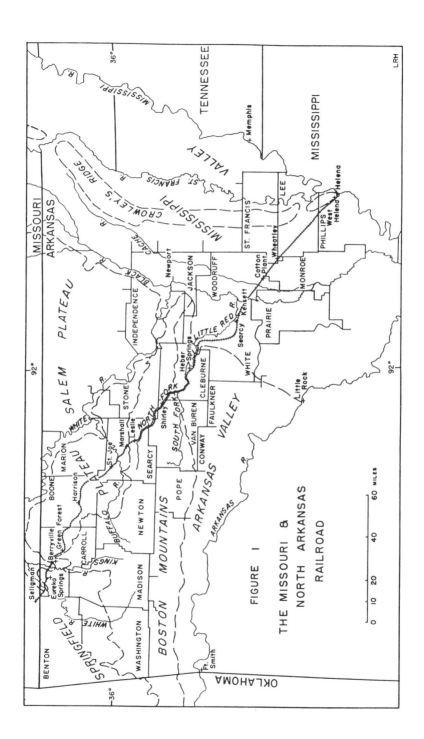

FIGURE I

THE MISSOURI &
NORTH ARKANSAS
RAILROAD

203

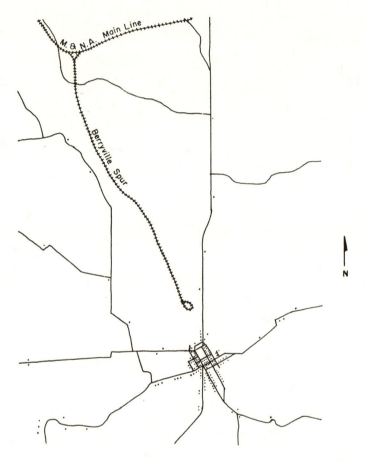

FIGURE   2

BERRYVILLE, 1901

FIGURE 3

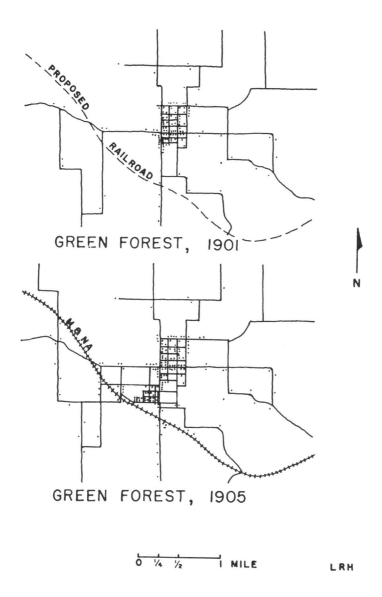

GREEN FOREST, 1901

GREEN FOREST, 1905

N

0   ¼   ½   1 MILE   LRH

FIGURE 4

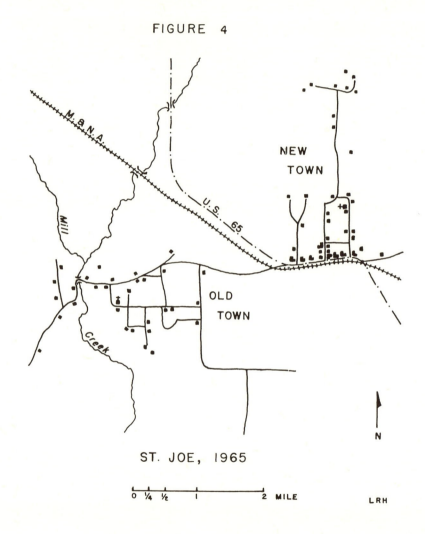

ST. JOE, 1965

0 ¼ ½ 1 2 MILE

LRH

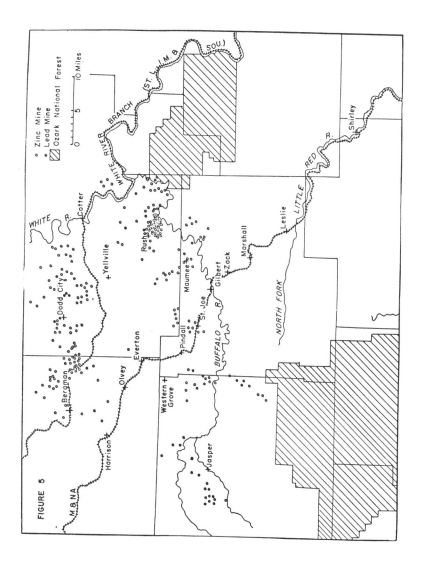

FIGURE 5

FIGURE 6

CARROLLTON AND ALPENA
1904

0   1/2   1 mile

Contour Interval   50 feet

FIGURE 7

·SETTLEMENT AND SHIRLEY,

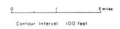

Contour Interval 100 feet

Belt Railroad,[44] the common name for the St. Louis and Southwestern Railroad.

Soon after the railroad went through J. R. Arnold and A. Brown, who owned Arnold & Brown, one of the largest general stores in Settlement, purchased eighty acres of land near the new station and platted it into the townsite of Shirley. They held a sale to dispose of their town lots, giving one lot free with the purchase of five or more.[45] Residents of Settlement were the principal buyers, and some of them bought whole blocks. In 1909 Arnold & Brown moved from Settlement to Shirley, locating their stock of merchandise into a large combination general store and post office building that stood on the hill above the present business section of Shirley.[46]

Thereafter Shirley grew rapidly. It incorporated in 1911, and by 1920 had a population of 349. By 1911 four of the five general stores, the drug store, and the bank had moved from Settlement to Shirley. That year a list of Shirley businesses included four general stores, the Arnold & Brown Hotel, Settlement Bank, C. H. Smith Tie Company, Western Tie Company, W. R. Lee Wagon Hub Mill, Humphries & Bucklew Lumber Company, Gorwich & Waller Lumber Company, and the Clinton-Shirley Mercantile Company. That same year a Clinton newspaperman, S. A. Myover, packed up his printing press at the county seat and moved to Shirley because he was convinced it would become one of the larger cities of the Ozark region at the rate it was growing.[47]

Shirley did, of course, promise to overshadow the county seat town of Clinton, which had no railroad and which now became dependent upon freight wagons hauling between Shirley and Clinton. As one writer put it during the heyday of the M&NA: "All roads in Van Buren County lead to Shirley and to the railroad yards, for this was the market."[48]

[44]*Ibid.*
[45]*Ibid.*, 11.
[46]*Ibid.*
[47]*Ibid.*, 12.
[48]*Ibid.*, 12, 36.

As in other towns on the M&NA, timber was the principal basis for Shirley's temporary prosperity. Ten to fifteen sawmills operated in the area about Shirley,[49] and wagons, hauling the rough lumber, came from as far north as Timbo in Stone County and they poured into Shirley from all parts of Van Buren County. Yet timber exploitation around Shirley, while it continued to some extent through World War II, was a fleeting proposition. The best stands of timber were cut over in a few years. For example, Gorwich & Waller Lumber Company was the largest in Shirley, but it operated only a few years. Indeed, in 1920 only three of the companies that began operations at the town in 1911 were still present, although other companies had moved into the area. The Goblebe Lumber Company and the National Cooperage and Wood Working Company were the two prime concerns in 1920.[50]

Mt. Pisgah in northwestern White County no longer exists. Established in the mid-nineteenth century, it was situated at the foot of the escarpment that divides the Arkansas Valley from the physiographic provinces of the coastal plain. It developed as an agricultural trade center at a crossroads location, and at the turn of the century was a thriving rural village of several general stores (one of which sold coffins),[51] a post office, a drug store, a photographer's studio, a cotton gin, and two churches. In 1908 the M&NA built through the area, missing Mt. Pisgah by a half mile. Business interests in Mt. Pisgah promptly began moving south to the tracks, where a new townsite called New Mt. Pisgah came into being. The school moved also, leaving only the two churches and the cemetery at Old Mt. Pisgah. The new town never had a depot but was only a flag stop for passengers on the railroad. The cotton gin and several sawmills at New Mt. Pisgah were, however, served by a siding for loading of cars. During the railroad strike and

---

[49]Ronnie Maxwell, "The Bridge Over the Little Red," *Arkansas Gazette*, Apr. 16, 1972, p. 4E

[50]Hackett, "Early History of Shirley, Arkansas," 32.

[51]Mrs. W. C. Welch, "Mt. Pisgah Community," *White County Heritage*, III (July 1965), 17.

shutdown of 1921-1923 and on through the Great Depression, the town began to disappear. "When the train stopped running in 1946 that was the end of Mt. Pisgah."[52]

The towns of Alpena, Gilbert, Shirley, and New Mt. Pisgah are but a few examples of settlements that developed as a result of the M&NA's penetration of north central Arkansas. Indeed, a total of thirty-three communities owed their origin to the coming of the rails. Many of these began as sites of railroad construction camps, such as: Alpena and Urbanette (Carroll County); Gilbert and Arlberg (Stone County); Shirley (Van Buren County); and McClelland (Woodruff County). The last town originated as a construction camp for the building of the White River bridge. Others originated as shipping points for local products, such as: Everton (Boone County) for farm and lumber products; Coin (Carroll County) for fruit; Pindall and Zack (Searcy County) for zinc ore; Rumley (Searcy County), Elba (Van Buren County), Lydalisk (Stone County) and Letona (White County) all for timber products; Enright (White County), Dixie, and Daggett (Woodruff County), and Aubrey, and Rondo (Lee County) all for cotton; and Little Prairie (Lee County) for rice. And several communities developed for other reasons, such as: Walden and Cisco, (Carroll County) and Betts (Stone County) as water stops; Freeman and Junction (Carroll County) as switching points for spur lines; Elk Ranch (Carroll County) and Armstrong Springs (White County) as tourist sites for viewing a herd of elk and for visiting mineral springs respectively; and Fargo (Monroe County) as the crossing of the M&NA and the Cotton Belt tracks.

If most of these little communities originated rather haphazardly, say, at sawmill locations, at a siding for loading zinc ore, or at a cotton gin's or plantation's platform for loading cotton, the three communities of Everton, Edgemont, and West Helena were actually planned communities, promoted and backed with enough capital to become fairly important commercial centers.

---

[52]*Ibid.*, 19.

Everton, in the southeast corner of Boone County, was the brainchild of Dr. J. L. Rush.[53] In 1902, immediately after the railroad built through, he bought seventy-five acres of land at the junction of Clear and Hog creeks and platted his town. Surveyed on a rectangular grid pattern with the railroad passing diagonally through the town, Everton was well layed out. It was surrounded by some of the best fruit and farm lands in the region and the original economic life of the new town focused on a canning plant and on the shipping of fresh fruit. Within four years it had two general stores, a drugstore, a furniture store, a grist mill, a sawmill, a cotton gin, and the Everton Hotel.[54] Eventually Everton also became a shipping point for railroad ties, posts, lumber, lead and zinc ore, glass sand, and other products from the four-county area of Boone, Marion, Searcy, and Newton.

Edgemont, in Cleburne County, was located along the railroad halfway between Shirley and Heber Springs. E. W. Stanfield, from Indiana, formed the Edgemont Improvement Company August 14, 1908,[55] which bought 1,200 acres of land and laid out 200 acres of it for a townsite.[56] To perpetuate the town, the Globe Cooperage and Lumber Company moved four large mills from Indiana and established a bank and other commercial enterprises. With the lumber mills employing about 150 men, the town grew rapidly, so that by 1910 it had a population of nearly 350. However, with an economic base that depended solely upon the timber of the area and the railroad, the town began to diminish in people and prosperity with the M&NA strike of 1921-1923, and with the depletion of usable timber. The original town of Edgemont lies beneath the waters of Greers Ferry Lake, but the name is perpetuated by the community on the north shore of the lake opposite the town of Greers Ferry.

---

[53]*Harrison Times*, Apr. 12, 1902.

[54]*Ibid.*, June 2, 1906.

[55]Letter received from Kelly Bryant, Secretary of State, Little Rock, Arkansas, Aug. 8, 1972.

[56]Hull, *Shortline Railroads of Arkansas*, 71.

West Helena grew out of the city of Helena's need for additional level land on which to build industries. The "Hardwood Capital of the World,"[57] as the rivertown called itself, fitted snugly between the Mississippi River on the east and Crowley's Ridge on the west. An attempt in 1900 to locate industry southward at Barton had failed because of lack of interest. In 1907, when the M&NA built around the end of Crowley's Ridge and northwestward from Helena to Cotton Plant, the tracks west of Crowley's Ridge crossed the 2,300-acre Hoggatt Clopton plantation belonging at that time to James R. Bush. In 1909 Bush sold his holdings to Edward Chaffin Hornor and John Sidney Hornor, who were cousins.[58]

The two men, with James Tappan Hornor, organized the West Helena Company and platted in 1910 the new townsite of West Helena. They laid out industrial sites of about ten acres each along the east side of the M&NA tracks and residential areas were developed adjacent to the industrial area. Plans for water, sewer, and other utilities were included, and the main east-west street, Plaza, was laid out 117 feet wide to provide for electric interurban streetcar service between West Helena and Helena. A spur track of the Missouri Pacific Railroad traversed the center of the industrial sites, so that the new town was served by two railroads.[59]

Several of the industrial sites were snapped up even before the town survey was completed and construction of mills and factories begun. The first companies, which bore testimony to Helena's importance as a hardwood center, were Helena Veneer Company, Ong Chair Company, Southwest Wagon Company, and Dennison Sawmill. It was the rapid growth of West Helena's wood mills that gave impetus

---

[57]E. G. Green, "A Brief History of West Helena," *Phillips County Historical Quarterly*, III (June 1965), 20.

[58]*Ibid.*; L. R. Parmelee, "Helena and West Helena: A Civil Engineer's Reminiscences," *Phillips County Historical Quarterly*, I (1962), No. 2, p. 2.

[59]Green, "A Brief History of West Helena," 20.

to the sprouting of sawmills up the M&NA tracks at Wheatley, Moro, Aubrey, and Rondo.[60]

West Helena grew·rapidly. It incorporated as a city on June 17, 1917, and in 1920 had a population of over 6,000. Like Helena, its economic base was grounded upon the milling of hardwood and the manufacture of wood products. The Chicago Mill & Lumber Company, employing 1,600 workers and using 3,500,000 board feet of lumber monthly, was the largest of the fourteen lumber and wood products industries in 1920.[61] West Helena also came to rival its mother city as a center for cotton. Cotton gins, cottonseed oil mills, cotton warehouses, and the St. Francis Cotton Mill—the latter had removed from Barton to West Helena in about 1913—all located at the new town. Today there are still several wood products mills along the old M&NA tracks.[62]

---

[60]*Ibid.,* 21.

[61]"Arkansas on Wheels," Little Rock *Arkansas Democrat,* Oct. 16, 1916.

[62]The old M&NA tracks are used by the Missouri Pacific Railroad to give access to many of the industries of West Helena.

Street Railways in Grand Forks, North Dakota: 1887-1935

by Colleen A. Oihus

The history of the street railway in Grand Forks, North Dakota, reflects the national street railway era. The local system arose at roughly the same time as the national network, emerged for similar reasons, exemplified similar growth patterns, experienced like problems, and declined at approximately the same time for comparable reasons. The street railway system on both the local and national levels represented a stage in the evolution of public transportation.

The use of the electric streetcar as a means of public transportation resulted from the inadequacy of the horsecar, the principle mode of public transportation prior to the electric vehicle.[1] This type of transportation traveled under six miles per hour, frequently derailed, halted service under conditions of extreme snowfall, and could not be used on steep grades.[2] Moreover, a horse utilized by a transit company averaged only five hours of service a day, consumed approximately 30 pounds of hay and other grains daily, required the service of veterinarians, blacksmiths and hostlers, and sustained a short life expectancy.[3] Finally, a transit company purchased approximately seven times as many horses as cars.[4]

Until the advent of the streetcar, several alternatives to the horsecar were unsuccessfully tried in American cities. For instance, high cost and dirtiness led to the failure of the steam dummy engine as a method of public transportation;[5]

the development of the tram, "operated by compressed air and internal-combustion engines," never reached a level of efficiency,[6] and battery cars were simply too slow. One alternative which did receive limited use, the cable car, necessitated massive expenditures for underground construction that rapidly deteriorated.[7] These alternatives all succumbed to the arrival of the electric streetcar.

[1] *Trolley Car Treasury,* as cited in George W. Hilton and John F. Due, *The Electric Interurban Railways in America* (Stanford, Cal.: Stanford University Press, 1964), 4.
[2] *Ibid.*
[3] *Ibid.*
[4] *Ibid.*
[5] George W. Hilton and John F. Due, *The Electric Interurban Railways in America* (Stanford, Cal.: Stanford University Press, 1964), 5.
[6] *Ibid.*
[7] *Ibid.*

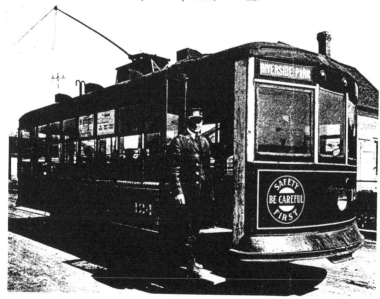

Conductor George Glass and car #124 of the Grand Forks Street Railway Company's Riverside Park Line. The photo dates from about 1921, immediately after the company purchased six of the Birney-type streetcars.

— *Courtesy Orin G. Libby Manuscript Collection*

12

The national street railway system emerged during the early 1890s after the use of the electric traction proved economically feasible.[8] In 1887, only 29 miles of electric street railway existed in the country.[9] By 1897, however, the total figure jumped to 13,765 miles,[10] and another decade increased the total approximately 34,000 miles.[11]

From 1900 to 1917, the national street railway system expanded greatly. The number of miles of line operated increased from 22,576.99 in 1902 to 44,835.37 in 1917. The number of passengers increased from 4,774,211,904 in 1902 to a surprising 12,666,557,754 in 1917.[12]

Together with this period of growth and expansion came two basic problems. First, franchises from city councils granted a company the right to construct a street railway system, but also required the company to pave and maintain areas where tracks were laid.[13] As time went on, this paving requirement became quite a financial burden. A second franchise requirement that companies encountered was the fixed five-cent fare.[14] This fixed fare failed to produce a net operating revenue and often created indebtedness, which led to receivership.

The year 1917 marked the peak of the national street railway system, and its prosperity continued until approximately 1927. However, by 1932 the number of street railway companies had decreased to a total below that for 1902.[15] The advent of the bus and the increased use of the private automobile spelled the end of an era in American public transportation.

In Grand Forks, the street railway system was the response to the demand for adequate public transportation by a growing population in the late 19th Century.[16] Horsecabs, the existing mode of transportation, were too expensive; the average fare was fifty cents per ride at a time when "a dollar would buy three meals and a room for a day in a hotel."[17] This need led to the first attempt to create a street railway system for the city.

On April 20, 1887, the Grand Forks City Council approved its first franchise for the construction of a street railway. The franchise granted to William O'Mulcahy and M.L. McCormack of New York City the right "to lay tracks and operate streetcars on Belmont, Division, International Avenues, and Third Street."[18] The franchise stipulated that construction commence within 90 days and that one mile of track be completed within one year's time. Construction never began, however, and a street railway in Grand Forks never materialized in 1887.[19]

In 1892, the "sprawling city of some 5,500 inhabitants became real excited over the prospect of a real electric street-car line, which would run through its unpaved streets to the far reaches of the city."[20] On May 2, the city council granted a second street railway franchise to the "projectors" of the community and several St. Paul businessmen. The franchise gave,

> the authority, right and privilege to build, equip, maintain and operate electric street railway lines, with double or single tracks, with all necessary side tracks and switches, poles, wires, conduits and appliances, over, along and upon the streets and avenues in the city of Grand Forks.[21]

On August 2, 1892, W.C. Merryman of the St. Paul Phillips and Merryman Engineering Firm reached the city and began to survey projected track lines.[22] Along with the surveying came a series of problems and delays. The ordinance required the erection of guard wires above the trolley wires for protection against contact with telephone or telegraph lines. The "projectors," however, thought these wires both needless and unsightly, and therefore objected. The franchise also required paving both between and one foot to each side of the tracks. The "projectors" agreed to this, but desired a provision in the ordinance exempting them from the costs of relaying pavement due to sewer, water or gas line construction.[23]

Consequently, the city council met on August 15, and amended several sections of the original franchise. With construction delayed, probably as a result of this process, C.F. Arrall came to the city to expedite the work. Regardless of these efforts, the St. Paul firm never laid any track.[24] The actual completion of a street railway line in Grand Forks took place over 15 years later.

In 1904, another attempt to effect a street railway system in Grand Forks began under the direction of Webster Merrifield, President of the University of North Dakota. At this time, the distance between the University and the city, well known by professor and student alike, effectively stimulated the final emergence of a public transportation system to bridge two "bleak and dismal miles . . .where the sun beat down in summer and arctic winds blew in winter."[25] Thus, "feeling the need for a connecting link between the Sioux Institution and the city," several professors and businessmen founded the

[8] John H. Hanna, "Evolution of Community Transportation," Electric Railway Journal, 75 (September 15, 1931), 498.

[9] Ibid.

[10] Ibid.

[11] Ibid.

[12] U.S. Bureau of the Census, Electric Railways and Affiliated Motor Bus Lines (Washington: U.S. Government Printing Office, 1931), 7.

[13] Henry W. Blake, "There Is Little Warrant for the Paving Tax as at Present Assessed," Electric Railway Journal, 60 (November, 1922), 839.

[14] Ibid., "The 5-cent Fare Level Cannot Come Back Generally," Electric Railway Journal, 61 (January, 1923), 152.

[15] U.S. Bureau of the Census, Electric Railways and Affiliated Motor Bus Lines: 1932 (Washington: U.S. Government Printing Office, 1934), 4.

[16] Untitled article on the street railways in Grand Forks located in the clipping-pamphlet file, "Grand Forks-Transportation," Dakota Room, Chester Fritz Library, University of North Dakota.

[17] "Trolley Line Was Great Day For City, U," Grand Forks Herald, February 23, 1958, sec. A, 16.

[18] Robert S. Anderson, "A Social History of Grand Forks, North Dakota 1880-1914," (Unpublished MA Thesis, University of North Dakota, 1951), 121.

[19] Grand Forks City Ordinance, No. 6, April 20, 1887, as cited in Anderson, "A Social History of Grand Forks, North Dakota 1880-1914," 121.

[20] "Old City Trolley Had Final Trip in 1934," Grand Forks-Transportation File, Chester Fritz Library, University of North Dakota.

[21] Grand Forks City Ordinance, No. 251, September 2, 1908.

[22] "Trolley Final Trip," clipping-pamphlet file.

[23] Ibid.

[24] Ibid.

[25] "Trolley Line," Grand Forks Herald, February 23, 1958, sec. A, 16.

13

A Grand Forks Street Railway Gallery

William Budge

Elwyn F. Chandler

David H. Beecher

Adison I. Hunter

Edward J. Lander

Earle J. Babcock

Jeremiah D. Bacon

James A. Dinnie

Thomas Roycraft, mgr.

Robert B. Griffith

— *State Historical Society of North Dakota Collection*

Grand Forks Transit Company.[26] Webster Merrifield directed the enterprise; Dean Elwyn F. Chandler, Robert B. Griffith, Edward J. Lander, David H. Beecher and Oscar S. Hanson[27] were also involved.

In his annual report to the University Board of Trustees, Merrifield suggested the great value of a streetcar line:

> The new motor line will, it is believed, prove of almost incalculable value to the University in putting it in cheap and easy communication with the city. The distance of the University from Grand Forks and the great difficulty during the greater portion of the year of getting back and forth has proved, heretofore, a serious handicap to the institution. Henceforth, the University will enjoy all the advantages of isolation with the added advantage of being practically in the heart of a considerable city. With the construction of the new railway the town will doubtless grow toward the University and there will, in a few years be an abundance of comfortable boarding places in the immediate neighborhood of the University for the accommodations of students. The lack of such accommodations, heretofore, has been one of the most serious difficulties in the way of building up the University . . . Not the least advantage of the line will be the possibility of bringing the towns people to lectures of a literary or scholarly character, convocation addresses, class plays, receptions and other forms of entertainment at the University as well as making similar attractions of an instructive and social character in Grand Forks more easily accessible to the students at the University.[28]

On April 11, 1904, the city council awarded a franchise o the Grand Forks Transit Company. It granted:

> to Leslie Stinson, A.G. Schultheis, William Budge, E.J. Babcock and R.B. Griffith, and their assigns, permission to construct and operate a street railway along certain streets in the city of Grand Forks and to the University of North Dakota, and establishing regulations and conditions under which said street railway shall be constructed and operated.[29]

[26] "UND Alumni Review," (December, 1938), located in the clipping pamphlet file, 16.

[27] The six men were prominent civic leaders in the city. Webster Merrifield (1852-1916) served as President of the University of North Dakota from 1891-1909; Elwyn F. Chandler (1872-1946) was professor of Mathematics at the University from 1904-1914, North Dakota State Engineer from 1904-1905 and professor of Civil Engineering from 1914-1938; Robert B. Griffith (1856-1934) owned the Ontario Store, a Grand Forks department store now known as Griffith's; Edward J. Lander (1856-1953) founded and operated the E.J. Lander Real Estate firm and had many local investments; David H. Beecher (no dates available) was a banking entrepreneur who was President of the Union National Bank; Oliver S. Hanson (1862-1939) was President of the Scandinavian-American Bank.

[28] "Annual Report to the Members of the Board of Trustees of the University of North Dakota, June 1, 1904," Merrifield Papers, Orin G. Libby Manuscript Collection, University of North Dakota, 10-11. Interestingly, Merrifield's prediction about the growth of the city of Grand Forks has proven apt. The city indeed filled the gap between itself and the University and is currently expanding westward beyond the Sioux institution.

[29] Grand Forks City Ordinance, No. 166, April 11, 1904, as cited in Anderson, "Social History of Grand Forks." Leslie Stinson (1861-1944) owned a large farm implement, coal and wood, and carriage dealership in Grand Forks; Albert G. Schultheis (1860-1945) was secretary-treasurer of the Grand Forks Foundry and Machine Company; William ("Billy") Budge (1852-1948) was a Grand Forks pioneer businessman who is sometimes called the "father" of the University of North Dakota; Earle J. Babcock (1867-1925) was Dean of the College of Engineering at the University of North Dakota from 1916-1925, North Dakota State Geologist from 1897-1902 and one of the leading early researchers into the industrial uses of North Dakota lignite coal. For biographical information about Babcock, see William O. Beck, "Earle Jay Babcock and North Dakota Lignite," *North Dakota History*, 41-1 (Winter, 1974), 4-15.

14

By the early months of 1904, Merrifield and the Transit Company accumulated $15,000 through the sale of stock,[30] Therefore, with both the right to build and sufficient capital, the company began initial construction of the University Avenue line.

Actual construction under the supervision of Andrew Morrison, the University Registrar, began on July 13 after the directors finally decided to employ electric power. The directors approved the use of electricity after a second generator was installed in the University powerhouse.[31] The laying of steel rails began on University Avenue directly in front of the present day Gamma Phi sorority house, and reached the Great Northern Railroad crossing on October 6.[32] The switchboard arrived shortly thereafter. October 10 witnessed the operation of the first streetcar in the city; it ran from the University powerhouse to the railroad crossing.[33] Construction ended at the junction of First Avenue and Third Street on November 20, 1904, and shuttle service began on a half-hour schedule.[34] An estimated 300 people rode the streetcar daily ''and at times reached a peak of 800 a day.''[35]

Initiative for a second streetcar line came from Robert B. Griffith, and resulted from the fact that the University car operated almost solely to the advantage of the students while the growing metropolis of over 12,000 people remained without a means of public transit.[36] Griffith organized a streetcar committee within the Grand Forks Commercial Club; together they started amassing the necessary construction monies, but encountered great difficulty because other projects competed ''for the limited capital of the city.''[37] The streetcar committee also went before the city council and requested a franchise. On September 2, 1908, the franchise was granted to:

> E.J. Lander, John Dinnie, W.H. Kelsey, E.H. Kent, and O.A. Webster, their successors and assigns, the authority, right and privilege to build, equip, maintain and operate a street railway line or lines with single or double tracks together with all necessary side tracks, turnouts, switches, loops, poles, wyes, conduits, and appliances in connection there with in, over, accross [sic] and along the following streets, avenues, bridges and public places within the limits of the city of Grand Forks, North Dakota.[38]

The above mentioned streets included Skidmore (Gateway

Car #124 travels the Riverside Park Line about 1921.

[30] Allan Dearden, ''The Days of the Trolleys,'' *They Came To Stay* (Grand Forks, North Dakota: Grand Forks Centennial Corporation, 1974) 25.

[31] ''Street Car Line Work,'' *Grand Forks Herald*, June 14, 1904, 8; ''Electricity Will Move University Line Cars,'' *Grand Forks Herald*, June 11, 1904, 6.

[32] Telephone Interview with Mrs. John C. McLaughlin, October 29, 1975. Mrs. McLaughlin is the daughter of the last General Manager of the Street Railway Company, E.O. Odegard.

[33] ''Street Car Line Ready,'' *Grand Forks Herald*, October 11, 1904, 8.

[34] ''Grand Forks Street Railway, Grand Forks, N.D. — East Grand Forks, Minn.,'' unpublished manuscript, (no page numbers, no date), located in the Minnesota Transportation Museum, Bloomington, Minnesota. Hereafter cited as Unpublished manuscript from the Minnesota Transportation Museum. See also: E.O. Odegard, ''Grand Forks,'' *Bus Transportation*, 14 (February, 1935), 62.

[35] Dearden, 26.

[36] *Ibid.*, See also, Odegard, 62

[37] Dearden, 26.

[38] Grand Forks City Ordinance, No. 251, September 2, 1908. John Dinnie (1853-1910) acted as Mayor of Grand Forks from 1896-1904 and became prominent as a contractor and builder; William H. Kelsey (1858-1936) owned a real estate, loan and insurance firm and was involved in city politics as either alderman or city commissioner from 1914-1926; Edward H. Kent (no dates available) was President of Kent Realty and Investment Co.; Oscar A. Webster (1856-1947) was President of the Pioneer Insurance Agency.

[39] *Ibid.*

Drive), DeMers, International, Woodland, Minnesota, Belmont, Boulevard, L'Hiver, Second and Tenth Avenues and on Chestnut, Conkling, Third and Fifth Streets.[39]

October 3, 1908, marked the beginning of construction. On that day the citizenry and a band gathered on south Third Street and listened to an optimistic address given by Edward J. Lander, President of the local Commercial Club:

> Citizens of Grand Forks, I congratulate you on the important work that here and now is about to be undertaken. That is is one of supreme importance to the city of Grand Forks is without question. That it will prove to be the most important step toward the building of a greater Grand Forks I firmly believe. The city is larger than the men who comprise it. Every city is what the men who live in it make it. That's were [sic] our individual importance comes in. This movement has been undertaken and brought to its present initial stage by the united efforts of unselfish, patriotic citizens.
>
> I believe this to be a critical time in the history of this city. I believe that having and not having a street railway means whether Grand Forks shall spell c·i·t·y or spell v·i·l·l·a·g·e. Without it our claim for metropolitanism is unwarranted and unfounded. With it our opportunity to make good is here. I have travelled over every portion of the United States and I tell you here and now no city is surrounded by a superior agricultural society.
>
> Think, act, and work metropolitanism, and a greater Grand Forks and it will surely be yours.

15

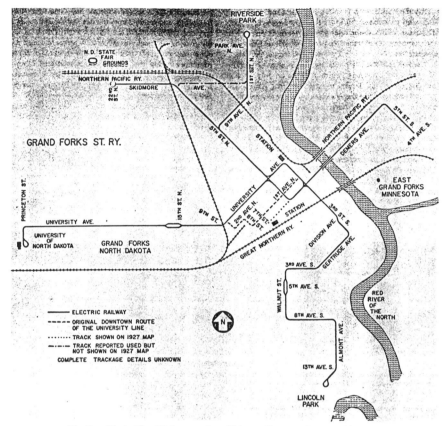

The Grand Forks Street Railway system. All known lines are shown on this map.

— *Courtesy Colleen A. Oihus*

With the conclusion of Lander's "boosting" address, Robert B. Griffith drove the first spike and inaugurated the construction of a true system of public transportation.[40] At 4:00 that afternoon, contractor McDonald and his crew began laying the steel rails.[41] Construction ended with the completion of the line to Lincoln Park on July 16, 1909.[42] This line offered public transit to a large residential section of the city.

While construction continued in the fall of 1908, the members of the streetcar committee formed a new company, the Grand Forks Street Railway Company, which incorporated on December 9, 1908. The Articles of Incorporation established an 11-person Board of Directors; they included

Edward J. Lander, Oscar A. Webster, Carlos F. Whitcomb, William H. Kelsey, James Dinnie, Jeremiah D. Bacon, Adison I. Hunter, David H. Beecher, William J. Murphy, Oliver S. Hanson and Oscar M. Hatcher. The Articles authorized capital stock of $150,000, divided into 1500

[40] "R.B. Griffith Drives Spike While Band Plays A Cheering Refrain," *Grand Forks Herald*, October 4, 1908, 8.

[41] Dearden, 27.

[42] Corporation Records of the Grand Forks Street Railway Company, unpublished manuscript consisting of four volumes located in the holdings of the E.J. Lander Realty Company, Vol. 1, 50. Hereafter be cited as Corporation Records.

16

shares at $100.00 per share.[43] Four men, Lander, Robert D. Campbell, Paul Griffith and Elwyn F. Chandler were the principal stockholders.[44]

Throughout 1909, the company consolidated its holdings. On June 18, it purchased the large Merrifield holdings in the old transit company. On July 31, director Kelsey "reported that he attended the meeting of the Grand Forks Transit Co., and that the purchase of all its assets by the Grand Forks Street Railway Company had been effected." Final incorporation of the old company into the newly-established one resulted shortly thereafter.[45] The company also secured power for the car service from the Grand Forks

Gas and Electric Company at a rate of two dollars per month per kilowatt.[46]

The year 1910 saw the extension of track across the Red River into East Grand Forks, Minnesota. This decision was made because Grand Forks, located in constitutionally "dry" North Dakota, simply offered less in the way of alcoholic beverages than did its "wide open" sister city.[47] With approval of the franchise,[48] work began from the Grand Forks Great Northern Railroad Depot, and eventually terminated at the corner of Fifth Street and Fourth Avenue in East Grand Forks. This line opened service in early 1911.[49]

This extension had immediate financial problems. According to one source, "it was thought that people from Grand Forks would like to ride by street car when going for their daily constitutional. However, this did not prove to be the case. It seemed that people thought they were too conspicuous when riding the cars to East Grand Forks and after being over there for a while they were unable to ride at all."[50] As revenue decreased during succeeding years, the directors finally decided on June 30, 1919, to discontinue this line.[51]

Since the State Fair had become quite a popular attraction during this era, the directors also extended the line to the State Fair Grounds north of the city. The tracks began from University Avenue, continued up north Fifth Street and ended on Skidmore Avenue (Gateway Drive) south of the fair grounds. Scheduled service began in 1911. In addition, the Riverside Park line, possibly an extension of the State Fair Grounds line, stretched from the corner of Fifth Street and Ninth Avenue to the junction of First Street and Park Avenue.[52] This line, completed either in 1911 or early 1912, offered public transit to the northern residential district. Reconstruction and relocation of the University line also took place in early 1913.[53]

By 1913, the Grand Forks Street Railway Company operated lines from the downtown district to the University, Lincoln Park, East Grand Forks, the State Fair Grounds and Riverside Park. These five lines constituted the entire amount

---

[43] *Ibid.*, 52. Carlos F. Whitcomb (no dates available) acted as Secretary, Treasurer and General Manager of the Kent Realty and Investment Corporation; James A. Dinnie (1863-1938) was President and General Manager of Dinnie Brothers, a construction firm, and Mayor of Grand Forks from 1914-1918; Jeremiah D. Bacon (1865-1933) was proprietor of the Dakotah Hotel, publisher of the *Grand Forks Herald* and owner of a well-known grain and stock farm near the city; Adison I. Hunter (1860-1936) was President of the First National Bank and a leader in the ill-fated Red River Valley Brick Corporation; William J. Murphy (1859-??) owned the Grand Forks Gas and Electric Corporation; Oscar M. Hatcher (no dates available) was President of the Hatcher Brothers Corporation, an investment and mortgage firm.

[44] Unpublished manuscript from the Minnesota Transportation Museum. Robert D. Campbell (1867-1961) was a very prominent physician and surgeon who was a bank director and at one time President of the North Dakota Medical Association; Paul B. Griffith (1887-1957), the son of Robert B. Griffith, became General Manager of his family's firm in 1934 and had substantial business interests in Grand Forks.

[45] Corporation Records, vol. 1, 50, 52.

[46] *Ibid.*, 43.

[47] Odegard, 62.

[48] Corporation Records, vol. 1, 76.

[49] Unpublished manuscript from the Minnesota Transportation Museum.

[50] Odegard, 62.

[51] Corporation Records, vol. 2, 179.

[52] Unpublished manuscript from the Minnesota Transportation Museum.

[53] Corporation Records, vol. 2, 27.

The street railway system pressed its special cars into service for events such as football games at the University of North Dakota.

— *Courtesy Orin G. Libby Manuscript Collection*

Car #102, one of the early, "double-end" models used in Grand Forks, turns for the car barns on Fifth Street. — *Courtesy Orin G. Libby Manuscript Collection*

operated throughout the lifetime of the Grand Forks street railway system and totaled approximately 8.5 miles.[54]

The company purchased and operated 16 streetcars during the system's existence. The first, "a small second-hand, double-end passenger motor car" purchased from the Duluth Street Railway Company, provided service on the first line constructed, the University line.[55] The term "double-end" meant that the car operated in both directions. When it reached the end of a line, the conductor simply switched the poles leading to the cable wire and thus reversed the direction of operation. The other man used in the streetcar operation, the motorman, ran control boards located at both ends of the car; when the car switched directions, the motorman switched control boards.[56]

The company bought additional cars as the various lines opened. Until 1920, it conducted service with seven cars purchased from the American Car Company as a "single truck, double end, deck roof semi-convertible" type construction.[57] The term "truck" referred to the number of sets of wheels carrying the car;[58] in this case, the cars used one truck with four wheels. The company painted the car bodies dark green and the roofs brown, and evenly numbered them from 102 through 108. Car number 110 resembled the other four cars, but was longer and had narrower windows. The last two of the seven cars, numbered 112 and 114, were second-hand and had a double-truck-type construction. The company also owned four trailers; these were open air structures with assigned numbers 2, 4, 6, and 8 that simply connected onto a regular car to provide extra seating for passengers when the need arose. The trailers operated almost exclusively on the State Fair Grounds line.[59]

On February 10, 1920, the Board of Directors approved the purchase of six Birney-type streetcars from the Safety Car Trust Corporation in St. Louis, Missouri.[60] These Birney cars were of single-end construction; in other words, they traveled in only one direction.[61] Therefore, the company built "turn-around" loops on the various lines in order to accommodate the new streetcars.[62] Construction of the loop on

University Avenue near Chandler Hall, took place even before the company secured approval from the University Board of Administration.[63]

The new streetcars began operation on the Riverside Park line on January 27, 1921, and on the University and Lincoln Park lines on January 30 of that same year.[64] The company acquired three additional Birney cars in 1930 from a firm in Oshkosh, Wisconsin, but these second-hand cars, numbered 128, 130 and 132, operated only on special occasions.[65]

The streetcar was a means of public transportation, but it also served as an object of mischief. If a sufficient number of people climbed on the back of the trolley, for instance, the front end rose up and the car slid off the track. Students found it quite amusing to hide the control lever from the conductor, and to knock the pole off the cable wire with snowballs and thereby immobilize the car.[66] The Birney car was also a source of complaints; they were light-weight and "swayed from side

[54] Unpublished manuscript from the Minnesota Transportation Museum.

[55] Ibid.

[56] Interview with William E. Thoms, held November 6, 1975, at the University of North Dakota.

[57] Unpublished manuscript from the Minnesota Transportation Museum.

[58] J.S. Dean, *The A B C of the Electric Car* (East Pittsburgh, Pa.: Westinghouse Technical Night School Press, 1924), 29.

[59] Unpublished manuscript from the Minnesota Transportation Museum; Thoms interview.

[60] Corporation Records, vol. 2, 199.

[61] Unpublished manuscript from the Minnesota Transportation Museum.

[62] Corporation Records, vol. 2, p. 209.

[63] "Report and Recommendations to the Board of Administration, November 12, 1920," (University Archives, University of North Dakota), 4.

[64] Corporation Records, vol. 2, 209.

[65] Unpublished manuscript from the Minnesota Transportation Museum.

[66] "Trolley Line," *Grand Forks Herald*, February 23, 1958, sec. A, 16.

During North Dakota winters, the snow sweeper was a very necessary piece of equipment for a streetcar line. The sweeper cleaned the tracks by means of large steel brushes attached to the car's undercarriage. — *Courtesy Orin G. Libby Manuscript Collection*

to side and bobbed up and down.'' Passengers referred to the Birneys as ''tin cans'' and ''jiggers.''[67]

The company occupied two buildings in Grand Forks for offices and housing maintenance of the streetcars. The location of the business office was 217 South Third Street. A building at 1008 North Fifth Street became the ''carbarns'' for service and upkeep of the cars.[68]

The amount charged for fares doubled throughout the existence of the streetcar system, even though the 1908 franchise limited the price of the fare to no more than five cents.[69] In September, 1918, however, fares increased to seven cents, with a special rate of 15 fares for one dollar and six fares for forty cents.[70] A second increase took place in January, 1929. At a special meeting of the Board, director Lander suggested increases to eight cents for cash fares, fifty cents for seven tokens, one dollar for 16 tokens, and $11.00 for 176 tokens. The directors approved and placed the new fare policy in operation.[71] By 1933, a fare sold for ten cents and four tokens sold for twenty-five cents.[72]

In 1921, the Street Railway Company entered a very critical period. The purchase of the six Birney cars, the cost of paving requirements, the issuance of bonds and the outstanding loans for operations put the company heavily into debt. By December 31, 1921, the company's debt statement included the following items:

Bonds (secured by first mortgage on all property owned by the company	$50,000.00
Banks and Trust Companies (money borrowed)	9,100.00
Unpaid balance purchase price of cars (secured by conditional sale contract)	19,005.08
Ties and other miscellaneous	1,280.00
North end pavement claims	8,260.60
Total	$88,695.68[73]

Committees from various civic bodies gathered to express their concern. They posed a number of questions in the form of a petition that evidenced how popular the street railway was:

1) The Street Railway Company furnishes cheap and convenient service to all of the parks, Lincoln, Central, University, and Riverside. How much would the value of these parks be diminished if there were no street car service?

[67] Unpublished manuscript from the Minnesota Transportation Museum.

[68] Ibid.

[69] Corporation Records, vol. 2, 163.

[70] Odegard, 63.

[71] Corporation Records, vol. 2, 165.

[72] Odegard, 63.

[73] ''Statement to the Citizens of Grand Forks in Reference to the Street Railway Situation,'' prepared by Committees from the Civic Bodies of Grand Forks, (no publisher, no date), 9.

[74] Ibid., 11.

[75] Ibid., 10.

[76] Corporation Records, vol. 3, 15-18.

[77] Ibid., 3.

[78] Ibid., 66-67, 79.

[79] Unpublished manuscript from the Minnesota Transportation Museum.

[80] Corporation Records, vol. 3, 82.

[81] Ibid., 108.

2) The Street Railway Company furnishes invaluable service during fair week. What would the State Fair amount to without street railway service?

3) We have had in Grand Forks during the past 13 years the best little street car system in existence. Shall we continue it?

4) Shall we continue to be a city, or shall we turn back and again be a country town?[74]

The several civic committees also formulated a policy in their petition for partial liquidation of the debt. They recommended:

that the Company be given a ten year vacation from the expense and cost of construction, repairing and maintaining pavement to begin January 1, 1921 and expire December 31, 1930. The company will pay for its own rails, ties and other material and the labor of installing its tracks, but the City will pay for all paving, including the north end pavement, the University avenue repairs and the south end pavement, together with such other construction, repairs and maintenance charges as may accrue during such period.[75]

This petition evidently never met with approval from either the general public or the city council because the company continued to pay for paving requirements and to request exemption from this requisition.[76] Too, the company evidently liquidated its debts because the street railway system endured until 1934.

Although no people gathered, no band played and no address was given, October, 12, 1926, was also an historic day for Grand Forks; it marked the beginning of the decline of the street railway system and its replacement by the bus. At a Board of Directors meeting, director Hunter suggested ''that it might be well for the company to consider the possibility of operating buses.'' Hunter thought the bus a better means of public transportation because it operated without tracks and therefore could provide service to areas not covered by the streetcar.[77]

Discussion of the change to bus service continued until 1930 when the directors finally decided in its favor. At a special meeting, held July 29, 1930, the Board of Directors instructed the General Manager to purchase two Reo buses with Eckland bodies from the Eckland Brothers firm in Minneapolis. On September 9, the directors approved the purchase of a third bus. These rubber-wheeled vehicles began total service on the Riverside Park line in June, 1931,[78] and on the Lincoln Park line later that year.[79] The 1931 also saw the purchase of two Mack buses at a total cost of $10,000.[80]

The complete transfer to buses took place in the spring and summer of 1934. The board met on April 10 and director Griffith reported the following conclusions:

1) The condition of University Avenue is so poor as to require large repairs immediately if street car service is to be continued long hereafter, perhaps [at a cost of] $1,500.00

2) If track is maintained on Fifth Street and DeMers Avenue after repaving, estimated cost would be $6,000.00.

3) Streetcar operations should therefore be abandoned permanently whenever the paving neccessitates [sic] it.

4) A fleet of five buses (2 large and 3 smaller) ought therefore to be purchased now.[81]

19

On July 1, 1934, the only remaining streetcar "clanged through the city for the last time." There was, however, a final "sentimental" run on July 15. Loaded with company directors, professors, city officials and businessmen, it traveled down University Avenue where track construction first began 30 years earlier. The company sold the old rails and cars to the public. Several farmers purchased the rails for supports in structures such as potato bins.[82] The old cars "became anything from farmers bunk houses to beer parlors, from living quarters to chicken houses."[83]

The establishment of the streetcar on the thoroughfares of the nation's cities took place between the years 1895 and 1905. In Grand Forks, it occurred later even though the intent had existed previously. The streetcar met a demand and a need for an adequate system of public transportation and it expanded rapidly until 1917. During this period, Grand Forks, like the other cities, increased its mileage of operation and thus increased service to the public. During this period, too, the various streetcar companies, like the Grand Forks company, encountered two problems, the paving requirements and the fixed fare.

Although the streetcar era of prosperity generally continued until roughly 1927, the national system started to decline thereafter. In Grand Forks, the decline began in 1930 with the purchase of two buses and ended in 1934 with the change to a total bus system. The street railway system in Grand Forks and in the nation's other cities, declined primarily because buses were more versatile. It also declined as a result of the financial burden created by the paving requirements and the increased usage of the private car.

The streetcar was a stage in the ontogeny of public transportation. Its use marked a phase between the outmoded horsecar and incoming bus. The streetcar emerged of necessity, grew because it was useful and declined when it became obsolete. The Grand Forks street railway system exemplifies one part of the history of American public transportation.

[82] Dearden, 27.
[83] *UND Alumni Review*, (December, 1938), 16.

Car #132 heads down Fifth Street in 1934, shortly before the complete switchover to buses.

— *Courtesy Orin G. Libby Manuscript Collection*

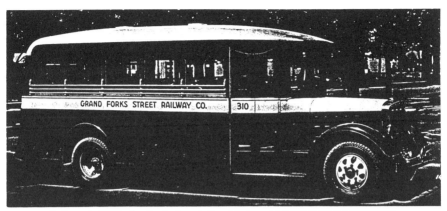

By the early 1930's, the street railway in Grand Forks was running largely on rubber tires.

— *Courtesy Orin G. Libby Manuscript Collection*

Car #114, the sole surviving streetcar from the Grand Forks railway network, was recently nominated to the National Register of Historic Places as an historic object and is presently being refurbished. The rebuilt car will be displayed in a local museum.

— *Photo by Dawn Maddox. State Historical Society of North Dakota Collection*

21

Lancaster's Streetcar Suburbs, 1890-1920

Gary R. Hovinen

We usually associate the movement of people to suburbs in search of lower density environments with the period after World War II and sometimes as well with the 1920s. But even before the automobile age, people may have preferred suburban locations where cheaper and more abundant land made possible larger building lots and where a less dirty and less noisy environment was found. Until the 1890s, the major limiting factors were the central location of most jobs and transportation technology that was insufficient to enable mass commuting. The electric streetcar, developed first in Richmond, Virginia, in 1887, was the most important transportation breakthrough before the automobile, since it enabled great numbers of people to commute longer distances at a relatively low cost. Lancaster, in 1890, was one of the first cities of its size to adopt the new innovation.

My purpose in this article is to examine the premise that the streetcar quickly encouraged first the subdivision of land and shortly thereafter the widespread building of houses in the suburbs outside the legal limits of Lancaster City. In the 1890s, much buildable land remained within the city, especially in the northeast and southwest sections. At the same time, the developed area of the city was very densely populated. Did the advantages of cheap, more abundant land outside the city, made accessible by the streetcar, encourage land speculators to become active and large numbers of people to buy houses there in newly subdivided areas? Furthermore, what classes of people located in Lancaster's streetcar suburbs? Were the new suburbanites middle class or were other classes also involved? Finally, how many suburbanites commuted daily to the city?

As a clue to the thinking of influential community leaders in the 1890s about the potential of the streetcar in leading to lower density development, we can examine the news media. An article in the Lancaster New Era of July 14, 1891, suggested:

227

Lots are far cheaper in the outskirts of the city than nearer its centre, and larger grounds can be secured the same money, while the quick and cheap communications with the business portion of the city renders the distance no objection whatever. Building operations will certainly extend themselves along the lines of street railway travel, for to most, some easy distance from the dust and din of the busy city is more desirable as well as healthier than the turmoil and heat of the thickly settled portion.

To discover the extent to which suburban land speculation and development resulted from the streetcar, various sources can be used. Newspapers and previous publications (Denney 1970, Shindle 1976, Cummings and Rohrbeck 1977) provide information on extension of the streetcar network from 1890 on. Early subdivisions outside the city are filed in the Register of Deeds office at the new county courthouse. At the same office, one can obtain information from the index of deeds on landownership transfers and then examine deedbooks in the old courthouse for more specific information. County atlases from 1875 and 1899 also provide information on landownership as well as on existing subdivisions and residential buildings.

Other sources provide information on class or socioeconomic status and commuting characteristics of early suburbanites. City and county directories list the names of residents and their type and place of employment. Tax assessment lists at the County Historical Society provide data on property values.

Extension of Electric Streetcar Network

Before the electric trolley suburban horsecar lines were constructed to Millersville in 1874 and to West End Park in 1888. A cable car ran along Old Philadelphia Pike to the Conestoga River in Lancaster Township beginning in 1888. John C. Hager, with real estate holdings in the West End of the city and in Lancaster Township, was president of the Millersville Street Railway Line. Hager was one of the principal promoters of the electrification of the city lines by September 1890 and of the West End and Millersville lines by June-July 1891.

By September 1891 streetcar promoters were proposing a new line out Marietta Pike to the West End Addition in Lancaster Township, where John Hager owned land, then south to join the existing Columbia Avenue line. The newspaper reported considerable speculation about suburban residential building activity that would occur once the streetcar line was finished.

Long-distance lines to Columbia and Lititz, completed in 1893 and 1895, made additional land to the west and north of the city ripe for subdivision and development.   In the latter case the Pennsylvania Railroad Company refused to grant permission for a trolley to cross its bridge just north of the city, so a horse-drawn omnibus had to transport passengers from North Duke Street in the city to a point in Manheim Township on the other side. Given the inconvenience to passengers and the loss of potential revenues, the trolley company decided to build an indirect line to Lititz via New Holland Avenue and Rossmere. The Rossmere Belt Line was completed in early 1895 and remained part of the long-distance Lititz line until 1907, when agreement with the railroad finally established a more direct line avoiding Rossmere. But the Rossmere Belt Line continued in use as a suburban line, since the trolley, as will be noted later, had given rise to suburban development.

In 1894, various electric lines in Lancaster County were merged into a single system owned by the Pennsylvania Traction Company, which developed grandiose plans for additional lines. But in 1896 the company went bankrupt owing to overcapitalization during a national financial recession. This halted additional building of electric lines until 1900, when the Conestoga Traction Company was formed. Between 1900 and 1913 new lines were built to communities throughout the county, and a direct line to Rocky Springs Amusement Park was finished by 1903.

Subdivision and Building Activity

To investigate in more detail the link between land subdivision and construction of houses, one can examine the suburbs where residential development took place between 1890 and 1920, the streetcar period. For purposes of this study, suburbs are defined simply as areas outside the city boundary but within possible range of places of employment within the city, where residential development may take place as a result.

Table 1 indicates the progression of subdivision and development activity as streetcar lines were extended outside the city. Table 2 shows the relation between dates of subdivision for selected suburban tracts and dates of trolley line openings. Examination of these two tables suggests that trolley line openings, or the anticipation of trolley line construction, often encouraged subdivision of suburban tracts in hopes of selling building lots. In many instances, such as the Herr Tract in Lancaster Township (subdivided by 1899), Fordney Road, Keller Tract, and Eden in Manheim Township, Sunnyside in West Lampeter Township, and Fairview in Lancaster Township (see map), these speculative ventures did not lead quickly to residential development. Apparently, abundant land closer to Penn Square, either within the city or just outside the city boundary, provided

TABLE 1. *Maximum radius of subdivision from Penn Square*

Maximum radius from Penn Square	Year
7/8 mile	1875
1 5/8 mile, Herr Tract, Lancaster Township	1900
2¼ miles, West Lancaster, Lancaster Township	1910
2 7/8 miles, Eden, Manheim Township	1920

Maximum radius of development from Penn Square

Corridor	1899	1920
East King Street	1½ miles	2 miles
Rossmere	1 3/8 miles	1½ miles
North Queen Street	1 mile	1 3/8 miles
Columbia Avenue West	1 1/8 miles	2¼ miles
Manor Avenue	7/8 mile	1 1/8 miles
South Queen Street	7/8 mile	1 mile

TABLE 2. *Subdivision and Trolley Line Opening Dates*

Subdivision	Date Subdivided	Trolley Line Opening
1. Clark's Eastern Addition, Lancaster Twp. (East)	1879	1890 (cable car by 1888)
2. Clark's Eastern Addition, by John C. Hager	1892	1890
3. Rohrer's Addition Lancaster Twp. (East)	1892	1890
4. Rossmere, Manheim Township	1895 (Nov.)	Early 1895
5. Real Estate & Improvement Co., Rossmere	Unknown	Early 1895
6. Columbia Ave. and Race; Lancaster Twp. (West)	1897	1891
7. West End Addition, Lancaster Twp. (West)	Unknown	1892
8. Keller Tract, Manheim Township	1895	1895
9. Fairview, Lancaster Twp (West)	1904	1891
10. West Lancaster, Manor Township	1906	1893
11. Sunnyside, West Lampeter Twp.	1912	1903

sufficient building opportunities to satisfy demand. But in some cases, subdividers succeeded in selling lots to people interested in land investment but not immediate building; the case of Fairview, now known as Bausman, will be documented later. In a few instances, mainly involving suburban tracts just outside the city, subdivision led rather quickly not only to sale of lots but also to development of a significant portion of those lots.

Case Studies of Streetcar Suburban Development

T hree indicators of the nature of Lancaster's early suburbs are occupational or socioeconomic status of residents, median assessed property values, and commuting characteristics. Using these indicators, one quickly sees that Lancaster's suburbs before 1920 were not simply homogeneous, middle-class areas and differed markedly from each other.

The West End of Lancaster Township, developed mainly during the streetcar period of 1890 to 1920, was the home of many high-status people who commuted to places of employment in the city. Some of the most prestigious suburban estates were found along Columbia Avenue, whereas others were located to the north. Promoters of the West End included John Hager and John Hager, Jr., and G. B. Willson, the owner of the Wheatland estate. The Hagers, as already noted, were active promoters of residential development along the street railway lines; they first subdivided their landholdings in the West End Addition of Lancaster Township and then sold building lots to many middle- as well as upper-status families. G.B. Willson was the person responsible for the subdivision of land and sale of lots along Marietta Avenue. He and Judge J. Hay Brown, who had common landholdings, exercised influence to have the streetcar and public utilities extended to the Marietta Avenue area, which became one of the most prestigious suburban residential districts outside Lancaster.

But the West End contained some lower-status people as well (see Table 3). Factory workers and day laborers inhabited the Rider Avenue area of the West End, not far from some of the most prestigious suburban estates, including the Hagers', along Columbia Avenue. The housing units along Rider Avenue were generally modest and semidetached and were rented, in contrast to the rest of the West End, where houses were generally owner-occupied. Overall, the West End had the lowest percentage of tenant occupants of any of the suburban areas studied — 33.8 percent of the heads of household in 1908.

By 1908, the West End had the highest median assessed property value for those properties containing housing ($2,000) of any of the suburbs studied. The rather mixed nature of the district was indicated by the range from high to low, from $400-$500 along Rider Avenue to a maximum of $25,000 along School Lane Avenue.

TABLE 3. *Socioeconomic Status of Streetcar Suburbs*

	West End, Lancaster Township*	Rossmere	East End, Lancaster Township	West Lancaster, Manor Township* *
Households	63	64	130	—
1908				
1918	93	74	205	9
Social group breakdown, 1908 (heads of household)				
Professional	8 (11.8%)	1 (1.6%)	7 (5.4%)	—
Owner, entre-preneur	15 (22.1%)	—	3 (2.3%)	—
Sub-managerial	6 (8.8%)	4 (6.3%)	5 (3.8%)	1 (11.1%)
Clerk	1 (1.5%)	4 (6.3%)	8 (6.1%)	1 (11.1%)
Agriculture, Self-employed	3 (4.4%)	1 (1.6%)* * *	2 (1.5%)	—
Shopkeeper	4 (5.9%)	2 (3.1%)	15 (11.5%)	—
Hawker, peddler	1 (1.5%)	—	—	—
Skilled worker	7 (10.3%)	32 (50.0%)	36 (27.7%)	6 (66.7%)
Semiskilled worker	—	1 (1.6%)	5 (3.8%)	—
Unskilled laborer	8 (11.8%)	11 (17.2%)	20 (15.4%)	—
Retired, no occupation, unknown	15 (22.1%)	8 (12.5%)	29 (22.3%)	1 (11.1%)

* Includes Columbia Avenue west to Abbeyville Road, south and east to Rider Avenue, north to Marietta Avenue
* * Social group breakdown data are for 1918.
* * * Agricultural laborer.

TABLE 4. *Commuting from streetcar suburbs*

Area	Year	No. commuting to central city	No. working locally	Retired or unknown
West End	1908	29	7	32
East End	1908	23	17	90
Rossmere	1908	2	24	30
W. Lancaster	1918	3	—	6

The East End in Lancaster Township, which developed mainly after 1890, also consisted of a mixture of classes but was somewhat lower in overall status than the West End. It was the largest of the streetcar suburban developments.

In the East End the median assessed property value was $1,400 in 1908, reflecting lower overall socioeconomic status. The range was also less than in the West End, from $500 to $5,000. The more expensive detached houses were along East King Street and on Cottage Avenue, where they tended to be owned by the heads of household. In general, however, the percentage of rental occupancy was much higher than in the West End (53.8 percent). On Clark Street, many inexpensive row houses were rented to people of low socioeconomic status. John C. Hager, Jr., one of the principal promoters of the West End, was also one of the principal sellers of lots for residential building in the Clark and Orange Street area of the East End.

Rossmere developed after 1894 as a basically working-class suburb (see Table 3) with little evidence of commuting to places of employment in the city; rather most people were employed in local factories, including the large Stehli Silk Mill, the Safety Buggy Company, and the Hubley Manufacturing Company. Promoters of Rossmere were John Hiemenz, a local realtor, and a company which he headed, the Real Estate and Improvement Company. After acquiring 80.9 acres of land in the area around the Pennsylvania Railroad Cutoff in 1893, the Real Estate and Improvement Company attracted the Hubley Manufacturing Company and the Safety Buggy Company to Rossmere the following year. Then, in early 1895, the Rossmere Belt Line was built, making the area a potential streetcar suburb. John Hiemenz, presumably encouraged by the streetcar line, in late 1895 purchased another tract of land which he immediately subdivided into sixty-one building lots and called Rossmere; he quickly succeeded in selling many of the lots.

But most of the houses in Rossmere were not built until 1899 to 1901, and the catalyst was the arrival of Rossmere's largest factory, the Stehli Silk Mill. The Lancaster Board of Trade had actively sought a silk mill for the Lancaster area, and in 1897 John Hiemenz donated the necessary land as well as foundation stones and bricks for construction of the mill in Rossmere and promised that he would not provide free any other land and bricks for a competing silk mill for at least five years. The Stehli Company of Zurich, Switzerland, built its mill in 1898, and many of Rossmere's residents after 1898 worked there; eventually the firm employed 1,200 workers at the Rossmere site, which developed into one of the largest silk mills in the world.

Incomplete evidence from the city and county directories suggests that most residents of Rossmere by 1908 worked locally, whereas both the West End and West End experienced considerable commuting to the central city by streetcar (see Table 4). Rossmere, although a streetcar suburb, was thus distinctive as a more self-contained area in an employment sense. The factories created Rossmere, since residential development followed their establishment. But the streetcar enabled outward commuting for those interested and, more importantly, allowed residents to obtain goods and services in the downtown or elsewhere in the city. It is likely that many people also commuted from the city of Lancaster to Rossmere to work at the local factories.

Most houses in Rossmere were modest rental units built by two builders. The median assessed property value for property with houses was $1,000 in 1908, with a range from $900 to $2,200, reflecting a lower overall socioeconomic status than the West End or East End. By 1908, 87.5 percent of Rossmere's resident heads of household were tenants, giving it by far the highest tenant proportion of the suburban areas being studied.

West Lancaster in Manor Township was the smallest of the streetcar suburban developments, having no residents in 1908 and only nine households in 1918. Served by the Columbia Avenue trolley line, it was subdivided in 1906 but did not experience building until shortly before World War I. Unlike the previous suburbs, it was not contiguous to Lancaster City, being located 1¼ miles from the boundary. Clearly, the streetcar encouraged the creation of West Lancaster, since there were no employment opportunities and the few employed residents by 1918 mainly commuted to work in the city. Subdivision by a local landowner followed the streetcar line opening by thirteen years. In 1918, the median assessed property value for property including housing was $1,450, with the range being $1,000 to $2,125. There were no rental occupants in 1918.

In the other three suburbs, discussed previously, additional development occurred between 1908 and 1918, the East End experiencing the greatest amount (see Table 3). But, unlike West Lancaster, Rossmere and the East End and West End were well established suburban districts by 1908 as a result of developments during the earlier part of the streetcar period. As noted above, they were all contiguous to Lancaster City, in contrast with West Lancaster, and thus provided building lots for those interested in a suburban location at an earlier date.

The four suburbs investigated thus differed to some extent in their socioeconomic composition and their economic relations with the central city. Although streetcars facilitated the outward movement of the middle class to detached houses in the suburbs, the three suburbs of the West End, the East End, and Rossmere each had by 1908 unskilled laborers as well as somewhat higher-status skilled factory workers as residents, usually occupying rental row houses or semidetached houses. Rossmere was a working-class suburb with considerable employment opportunities so that residents worked largely in local factories; the streetcar undoubtedly facilitated commuting to Rossmere from the city of Lancaster as well as making possible some commuting from Rossmere to the city. The East End and West End were more heterogeneous than Rossmere, since they consisted of higher-status, middle-class owner-occupier families in addition to differing numbers of lower-status renters. They also differed from Rossmere in the great extent of commuting to the central city and the relatively fewer local employment opportunities. Finally West Lancaster, without local em-

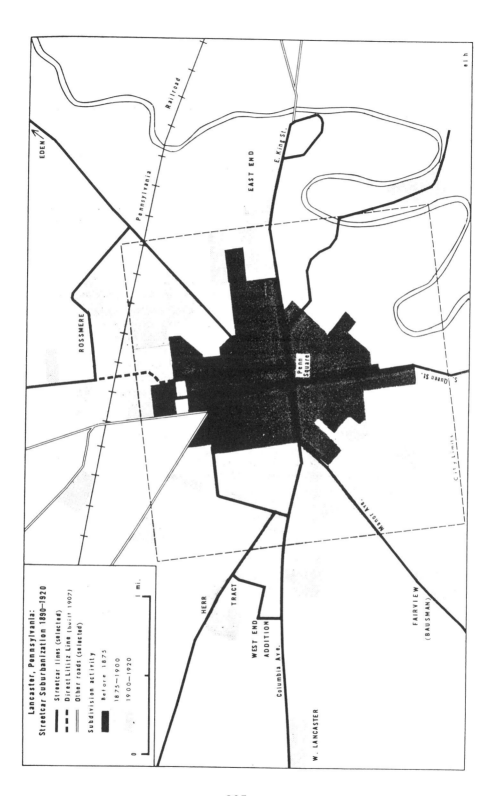

Lancaster, Pennsylvania:
Streetcar Suburbanization 1890—1920

Streetcar lines (selected)
Direct Lititz Line (built 1907)
Other roads (selected)

Subdivision activity
Before 1875
1875—1900
1900—1920

0   1 mi.

EDEN

ROSSMERE

Pennsylvania

Railroad

EAST END

E. King St.

Penn Square

S. Queen St.

City Limits

Manor Ave.

HERR
TRACT

WEST END
ADDITION

Columbia Ave.

W. LANCASTER

FAIRVIEW
(BAUSMAN)

e l h

ployment opportunities, was wholly a commuting suburb of very small size by 1918 where lower middle-class workers owned their own detached house.

In addition to commuting from the above suburbs, some commuting to the city of Lancaster apparently also took place from more distant county communities located on long-distance streetcar lines. A 1908 brochure of the Conestoga Traction Company claimed that hundreds of wage earners commuted daily from half a dozen county towns to the city of Lancaster and thus "earned town wages while having the comfort and advantages of rural life." Insufficient information in the county directories prevented the documentation of this commuting pattern for selected communities.

The streetcar also encouraged speculative subdivisions which did not experience development before 1920. One such area that was investigated in more detail was Fairview, located along Millersville Pike in Lancaster Township, an area that subsequently came to be called Bausman after development took place beginning in the 1920s. Fairview was subdivided by Edward Larter in 1904. The subdivision plan proclaimed that it was Lancaster's newest suburb and was situated along the Millersville trolley road. Larter had purchased thirty-two acres from the estate of a deceased person. Between 1906 and 1912 he succeeded in selling twelve subdivided acres of lots to people mainly interested in land speculation. Only one person had built on a lot by 1912. In the next year, the Lancaster Trust Company offered for sale an additional forty-eight acres of building lots in the Fairview area. By 1920, 252 lots had been purchased, including eighty-three lots by a man from Brooklyn, New York, but only one lot was developed. Apparently, enough building opportunities were found closer to Lancaster or within the city itself to satisfy existing demand by the beginning of the automobile period in 1920.

Conclusion

Subdivisions immediately outside the city boundary were the first to be developed, whereas those somewhat farther out generally remained undeveloped during the streetcar period of 1890 to 1920. The outlying subdivision of West Lancaster in Manor Township, as noted, experienced a small amount of development late in the period. Larger building lots and less polluted environments were not enough to overcome the disadvantage of longer commuting journeys and poorer services, particularly when buildable land existed within or just outside the city. At the same time, subdividers sometimes did succeed in selling large numbers of building lots for purposes of speculation, as the example of Fairview (Bausman) shows. Bausman and other outlying subdivisions were eventually developed, sometimes after a new subdivision plan was prepared, during the early automobile period after 1920.

Commuting characteristics varied widely among the streetcar suburban developments that did occur. The early suburbs except for West Lancaster in Manor Township were not simply bedroom communities. In Rossmere, most residents worked locally. The East End and West End in Lancaster Township had some employment opportunities, but most people commuted to jobs in the city.

Finally we have also seen that the class or socioeconomic status of the streetcar suburbs varied. The early suburbs were by no means simply middle-class communities. Rossmere was largely working class, with many factory or lower paid service workers inhabiting rental housing. The East End and West End also had lower paid people inhabiting rental housing, although middle-class families predominated and, particularly in the West End, there were some upperclass or elite inhabitants. West Lancaster in Manor Township, the smallest of the streetcar suburbs, was also the most uniformly middle class.

To conclude, we can see that the premise stated at the beginning of the article is only partially correct. Outside the city streetcars encouraged much more land subdivision than actual development.

References

Cummings, Luther, and Rohrbeck, Benson (1977). *Garden Spot Trolleys.* West Chester, Pa.: Ben Rohrbeck Traction Publications.

Denney, John, Jr. (1970). *Trolleys of the Pennsylvania Dutch Country.*

Shindle, Richard (1976). "The Conestoga Traction Company, 1899-1931." *Journal of the Lancaster County Historical Society* 80:31-56.   ☐

The Origins of the Decline of Urban Mass Transportation in the United States, 1890-1930

STANLEY MALLACH

From the 1890's to 1920 in virtually every city with a population of at least 10,000 electric traction dominated mass transit, and mass transit dominated urban passenger transportation. In the twenties the composition of mass transit changed, as buses became an important part of urban traffic. But throughout the decade in most of the nation's cities its health continued to depend on the health of street railway companies. Their trolley cars carried the bulk of mass transit users and by the end of the twenties they operated the majority of the buses in local service. The dependence of mass transit on the street railways was unfortunate, because everywhere firms were beset by problems. So much so that mass transit itself faced an uncertain future.

The slow, steady decline of the street railways and mass transportation in America's cities has often been blamed on the automobile. But the car was only one of their problems. When automobiles first seriously contested mass transit's predominance on urban roadways in the 1920's the financially shaky street railway industry would have had a hard time meeting the challenge under any circumstances. What made its task impossible and what made the twenties a crucial decade for urban mass transport were governmental policies and public attitudes that aided the spread of car travel and undermined the nation's street railways. The operating characteristics and financial standing of trolley companies and the harmfulness of governmental policies varied from place to place, so there

STANLEY MALLACH is Bibliographer of the From-kin Memorial Collection at The University of Wisconsin—Milwaukee Library.

were differences in the rate of decline in, for example, east coast and west coast and very small and very large cities. But everywhere, to a greater or lesser degree, mass transit's and the street railways' fall from dominance in the streets conformed to a pattern.

The electric traction industry's sorry condition in the twenties was largely the result of its pioneers' malpractices and the devastating impact of post-1914 economic trends. The industry was born at the end of the 1880's. For the next twenty-five years it mushroomed, as traction firms and cities of all sizes cooperated in a mad scramble to electrify horsecar and cable trackage and build new lines. By 1917, street railway companies were operating thirty-five times more miles of electrified single track than in 1890 and carrying billions more passengers.

To the lasting misfortune of the street railways, shabby and unwise practices pervaded this rapid growth. As often as not, the principal figures in a street railway scheme milked their properties by letting construction contracts to a company they themselves owned. Such companies frequently charged exorbitantly for their services and did cheapjack work besides. Another dubious practice was overbuilding. To be sure, during the industry's early years firms built needed and profitable lines through densely populated areas. But they also laid miles and miles of track in places too thinly populated to support the trackage, and they built lines simply to drive other operations out of business or to enhance the value of distant vacant lots owned by men with street railway connections or political influence. Too few of these lines ever turned a profit. In later years companies either had to abandon them or drain the profits of other lines to keep them in service.

1

William D. Middleton, *The Time of the Trolley* (Milwaukee: Kalmbach, 1967); Howard E. Johnson Collection

Double-decker streetcar of the Terre Haute (Ind.) Street Railway. Built in the late 1890's.

The exuberant financing of growth also harmed the street railways. Driven by a need for capital and a vision of the industry as "a veritable El Dorado," pioneer firms issued huge amounts of bonds and heavily watered stock to a hungry public of large and small investors. Their profligacy resulted in booming overcapitalization everywhere but in Massachusetts. There lines were merely slightly overcapitalized because the state more or less regulated street railway financial practices from the industry's first days.[2]

A substantial portion of the overcapitalization was in bonds, thereby saddling every company in the industry with a long-term burden of fixed charges. In later years, when the industry's prospects had become leaden, these charges depleted the earnings most companies had available for the maintenance and replacement of equipment and other necessities. Moreover, by the time many of the bonds matured the issuing companies had fallen on hard times and could not afford to retire them. Consequently, they either had to reduce their margin of profits over fixed charges further by refunding their debt at a higher interest

rate or default and possibly go out of business or collapse into receivership.

Like so many other early street railway practices the almost universal flat five-cent fare at first benefited companies, but in the long run did much harm. Street railway pioneers believed that such a fare structure would yield high profits indefinitely, so they strove zealously and often corruptly to have it included in their franchises. Their efforts succeeded in an overwhelming majority of the nation's cities, making the United States the only major country in the world where a flat fare predominated over some sort of zone charge. During the industry's first decade or so the nickel fare—in combination with urban population growth, low wages and long hours for street railway workers, and gross neglect of maintenance and depreciation—produced bonanza profits throughout the industry. At the same time, the fare, along with free or low-cost transfers, encouraged urban decentralization by making long trips as cheap as short ones. This led to an increase in the operating cost per passenger on numerous routes, especially

2

240

TABLE 1

MILES OF SINGLE TRACK OPERATED BY ELECTRICITY
AND NUMBER OF REVENUE PASSENGERS
CARRIED BY ELECTRIC RAILWAYS,[1] 1890-1932

Year	Miles of Single Track Operated by Electricity	Revenue Passengers (Including Pay-Transfer)[2]
1890	1,262	2,023,010,202
1902	21,902	4,774,211,904
1907	34,038	7,441,114,508
1912	40,808	9,545,554,667
1917	44,677	11,304,660,462
1922	43,789	12,666,557,734
1927	40,585	12,174,592,333
1932	31,432	7,955,980,642

U.S. Bureau of the Census, *Census of Electrical Industries, 1922: Electric Railways* (Washington, 1925), 8; U.S. Bureau of the Census, *Census of Electrical Industries, 1932: Electric Railways and Motor-Bus Operations of Affiliates and Successors* (Washington, 1934), 4.

1. The electric railway industry consisted of urban, suburban, and interurban trackage. The street railways ran intracity lines. However, the interurbans also carried some suburban and intracity traffic. The Census Bureau never adequately distinguished between street railway and interurban operations, nor did it attempt to separate interurbans' strictly intercity operations from their urban and suburban service. Therefore, the data contain an indeterminate amount of intercity trackage and traffic.

2. The number of revenue passengers includes traffic in cars operated by cable, animal traction, steam, gasoline engines, and gravity. Such traffic was a substantial portion of the total only in 1890. It was insignificant after 1902.

long-distance ones where population was sparse or unevenly distributed. Low earnings on such routes nullified some of the growth in earnings from increases in short-haul travel. Over time, as the amount of flat-fare, long-haul traffic swelled and short-haul traffic growth failed to keep pace, street railway earnings suffered. Companies were unable to adjust their fare structure swiftly to meet this situation because of inflexible franchise provisions, the gradual decline of their political power after 1900, and the rise of rate-controlling state public service commissions after 1907. When wartime inflation sent operating costs skyrocketing the once cherished flat nickel fare became a pernicious canker.

The nickel fare also caused widespread public hostility. Before 1910, critics of the industry incessantly protested that a nickel was too much for most trolley rides and that companies were fleecing the public to pay extravagant dividends on water-logged stock. Further, the five-cent fare stood as a major symbol of the street railways' corruption of governments in their ceaseless quest for favors. Whether they looked at San Francisco, where Boss Ruef funneled money from the United Railroads into the pockets of city councilmen; or at Milwaukee, where Henry Clay Payne was neck-deep in state and local politics to promote the interests of the Milwaukee Electric Railway and Light Company; or at Chicago, where Charles Tyson Yerkes built an empire of trackage and rolling stock on a foundation of bribes—everywhere people witnessed the street railways buying franchises and favors which cost the public in money and poor service. The anger over high fares and political corruption resulted in a lasting distrust of the industry and a widespread lack of sympathy for its authentic and serious problems. The abuses of electric traction pioneers also aroused reformers who in the 1890's began organizing in city after city to bring street railway operations and fares under strict regulation.

Had they been guilty only of monopoly, financial manipulation, corruption, and gouging the public, the street railways probably would have occupied a high place in the elaborate demonology of the Progressive Era. Their service assured them this dubious honor. Unlike other public utilities the street railways provided an intimate service which millions of people used and could evaluate every day. Too often, fairly or unfairly, they did not like the service they got. In various cities at various times they criticized long waits on windy corners, overcrowded and unheated cars, rock-hard seats, and the high accident rates throughout the industry which the traction barons callously ignored. Particularly irritating was the reluctance of companies in large and medium-sized cities to build extensions of lines or needed new ones. Urbanites observed trackage being laid into wildernesses to enhance the value of real estate speculations, but they could not obtain service in built-up territories, nor could they get crosstown east-west routes in cities where lines ran predominantly north-south. The haughty attitude of most traction pioneers toward complaints and the needs of cities left a running sore on the public's memory. In the days before effective public regulation, service problems were solved, if at all, only through prolonged and rancorous agitation against street railways or the use of threats or coercion by municipal governments.

On the eve of war in Europe incipient decay and widespread public mistrust plagued the nation's street railways. Many observers inside and outside the industry recognized the signs of trouble, and some traction men even dared to suggest that the industry had reached its zenith in 1910. But not even the most prescient people foresaw the calamities in the industry's immediate future.

3

The first signals of the coming crisis were rising wages and materials costs. Between 1900 and 1915, both had risen only modestly and fluctuated within narrow limits. Then, in 1916, wages began to soar and materials costs to rise erratically. By 1920 American Electric Railway Association indexes showed that both had more than doubled in a five-year period.

The wage spiral particularly troubled traction men, because aside from raising labor costs it exposed the problem of dispersed government regulation of the industry. During part of the wartime and postwar period wages increased in response to the decisions and influence of the National War Labor Board, which began work in April, 1918, and lasted until August, 1919. Its actions prefigured regulatory trends of the 1920's, when state public service commissions rigorously regulated the street railways but could not secure the conditions necessary to keep companies healthy. Its guidelines for union organization gave the Amalgamated Association of Street and Electric Railway Employees of America enough muscle to wring wage boosts from many companies on its own.[3] Fear of the Board and the threat of having the Amalgamated unionize its employees prompted others to accede to their workers' wage demands without a fight. Where wages failed to rise and conflict erupted the Board stepped in.[4] Using the then novel principle that workers were entitled to a living wage regardless

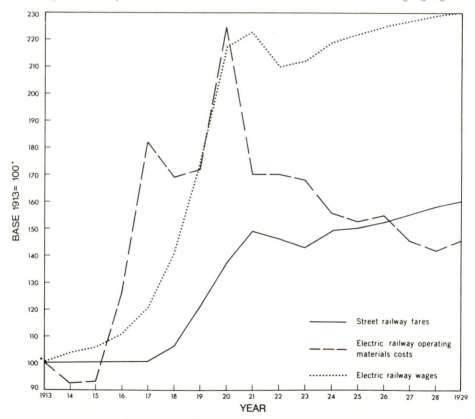

Clifford A. Faust, "Expenditures for Improvements Mount Upward," *Electric Railway Journal,* 74 (January, 1930), 12; UWM Cartography Lab.

Indexes of Street Railway Fares, Electric Railway Operating Materials, Costs and Wages, 1913-1929.

4

of the financial condition of an industry or a company, it awarded wage boosts in every street railway case it heard. The raises often ranged from 40% to 50%; in some cases they exceeded 100%. The Board realized that its decisions were increasing the costs of already hard-pressed companies. It therefore frequently urged state and local regulatory agencies to aid the street railways by immediately raising fares. When many failed to do so, the Board urged Congress to raise fares, but to no avail.

That state and local regulatory agencies failed to respond swiftly and positively to the mounting crisis was not entirely their fault. A United States Supreme Court decision in 1919 seriously hindered their ability to tamper with fares set by franchise provisions. And public protest against fare boosts was rabid. Urbanites remembered the past and were wary of claims that the street railways urgently needed higher fares. Joined in their protest by opportunistic politicians, street railway critics made the once vilified five-cent fare sacrosanct. Meanwhile, they ignored the forces which were undermining the service and solvency of the street railways, in the apparent belief that the trolleys would continue to run no matter what. Only the street railways' demonstrable and dire need for extra revenue finally induced many cities and state regulatory agencies to overcome all obstacles and begin raising fares.[5]

Unfortunately for many companies, their fare increases were not enough to offset spiraling operating costs. Even in cases where a raise was adequate, the higher fare did not always produce the higher revenue anticipated, and sometimes revenue actually declined for a time. For as the cost of riding went up, the number of passengers went down. Among the several reasons for this the most important was public resistance. In a few places mob violence against trolley cars and organized boycotts emptied the streetcars until the fare was rolled back or the public was forced to accept the higher charge. Elsewhere, traffic declined for a short or long time simply because thousands of individuals, including a large number of profitable short-haul passengers, made a personal decision to walk or use another means of transportation rather than pay more for a streetcar ride.[6]

By 1919, the magnitude of the street railway crisis was startlingly evident. So much so that in May, President Woodrow Wilson appointed a commission to investigate the problems of the nation's electric railways and to generate favorable publicity for them. The commission carried out a thorough investigation, but it failed utterly to arouse sympathy for the industry's plight. Meanwhile, in the years following the war through 1920 and into 1921, industry-wide, the critical ratio of operating expenses

to operating revenue crept relentlessly upward. This left many companies with little money for fixed charges and repair of their run-down properties, much less for expanding service, paying dividends, or retiring bonds.

Even more revealing of the electric traction industry's ravaged condition was the trend of companies collapsing into receivership. During the seven years between 1909 and the end of 1915 an average of 19 companies per year, a few of which were probably interurbans, passed into the hands of receivers. Between 1916 and 1920 the average rose to more than 26 firms per year. Furthermore, the companies of the 1916-1920 period were bigger, with a few servicing some of the nation's largest cities. In the 1909-1915 period an average of 572 miles of single track per year passed into the hands of receivers; the average bankrupt company operated 30 miles of track. During the 1916-1920 period the average rose to 1,680 miles, and the average mileage per company rose to almost 65. In the bleakest year of the crisis, 1919, receivership struck 48 companies, some probably interurbans, that operated 3,781 miles of single track and represented $634,174,458 in outstanding stocks and bonds. In the same year foreclosure sales were held for 29 companies, again, some probably interurbans, that operated 2,675 miles of single track and represented $169,772,438 of securities.[7]

News of the industry's desperate straits shot through investment circles. Investors had been concerned about the street railways' ailments even before the war. Wartime and postwar conditions so increased their concern that the softening market for street railway securities of 1914 had turned to mush by 1919. If companies were able to obtain badly needed new capital at all they had to pay a ruinous price. Their stocks sold at a discount and their bonds paid yields of up to 13%. Investors clearly were terrified by the wave of receiverships and foreclosures, and many doubted that even the strongest and most conservatively capitalized companies could earn enough to pay a return on investment competitive with other utilities and manufacturing firms. Also, they viewed the rising competition of the automobile, the war-spawned higher labor costs, the aggressiveness of the Amalgamated encouraged by the National War Labor Board, the failure of regulatory agencies to relieve the postwar crisis, public antagonism toward poor service and higher fares, ceaseless attacks by public officials, and the decrepit condition of innumerable properties, all as ominous clues to the industry's future.

Thus, the street railways entered the 1920's staggering wrecks. During the decade the industry revitalized itself remarkably, causing street railway advocates to speak with cautious optimism about the

5

future of electric traction and all forms of mass transport. But not too far removed from the signs of revitalization, many of the problems accumulated over thirty years persisted, and powerful forces, some old and some new, began to shake the foundations of mass transportation and inflict damage from which no form of urban mass transit would ever fully recover.

Much of the optimism of the 1920's was generated by trends toward better service, improved operating efficiency, peaceful labor relations, more elastic fare structures, better management, and a marked lessening of public hostility. Many of these tendencies had appeared before 1917 in a few cities where empire-building traction barons had given way to managers who were more sensitive to the desires of the public and the long-term needs of their companies. During the twenties such people controlled most street railway firms and brought about industry-wide progress.

The new managers' skillful performance enabled them to more than double the industry's net income during the 1920's, despite high taxes and fixed charges and other adversities. However, rising net income was a deceptive signpost to prosperity, because pointing in another direction stood the industry's operating finances, which were in a very mixed condition. The most encouraging aspect of operating finances was the stabilization of the industry-wide operating ratio. Yet, for all companies the ratio remained at high levels throughout the decade, and for most operations in cities of under 50,000 population it was perilously high. The state of the ratio reflected the movements of operating revenue and expenses. Industry-wide, fares rose a bit; nevertheless, after fluctuating throughout the decade, in 1929 operating revenue was less than in 1920 because of a drop in traffic in some cities and stagnation in others. Meanwhile, one basic expense, materials costs, declined sharply from wartime and postwar levels. However, another expense, wages, continued to climb. As during the wartime and postwar years, some of the wage increase was offset by productivity gains. Still, the increase absorbed a portion of the extra income from higher fares and part of the savings from lower materials costs and improved management. For the future the sum of these mixed trends was bad news: the industry had stagnated and was in incipient decline.[8]

The industry's record in the securities market was mixed and, on the whole, discouraging. During the decade stock prices generally rose and bond yields fell. But the only companies that were able to sell their securities easily or at good prices were the few that were reasonably sound financially or ones which gained strength from their connection with

commercial electric power generation. Only around half the companies in the industry fell into these categories. They tended to be large operations located mainly in the southern and central states. The other half of the industry was more or less shut off from reasonably priced new capital needed to improve operating efficiency and service. Investors were reluctant to put their money into companies plagued by current financial problems and harmful franchises and leases, some of which had only a short time to run. The industry's low return on investment and the defaults on bond issues during the decade did nothing to change their attitude.

The most significant and hopeful trend of the 1920's, both for the street railways and for mass transit, was the expansion of bus service. The first motorized buses in the United States appeared in 1905 plying New York City's Fifth Avenue. They proved so popular that within the next three years the Fifth Avenue Coach Company retired all its horse-drawn vehicles in favor of a fully motorized fleet. The first use of buses by a street railway firm occurred in 1912 in Cleveland, where the local company employed buses in outlying areas as feeders. During the next few years, however, only a handful of street railways imitated this example.

American City, 12 (1915)

Tiffin type of Motor Bus. In service around 1915.

To most electric traction men buses were monsters. Independent operators began running them in significant numbers after 1910 when the shortcomings of the industry's development were becoming pernicious. Then, bus service spread during the wartime and postwar years when few street railway companies could afford to buy a fleet of buses out of earnings or obtain the money at reasonable prices in the securities market. However, even if most companies had been able to buy buses, they would not have, because buses were considerably more expensive to operate than streetcars and had other disadvantages.

But as independent bus operations burgeoned

6

and lax government regulation allowed some to run parallel to street railway routes and skim the cream of the short-haul traffic, the street railways felt compelled to move against this threat to their survival. Companies enlisted the aid of government regulatory agencies, which swept directly competitive independent operators from the streets. At the same time, they raced to inaugurate or expand their own bus service. In 1920, only 21 electric railway firms ran buses, and they carried but 8.3 million riders. By the end of 1929, 361 companies were operating buses and carrying 1.3 billion riders. As the street railways rushed into the bus business independent operators were forced out of business were relegated to servicing mainly small cities. Between 1922 and 1929, the number of independent bus companies in cities of over 10,000 population plummeted from 2,871 to 813, and their share of bus traffic shrank from 94% to 27%.

TABLE 2

BUS OPERATIONS OF ELECTRIC RAILWAY
COMPANIES AND SUBSIDIARIES, 1920-1929

Year	Number of Companies and Subsidiaries Operating Buses	Number of Passengers Carried by Buses	Number of Buses Operated
	(1)	(2)	(3)
1920	21	8,270,000	110
1921	48	18,795,000	250
1922	72	40,447,000	538
1923	123	92,471,000	1,230
1924	171	225,000,000	2,660
1925	251	520,000,000	4,452
1926	339	845,000,000	6,556
1927	369	991,000,000	8,387
1928	361	1,126,400,000	9,669
1929	361	1,280,700,000	11,860

Columns 1 and 3 are from Kenneth L. McKee, "Electric Railway Bus Operations," Aera, 20 (November, 1929), 671. Column 2 is from "More Riding and More Revenue in 1935," Transit Journal, 80 (January, 1936), 6. The figures in columns 1 and 3 from 1920 through 1924 are for December of each year; the figures from 1925 through 1929 are for September of each year. All columns, especially column 1, contain a small but unknown amount of data on interurban bus service; some of this service was probably commuter service.

Street railway companies went into the bus business initially to maintain their mass transit monopolies. But once in it they gradually learned that buses had some advantages over trolley cars. The most important was the public's relish for them. Like streetcars three decades before, buses appealed to urbanites'

taste for novelty, and they called up no memories of real or imagined past discomforts and poor service. Further, buses required no obstructive and ugly poles, wires, and tracks, and they could be loaded at the curb rather than in so-called safety zones in the middle of streets. Other advantages of buses were financial: the capital costs of providing bus service were less than building new trolley lines or extensively repairing old ones; labor costs were less on buses than on trolley cars with two-man crews; and buses were neither taxed nor regulated nearly as stringently as streetcars. However, buses were not totally advantageous financially. Depreciation and maintenance costs were higher than for streetcars and operating costs also were higher, though they fell steadily during the 1920's as manufacturers produced vehicles with larger seating capacities and as gas, oil, and tire prices declined.

Street railways employed buses in various ways. Everywhere companies ran them in newly developed districts to test the market for mass transit. In already settled sections of cities buses were used on newly created crosstown routes, thereby eliminating an ancient complaint against the street railways. In many large and medium-sized cities companies coordinated bus and trolley service, with each form of transportation serving the functions for which it was best fitted. On some lightly travelled routes buses replaced trolley cars altogether. And finally, in some small cities, where street railway service made little or no profit and the increasing use of automobiles was depleting the small pool of potential riders, companies gave up on electric traction completely and turned to all-bus service.

The bus helped mass transit a good deal. In scores of places it enabled the extension of service into areas where the street railways could not afford to build new trolley lines, and on routes and in entire cities where streetcar service had become unbearably unprofitable the bus allowed mass transit service to survive. In some cities where streetcars were losing passengers bus traffic made up most or all of the loss of revenue.[9]

But alone buses could not halt the decline either of the electric traction industry or of mass transit, for the bus was a mere technological solution to the multiplicity of problems facing both. These problems were made more complex and difficult to solve by changing patterns of transit use, which in turn reflected some changing patterns of urban life. After about 1915, and especially after 1920, the volume of long-distance morning and evening rush hour riders increased sharply. Meanwhile, the number of daily off-peak riders declined and the once booming Sunday and holiday traffic to street-railway-owned amusement parks almost vanished. The increase in rush hour

7

traffic forced street railway companies to employ a great deal of labor and equipment during only a few hours of the day, while the decrease in off-peak traffic imposed a costly idleness on much of the same labor and equipment during the other hours and on Sundays.

There were several reasons for the changing patterns of transit use. Cities were spreading out more and more, which swelled the number of long-distance commuters. Shopping and leisure facilities were sprouting around new residential areas and in places mass transit could not afford to service, thereby reducing the number of shopping and leisure-time riders. But most of all, the automobile was changing people's ideas about urban travel. Once they had been willing to walk a few blocks to transportation lines and to adjust their activities to mass transit schedules. Now, the existence of the automobile fostered a desire for a more personalized transportation service. The efforts of car manufacturers during the twenties allowed many to realize their desire.

Before the 1920's, when the industry had only begun to produce cars for the millions and not just millionaires, the privately operated automobile was to the street railways merely a pest with an increasingly sharp sting. Even so, its popularity was already widespread and it had proven its usefulness to those who owned one and to those who longed to own one. Far more than any previous form of transportation the automobile added a desired garnish of variety to people's lives by giving rise to new choices as to where they might make their homes, where and when they might shop, and where and when and how they might spend their leisure hours and vacations. During the twenties the car became an inexorable threat to all forms of mass transportation. The industry's volume of production swelled, its cars became more attractive and better engineered, and its marketing and managerial methods became the envy of businessmen around the world. The industry's strength, along with low car prices, declining operating costs, and rising real income for millions of Americans, enabled the car to become a securely established and hugely popular utility of urban life.

Because the automobile industry was robust while the street railway industry was sick, the car held an important competitive edge over mass transit—but not enough to decimate it. What decisively tipped the existing unbalance against mass transit were the actions of local and state governments and the federal government. During the twenties old governmental policies injured public carriers more than in the past and meshed with new ones to form a ruinous pattern of discrimination against mass transit. By their policies of taxation and regulation of street railway companies, by their failure to control land uses ade-

quately, and by various kinds of assistance to car travel governments inflicted fateful wounds on urban mass transportation.

The most visible and conscious aid to automobile travel was the gargantuan investment in road construction and improvement by every level of government.[10] Federal and state governments spent their funds on so-called rural roads, which were highways outside incorporated communities or delimited places of at least 1,000 population. Some of the money financed work in authentically rural areas, and some paid for road work that connected emerging urban fringes with central cities. Federal funding for state-owned rural roads began in 1917 with an expenditure of less than $50,000. In the twenties it reached a high of over 100 million dollars in 1925. State governments had funded work on rural roads for decades. By 1914, the first year for which reasonably reliable nationwide statistics are available, expenditures were more than 75 million dollars. By 1920, they had jumped to over 221 million, and by 1930, to over one billion dollars.

Within the legal boundaries of cities, local governments did all the road work, almost always without any financial help from higher levels of government. The statistics of municipal road work are exceedingly poor. Comprehensive mileage statistics do not exist, nor do comprehensive financial statistics for the period before 1921. The available data show that, like state expenditures, those of all urban places increased almost every year of the decade, as spending climbed from 337 million dollars in 1921 to 739 million in 1929.

What all this money bought is as important as the sums spent. Governments constructed some new streets and highways, and in Chicago, Los Angeles, Detroit, and a few other large cities several types of superhighways were built expressly to facilitate suburban travel by car. However, the nation's total road mileage increased but little during the 1920's. Instead, a large portion of the governmental expenditures financed the improvement of existing roads. In and around cities of all sizes roadways were widened and surfaced or re-surfaced, alignments were adjusted, railroad grade crossings were eliminated, and already existing thoroughfares were connected. Often such improvements aided travellers more than new roads, because for most motorists the improved streets and highways were already the main arteries of intraurban and suburban travel; now, travelling on them was safer, faster, cheaper, and more comfortable.

Street railways only rarely objected to the governmental road policies of the twenties. Their buses benefited from the new and improved streets and highways, and the hope lingered throughout the decade that automobile ownership would soon reach

TABLE 3
ROAD EXPENDITURES[1] BY FEDERAL AND
STATE GOVERNMENTS AND BY
INCORPORATED AND OTHER URBAN PLACES,
1914-1929[2]

Year	Federal Government Expenditures (thousands of dollars)	State Government Expenditures (thousands of dollars)	Expenditures by Incorporated and Other Urban Places (thousands of dollars)
1914		75,423	
1915		90,694	
1916		87,217	
1917		116,469	
1918		139,730	
1919		221,260	
1920		358,145	
1921	95,000[3]	444,413	337,000
1922	80,000	492,736	376,000
1923	57,000	493,317	403,000
1924	93,000	691,963	482,000
1925	100,000	761,914	582,000
1926	93,000	747,141	630,000
1927	84,000	847,803	734,000
1928	83,000	983,924	728,000
1929	80,000	1,089,411	739,000

U. S. Bureau of the Census, *Historical Statistics of the United States, Colonial Times to 1957* (Washington, 1960), 458, 459, 461.

1. These figures include expenditures for maintenance and repair, administration, and capital outlays.
2. These figures do not include expenditures by county governments, which spent almost as much money on rural roads during the 1920's as incorporated and other urban places.
3. Expenditures for 1917-1921.

a saturation point. Occasionally companies actually led fights for certain improvements, especially street widenings, in order to relieve the traffic congestion which slowed down their trolleys and buses. When street railways did object to road work, they complained not so much that private car travel benefited from more and better roads, but that mass transit operations did not benefit enough. Preoccupied with gaining momentary benefits from specific improvements, most street railway spokesmen seemed unaware of the possible long-term consequences of the huge public subsidies car travel was receiving.

The infrequent and mild attacks on road-work expenditures contrasted sharply with the repeated and vehement assaults on damaging governmental tax policies. Throughout the twenties street railways paid substantially higher taxes than other public utilities and manufacturing and mercantile establishments. Across the country cities and states exacted

some combination of normal business taxes like license fees and taxes on property and receipts. Beside this, most cities levied unusual imposts. In various places companies were required to pave and maintain, often from curb to curb, the streets they traversed; to clean and sprinkle these thoroughfares and others; to make special contributions to the wages of traffic officers; and to carry public employees, children, and other groups free. The requirement to pave and maintain streets was the most widely condemned and expensive of the special imposts. Every time a city decided to improve a tracked street the local streetcar company had to tear up and then rebuild its tracks at its own expense and then shoulder a share or all of the cost of paving the new roadway. Quite apart from being expensive, the paving tax forced the street railways to provide the sort of roads which facilitated the spread of automobile travel.

Cities and states levied some of these taxes on the street railways strictly to raise revenue. Others, particularly the special imposts, were relics of the industry's early days when franchise-hungry traction companies had accepted even the most outrageous obligations and when cities had imposed normal and special taxes to regulate the street railways and to punish them for their bad service and arrogance. In Swarthmore, Pa., for example, to get permission to lay track on a certain street, the street railway company had agreed, among other things, to build a pine picket fence around the property of a Mrs. Johnson and other residents along the thoroughfare and to build and maintain a fence around a public school. In Pittsburgh, when the city and Allegheny County bought the toll bridges around the city, they gave free passage to all but the street railway company, which had to pay the same tolls it had paid private owners. By the twenties exactions like these had outlived their original purpose and were just the most excrescent part of an onerous tax burden.

During the twenties the tax overload was assailed in trade journals and other publications and even in public service commission decisions. Employing arguments already a decade or more old, mass transit advocates denounced the excessive taxation as discriminatory and demonstrated that high taxes were helping undermine many companies' financial stability and capacity to provide adequate service. They further claimed that high taxes were pushing up fares to levels which discouraged people from using mass transit. Despite such arguments from a broad range of interests street railway companies received minimal tax relief during the decade.

Another problematic aspect of government policy was the work of state public service commissions. States began regulating street railways—along with other public utilities—after cities had completely

9

failed to control them through franchise provisions, taxation, publicity of their abuses, and other devices. The first modern commissions appeared in New York and Wisconsin in 1907. By the mid-twenties forty-one other states had followed their lead.

The original purported purpose of these commissions was to protect the public against street railway abuses. To accomplish this, they received power, which varied in kind and potency from state to state, to control service, securities issues, consolidations, mergers, leases, valuations of properties, accounting practices, and fares. In the years before World War I, many commissions served the public admirably, chiefly by holding fares steady and compelling companies to improve their service and safety standards. In the course of their work commissions also helped the street railways. They assisted companies directly by halting unwise overbuilding and putting an end to destructive financial and consolidation practices, and they aided them indirectly by scotching unjustifiable demands for increased service and lower fares and by deflecting to themselves some of the agitation against the street railways.

Along with such achievements, however, commissions' actions created problems. Their regulations measurably increased companies' operating, maintenance, and depreciation costs. When hard times struck the street railways they seldom relaxed their regulations, and they lacked the jurisdiction to deal with many of the forces that were rotting the industry. Consequently, by the mid-1910's everywhere commissions faced a dilemma: how to fulfill their original assignment of protecting the public against street railway abuses and assuring good service at low fares while keeping the hard-pressed street railways solvent and operating. As early as 1910, many observers inside and outside the industry saw the dilemma developing, and critics soon began scoring the ways commissions were resolving it. Muted at first, the criticism grew constantly louder and more diverse as the crisis of the street railways unfolded. By the twenties commissions were being assailed by a noisy babel of tongues. Most critics believed commissions were bungling their basic function, which different people defined differently. Some thought they were not helping the street railways enough in their battle for survival; others, that they were being too lenient with companies and not fully serving the public's service and fare interests. Naturally, the critics also disagreed on why the agencies were not doing their job properly. Some charged that their powers and jurisdiction were too limited; others, that provisions in local franchises worked at cross-purposes with commissions' policies. Yet others claimed that the federal courts consistently acted as "company tribunals" by nullifying

commissions' decisions against street railways, while still others argued that the agencies themselves were mere tools of the industry. Critics saved some of their harshest words for regulated fares. Some declared them too high and some too low; however, all agreed that commissions decided rate cases, and all cases for that matter, too slowly either for the health of the industry or public peace of mind. Despite the blizzard of criticism during the decade no state abolished its regulatory agency, nor did any commission significantly change its ways.

From a mass transit point of view much of the criticism of public service commissions was whistling in the dark. Critics dealt only with specific problems of regulation rather than the flawed principles on which it was based. The first agencies were created when few people owned a private means of travel and street railways monopolized mass transportation in almost every city in the nation. In most places if a person wanted to go somewhere, he either had to use the streetcar or walk. The specific task assigned the regulatory agencies was to protect the public from avaricious street railway companies. Given the circumstances, before 1910 this meant bringing a large portion of all urban passenger transportation under regulation.

By the twenties commissions faced a changed situation. Though throughout the decade the street railway remained the principal form of mass transit, everywhere the bus and the privately operated car were shattering its dominance in urban transport. Regulatory agencies' rulings did not fully recognize this new condition. Meanwhile, states and cities regulated competing forms of transportation less rigorously, if at all, through agencies which did not significantly coordinate their policies with those of public service commissions. The diffusion of regulatory power prevented the formulation and enforcement of a general urban transportation policy, one which would have assured a mix of various kinds of public and private means of travel, each having a large enough pool of riders to remain economically viable.

Some of the more astute members of the industry and mass transit advocates realized the anachronisms and shortsightedness of public policies. They believed that if governments were going to treat the street railways like monopolies they should hold competing forms of transportation in check. However, these observers usually held a narrow view of competition, one encompassing only other public carriers. They felt that such carriers should be coordinated with each other and with streetcar service. In such a scheme each type of transit would serve the functions for which it was best suited, and ideally one company would operate all of them. Never did these observers recommend legislation stringent enough to discourage

10

people from using their cars, and only the most advanced theorists proposed methods of integrating public and private carriers into a general system of urban transportation. It was well known that legislators were not prepared to place severe restrictions on automobile travel, and advocates of mass transit and public service commissions lacked the power to do so, even if most of them had wanted to. Insofar as coordination of various forms of mass transit occurred anywhere, it was usually because one company owned them all. But coordination was no solution to the problem of inadequate governmental regulation of competing forms of transportation, notably the jitney and the privately operated automobile.

"Jitney" was a term which before about 1920 many people indiscriminately applied to all motorized forms of mass transit. Often, "auto buses" which ran regular routes far away from streetcar lines were wrongly included under this rubric. The true jitney generally was a plain, large automobile which provided a wildcat transportation service in direct competition with street railways. The first ones seem to have appeared on the streets of Phoenix, Arizona in 1913 during a street railway strike, whence they spread across the country with the encouragement of automobile interests and the most rabid critics of the street railways.

Usually jitneys ran only during rush hours; they invariably operated along the most heavily travelled sections of street railway lines; they had no fixed stops; and they charged the same fare or less than the trolley cars—all of which practices enabled them to spirit away a large number of the profitable short-haul riders that streetcar companies needed so badly. In numerous cities they had a devastating impact on street railway traffic. In Bridgeport, Connecticut, for instance, by the spring of 1919, jitneys, some of which were probably regularly scheduled buses, were carrying two-thirds of all mass transit passengers.

A substantial number of cities did not immediately regulate jitneys as public carriers, and the ones that did almost always regulated them less rigorously than the street railways. In their quotidian operations jitney drivers took advantage of this and unsafely packed their cars, drove recklessly, and generally disregarded public convenience in their scheduling and other aspects of their service. Nor did cities impose on jitneys the multitude of financial obligations borne by the street railways. This economic advantage was vital to jitney operators, because with their high operating costs they could not have afforded such expenses and still turned even a meager profit. After a time, the irresponsibility of jitney operators and the anguished cries of the street railways and their friends prompted laggard cities and states to act.

By the mid-twenties almost everywhere jitneys had been placed under strict regulation and had disappeared from the streets. But not before they had dealt another stunning blow to the street railways and helped expose the problem of slow and discriminatory regulation during the crisis-ridden wartime and postwar years.

The automobile proved to be a far greater threat to the street railways than the jitney. In dealing with the car governments effectively discriminated against the industry, and in the process initiated lasting forms of discrimination against all kinds of mass transit. For mass transit, the kernel of the automobile problem was, of course, that people deserted public carriers for their cars. Governments never directly encouraged or discouraged the use of any particular form of transportation. However, in their efforts to facilitate the movement and parking of automobiles they indirectly aided private car travel enormously.

One kind of assistance was their solutions to the problem of urban traffic. During the 1920's the stupendous increase in the number of cars roaming the streets and highways created the most glutted and homicidal traffic conditions American governments had yet seen. This hurt motorists, pedestrians, and merchants, but no group suffered more than the street railways. Cars ran and parked on their tracks and so congested roadways that trolleys were reduced to painfully slow speeds, especially during rush hours. Not only did the slow speeds raise operating costs, but they also made the streetcar less attractive as a means of travel.

Cities were under constant pressure to unsnarl the tangle of trolleys, cars, buses, pedestrians, and horse-drawn vehicles in their narrow, often winding streets. Ideas on how to relieve the congestion poured from drivers and automobile interests, street railway companies, and local merchants, as well as from rising new professionals in traffic engineering who were beginning to insinuate themselves into government service. In response to the pressure, cities across the country resorted to various procedures suggested by traffic engineers and city planners to help both public and private transport. They obtained traffic surveys which defined the nature, scope, and location of problems; improved their horse-and-buggy street plans; prohibited left-hand turns; created one-way streets; installed traffic lights; re-routed streetcars; and urged businesses to stagger their work hours and make deliveries in the central city at night. The federal government, too, under the leadership of Secretary of Commerce and then President Herbert Hoover, worked to ease congestion and improve road safety by inducing numerous states and cities to adopt uniform traffic codes and to install uniform traffic signs and appliances.

11

William D. Middleton, *The Time of the Trolley* (Milwaukee: Kalmbach, 1967); Duke Middleton Collection.

Traffic congestion at Newark's famous "Four Corners," Broad and Market Streets about 1920. Once claimed the world's busiest trolley intersection.

The strenuous efforts to control traffic produced disappointing and ultimately harmful results for mass transit. Governments concentrated on facilitating car travel, in accordance with the widely accepted theory that this was the key to relieving congestion and aiding all forms of transportation. Unfortunately, governments did not always plan and execute their relief schemes well, and even when they did, too often the improvements in traffic conditions turned out to be transitory. For relieving congestion in an area set in motion a treadmill effect: the more the area was relieved, the more automobiles used it, thereby causing some re-congestion and the need for new relief efforts.

Though even partial improvements provided trolleys and buses with additional room in the streets and enabled them to achieve faster running times, car travel was speeded up far more, especially in the

rapidly growing residential zones inside and outside city limits. By the middle of the decade most authorities agreed that people who had a choice would use mass transit only if it were as fast as automobile travel. Though efforts at traffic control increased the ease of movement and speed of both, by increasing automobile speed far more governments enhanced the utility and popularity of cars and contributed significantly to the decline of mass transit.

Cities further aided automobile travel by their solutions to the parking problem. The proliferation of cars during the twenties created an unprecented demand for parking space. As motorists seized the most readily available territory for their vehicles, the area next to the curb, the amount of room for moving traffic shrank. This harmed the street railways. They desperately wanted the amount of curbside parking reduced to speed up traffic and to stop stand-

12

250

ing vehicles from encroaching on their tracks. Some cities accommodated the street railways by prohibiting curbside parking on certain tracked thoroughfares and by imposing other restrictions. But most, in response to pressure from motorists and automobile interests and merchants afraid of losing car-borne business, refused to limit or significantly regulate street parking.

While cities were divided in their handling of street parking, all municipal officials, as well as many authorities on urban problems, were in complete agreement that the best way to solve the parking problem was to increase the total amount of off-street space. Cities therefore opened free or low-cost parking areas on public land and permitted a constantly increasing number of private lots to open. Few champions of mass transit opposed these measures, because they either did not recognize or they chose to ignore the relationship between increases in parking space and increases in car use and the exacerbation of congestion. The street railways paid for their neglect. By the end of the twenties it was evident that cities' parking policies were contributing to the decline of mass transit. It was also clear that cities were as far away as ever from solving the parking problem. The more space they provided, the more motorists flocked to use it. This kept roadways filled with standing vehicles and people seeking a place to park and created a demand for yet more space.

In dealing with the automobile, governmental actions hurt mass transit. In dealing with land subdivision, governmental inaction was the problem: the general failure of cities and states to prevent urban sprawl within the legal boundaries of cities, especially in recently annexed areas, and to regulate the location of new subdivisions outside city limits. The spreading out of cities was not new in the 1920's. During the preceding half-century the steam railroad, the animal-drawn streetcar, the electrically-powered streetcar, and the automobile all had created more extensive urban areas and nurtured the growth of suburbs.

In the twenties, however, decentralization became almost universally accepted as "a magic theory" for curing the physical defects and social ills of cities, and the magnitude of population deconcentration increased dramatically. During the decade, for the first time in American history, the population grew statistically faster in urban rings than in central cities. By 1930, more people lived farther from the center of cities than ever before. The pace, character, and extension of decentralization varied from place to place, but all cities, large and small, were touched by it to some degree.

Decentralization would have harmed mass transit under any circumstances. The pattern of land development that occurred in the spreading cities made the damage greater than necessary. Before 1920, under the regime of fixed-rail forms of transport, most new land subdivision and building had taken place along existing or planned or hoped-for spines of track. Some scatteration had marred the regularity of development, but compared to that of the twenties it was insignificant.

The fast and flexible bus and automobile smashed the locational discipline imposed by fixed-rail transport. They allowed subdividers and builders to reach out over an unprecedentedly large area and begin creating a leopard-spot pattern of development on the land. Main traffic arteries and roads accessible to them did restrict where real estate operators could locate their enterprises, but not enough to prevent scatteration. For primary and secondary streets and highways were widely distributed in most large and medium-sized cities, and governments were spending a lot of money on building new roads and improving old ones, a massive cost subdividers and builders did not have to bear to provide their developments with transportation facilities.

During the twenties some cities and states successfully controlled certain aspects of real estate operations, such as platting, with a panoply of recently passed and, to observers of the day, revolutionary land use laws.[11] And a few government officials and shrewd commentators on urban affairs worried about scatteration. But cities and states had never tried very hard to control the location of new subdivisions or the amount of land to be subdivided. This tradition held sway throughout the decade.

Other characteristics of the real estate boom also affected mass transit. A good deal of suburban building was one- and two-family dwellings, giving many of the new areas a low population density. Then, large numbers of the new suburbanites, especially those who owned single-family dwellings, were not from the ranks of the urban poor; therefore, they often had at least one car and did not have to rely on mass transit for work or leisure trips. If they did use mass transit, it was usually to join the rush hour crush that was not helping the operating efficiency of the nation's street railways. Finally, as the population moved, businesses slowly began to follow, huddling together in conveniently located shopping areas. These commercial centers provided many of the services and kinds of merchandise people had once travelled by streetcar to find; now they did not have to.

During the 1920's few people made much of the impact of decentralization on mass transit, but the effects were real enough. First of all, the spreading out of the population decreased the lucrative short-haul traffic and increased the often unprofitable

13

251

long-haul and transfer traffic. Insofar as people began driving to and from work mass transit lost passengers permanently and more cars were added to the rush hour traffic tangle. Second, most companies could not afford to extend streetcar service to thinly populated, scattered new subdivisions or to provide all of them with adequate bus service. Therefore, some territories had only poor mass transit service and others had none at all. Residents of these districts might have chosen to use their cars for work, shopping, and leisure trips even if they had had first-class service, but under the circumstances they had no choice, and for reasons mainly beyond its control mass transit lost some of the fastest growing parts of the urban transportation market.

By the end of the twenties the fate of the street railway industry was sealed and with it the fate of mass transit. Only during World War II, when government rationing of gas and oil and rubber sent hun-

dreds of thousands of people scurrying back to streetcars and buses, did mass transit ever again flourish in most of the nation's cities. The street railways never completely recovered from the financial buccaneering and reckless management of their pioneers and the smashing impact of World War I and its aftermath. The emergence of buses and other signs of revitalization helped the industry in the 1920's, but they could not restore it to full health. The growing popularity and use of the car and the robustness of the automobile industry exacerbated the problems of mass transit. But most decisive in determining its fall from supremacy in urban transport were the actions of all levels of government. The weave of governmental policies which first became apparent in the 1920's effectively sponsored the spread of automobile travel and formed a noose around the neck of mass transit that has been tightening ever since. □

NOTES

1. The author wishes to acknowledge the helpful criticism of Ira Berlin, Thomas D. Phillips, George H. Roeder, and William F. Thompson of a fully footnoted, substantially identical version of this essay written in 1973.

2. During the first two decades of the industry's history a major reason for overcapitalization and heavy bonded indebtedness was the financial manipulations which accompanied the consolidation of lines and companies in virtually every city in the nation. These consolidations produced other unfortunate results: they stimulated the wrath of anti-monopolists and reformers; they undermined profitable lines by attaching unprofitable ones to them; and they brought on the street railways constant pressure to issue free or low-cost system-wide transfers, which considerably reduced the operating revenue of many companies and converted many marginally profitable trips into completely unprofitable ones.

3. The Board was stepping into one of the most vexatious aspects of street railway operations. The industry had long been plagued by terrible labor-management relations and bitter strikes over wages, unionization, and working conditions. The Board's guidelines, which foreshadowed many of the provisions of the National Labor Relations Act of 1935, had much the same effect on the Amalgamated during and after the war that the Wagner Act had on all unions in the late 1930's.

4. The Board had no enforcement powers. However, the President had the power to seize any industry which failed to comply with its rulings. Since it was well known that Wilson would use this power in order to keep the war effort functioning smoothly, the Board had little trouble inducing employers and employees to carry out its edicts.

5. In 1917, of the 303 systems in cities of over 25,000 population, 267 charged the original five-cent fare, 25 charged six cents cash, and the rest had some sort of zone fare or fare which included a charge for transfers. By the end of 1921, only 22 systems were still charging the original nickel fare, 45 were charging six cents, 67 were charging seven cents, 65, eight cents, 4, nine cents, and 61 were charging ten cents; the rest had some sort of zone charge or other complex kind of fare structure.

6. The case of the Scranton Railway Company illustrates the problems of many companies. In September, 1917, the company applied to the public service commission of Pennsylvania for an increase in its basic fare from five cents to six. The city of Scranton promptly challenged the power of the commission to override the five-cent fare provision in the company's franchise. The company, probably wishing to avoid a donnybrook with the city, agreed not to increase its fare until the legal problem of jurisdiction was settled. In March, 1918 the courts upheld the power of the public service commission to contravene the company's franchise and the following month the company raised its fare. The public did not resist this increase very stiffly, traffic declined only slightly, and the company's income rose. Then, in August, 1918, the War Labor Board granted the company's employees a 50% wage increase, which completely wiped out the benefits of the fare increase. In September, the company received permission to raise its fare again, this time to eight cents. On this occasion the public resisted, and in December, 1918 the national flu epidemic hit the city. The result was a 20% decline in traffic. However, gradually riders returned and the company's income rose in 1919 even though the public

14

service commission, under great pressure, cut the company's fare to seven cents.

7. This discussion of industry-wide trends and others in this essay obscure the differences in the financial condition of operations in cities of various sizes. The rule of thumb was, the smaller the size of the operation and the city it serviced, the poorer was that operation's financial condition. However, it should be noted that *independent* small operations in small cities (many, perhaps most, operations in small cities were part of large consolidated systems) constituted only a small portion of the industry and therefore counted for little in the industry-wide statistics used in this paper.

8. A highly controversial problem of the decade was the abandonment of lines and street railway systems. Between 1915 and 1929, 3,819 miles of single track were abandoned. This included entire systems, usually small ones of ten miles or less which serviced communities of under 20,000 population; these systems had never been any more than marginally profitable, if that much, and were ruined by automobile competition. Lines or parts of lines were abandoned in cities of all sizes; these usually ran through sparsely settled areas or needlessly paralleled profitable routes. From the public's point of view, line abandonments were bad because they reduced the amount of service available, and abandonments of systems eliminated service altogether. On the other hand, from a street railway point of view, such abandonments bolstered companies financially by eliminating unprofitable service. Companies constantly pressed regulatory agencies to allow abandonments, with more and more success as the twenties passed. Buses mitigated the impact of abandonments on mass

transit service by replacing streetcars on approximately two-thirds of the abandoned route mileage. However, on the other one-third traffic was so light it could not support any form of mass transportation.

9. At hearings in 1973 before the Subcommittee on Antitrust and Monopoly of the Senate Judiciary Committee on the Industrial Reorganization Act a controversy arose over whether General Motors, through its market power and industrial and political connections, was a major cause of the decline of electric traction in American cities. Critics portrayed the corporation as working aggressively and sometimes nefariously from the mid-1920's on to replace fixed-rail transport with rubber-wheeled vehicles. General Motors responded that traction companies had already begun using buses before the company entered the bus business in 1925 and that it was the negative economics of electric traction and public preferences, not the machinations of the corporation, that led to the almost complete substitution of buses for trolleys in the nation's cities by the late 1950's.

10. In contrast to the situation in the United States, in England the lack of a vigorous roadbuilding program—along with hostility toward the automobile among government officials and the public, high taxation of cars and gas, and other factors—retarded the progress of automobile travel.

11. The most widely employed land use control of the decade was zoning, which had an adverse effect on mass transit. It promoted low population densities in many suburban areas while promoting high building densities in central cities. The results were a scattered pool of riders in the suburbs and heavy traffic congestion in central cities.

REFERENCES AND FURTHER READING

This essay is based primarily on dozens of articles in the following periodicals: *Street Railway Journal; Electric Railway Journal; Aera; Proceedings of the American Electric Railway Association; Public Utilities Fortnightly; Municipal Affairs; Quarterly Journal of Economics; National Municipal Review; Municipal Engineering; American City; Journal of Land and Public Utility Economics; Bus Transportation; Proceedings of the National Conference on City Planning;* and the May, 1908, January, 1911, November, 1924, and September, 1927 numbers of the *Annals of the American Academy of Political and Social Science.*

Of works on the street railway industry there is no end. By far the best primary sources for the period before 1920 are U.S. Federal Electric Railways Commission, *Proceedings of the Federal Electric Railways Commission . . .,* 3 vols. (Washington, 1920) and Delos F. Wilcox, *Analysis of the Electric Railway Problem . . .* (New York, 1921). Other noteworthy contemporary writings that deal wholly or in part

with street railways are Edward W. Bemis, ed., *Municipal Monopolies: A Collection of Papers by American Economists and Specialists,* 4th ed., rev. (New York, 1904); F. W. Doolittle, *Studies in the Cost of Urban Transportation Service* (New York, 1916); Henry W. Blake and Walter Jackson, *Electric Railway Transportation* (New York, 1917); Herbert B. Dorau, ed., *Materials for the Study of Public Utility Economics* (New York, 1930); Chester T. Crowell, "Consider the Street Car," *Saturday Evening Post,* 198 (August 22, 1925), 45-48; and Raymond S. Tompkins, "The Troubled Trolley," *American Mercury,* 13 (April, 1928), 400-408. An influential series of articles that epitomized public anger toward the traction barons is Burton J. Hendrick, "Great American Fortunes and Their Making—Street Railway Financiers," *McClure's Magazine,* 30 (November, 1907), 33-47, (December, 1907), 236-250, and (January, 1908), 323-328.

Two secondary sources of importance on street railways are Edward S. Mason, *The Street Railway in Massachusetts: The Rise and Decline of an Industry*

15

(Cambridge, Massachusetts, 1932) and Emerson P. Schmidt, *Industrial Relations in Urban Transportation* (Minneapolis, 1937) Both are broader in scope than their titles suggest. William D. Middleton, *The Time of the Trolley* (Milwaukee, 1967) is an eye-delighting picture book with a worthwhile text. Clay McShane, *Technology and Reform: Street Railways and the Growth of Milwaukee, 1887-1900* (Madison, Wisconsin, 1974) is a case study that contains two superlative chapters on the first decade of the industry's history nationwide. Paul Barrett, "Public Policy and Private Choice: Mass Transit and the Automobile in Chicago Between the Wars," *Business History Review*, 49 (Winter, 1975), 473-497 makes many of the same points as this essay, while Larry J. Saylor, "Street Railroads in Columbus, Ohio, 1862-1920," *Old Northwest*, 1 (September, 1975) 291-315 takes an interesting legalistic approach to the subject. Donald N. Dewees, "The Decline of American Street Railways," *Traffic Quarterly*, 24 (October 1970), 563-581 tells the story of the industry in the twenties trenchantly and carries it into the 1950's. Robert E. Ziegler, "The Limits of Power: The Amalgamated Association of Street Railway Employees in Houston, Texas, 1897-1905," *Labor History*, 18 (Winter, 1977), 71-90 shows the kind of troubled labor relations that prevailed throughout the street railway industry during its first three decades. Glen E. Holt, "The Changing Perception of Urban Pathology: An Essay on the Development of Mass Transit in the United States," in Kenneth T. Jackson and Stanley K. Schultz, eds., *Cities in American History* (New York, 1972), 324-343 is an ambitious brief essay. G. Lloyd Wilson, James M. Herring, and Roland B. Eutsler, *Public Utility Industries* (New York, 1936) is useful on a variety of types of mass transportation in the 1930's. This last work and those of Holt and McShane contain good bibliographic guides to the literature on street railways. George W. Hilton and John F. Due, *The Electric Interurban Railways of America* (Palo Alto, California, 1960) is a standard work on an industry that supplemented street railway service in many cities.

Numerous urban histories and topical and general studies of individual cities contain material pertinent to the various subjects of this essay. Three of special interest are Robert M. Fogelson, *The Fragmented Metropolis: Los Angeles, 1850-1930* (Cambridge, Massachusetts, 1967); Harold M. Mayer and Richard Wade, *Chicago: Growth of a Metropolis* (Chicago, 1969); and Blaine A. Brownell, *The Urban Ethos in the South, 1920-1930* (Baton Rouge, Louisiana, 1975).

On the rise of the bus in urban transportation the principal sources are electric railway trade publications and *American City*. Also, though they deal more with interstate than with local bus service, *Bus Transportation* and a statistical publication, *Bus Facts*, are useful. *The Attitude of State Regulatory Commissions Toward Motor Bus Competition With Railways* (New York, 1924), a mimeographed publication of the American Electric Railway Association, is of interest. There is no historical writing of any consequence on bus transportation in the 1920's.

Topical and general histories on the rise of the automobile are superabundant. A useful general study is John B. Rae, *The American Automobile: A Brief History* (Chicago, 1965). No one interested in the history of the car should miss Allan Nevins' magisterial three-volume biography of Henry Ford. The volumes pertinent to this essay are Nevins, *Ford: The Times, the Man, the Company* (New York, 1954) and Allan Nevins and Frank Ernest Hill, *Ford: Expansion and Challenge, 1915-1933* (New York, 1957). For comparing governmental policies in the United States with those of a nation where the car was not allowed to undermine public transport, William Plowden, *The Motor Car and Politics, 1896-1970* (London, 1971) and James A. Dunn, "The Importance of Being Earmarked: Transport Policy and Highway Finance in Great Britain and the United States," *Comparative Studies in History and Society*, 20 (January, 1978), 29-53 are instructive.

On the jitney the best sources of information are trade journals, *American City*, the *Proceedings of the Federal Electric Railways Commission*, the books by Wilcox and Blake and Jackson, and F. W. Doolittle, "The Economics of Jitney Bus Operation," *Journal of Political Economy*, 23 (July, 1915), 666-695. The only historical work of any significance is Blaine A. Brownell, "The Notorious Jitney and the Urban Transportation Crisis in Birmingham in the 1920's," *Alabama Review*, 25 (April, 1972), 105-118.

Only a few historians have discussed the traffic and parking problems of the 1920's, and then only *en passant*, but contemporary sources are voluminous. Good starting points are electric railway trade publications; *American City*; the September, 1927 number of the *Annals of the American Academy of Political and Social Science;* Miller McClintock, *Street Traffic Control* (New York, 1925); Theodora Kimball Hubbard and Henry Vincent Hubbard, *Our Cities Today and To-morrow: A Survey of Planning and Zoning Progress in the United States* (Cambridge, Massachusetts, 1929), which is a cornucopia of information on many aspects of urban life during the twenties; William F. Sturm, "The Traffic Tangle," *Liberty*, 5 (January 29, 1928), 41-44, (February 4, 1928), 59-62, and (February 11, 1928), 73-76; and Harvard University, Erskine Bureau for Street Traffic Research, *Street Traffic Bibliography* (Cambridge, Massachusetts, 1933). On the important subject of the rise of

16

the traffic engineer Burton W. Marsh, "Municipal Traffic Engineering—Recent Developments in Their Appointments and Activities," *American City*, 40 (May, 1929), 85-87 is suggestive. Contrary to some recent writings, traffic engineers were a different professional group from city planners, and during the twenties they were far more influential.

General information on governmental street and highway expenditures tends to appear in a wide variety of places. A convenient statistical source is U.S. Bureau of the Census, *Historical Statistics of the United States, Colonial Times to 1957* (Washington, 1960), though the quality of the statistics for the period covered by this essay leaves much to be desired. *American City* contains scores of articles on urban road work during the twenties, and the book by Hubbard and Hubbard synopsizes a decade of progress. Also directly or indirectly relevant to the subjects of this essay are: Charles L. Dearing, *American Highway Policy* (Washington, 1941); John C. Burnham, "The Gasoline Tax and the Automobile Revolution," *Mississippi Valley Historical Review*, 48 (December, 1961), 435-459; Malcolm M. Willey and Stuart A. Rice, *Communication Agencies and Social Life* (New York, 1933); and the report on urban road finance by Jacob Viner in Charles M. Upham and S. S. Steinberg, eds., *Proceedings of the Fifth Annual Meeting of the Highway Research Board, Part I: Reports of the Research Committees and of Special Investigations* (Washington, 1926).

Works on public utilities regulation number in the thousands, but outside trade publications few published after 1920 deal extensively with urban transport. A useful bibliography for the period before 1918 is Don Lorenzo Stevens, *A Bibliography of Municipal Regulation and Municipal Ownership* (Cambridge, Massachusetts, 1918). Many of the primary and secondary works on street railways mentioned above are useful for this subject. Other contemporary sources of interest are the January, 1906, May, 1914, and January, 1915 numbers of the *Annals of the American Academy of Political and Social Science;* Clyde Lyndon King, ed., *The Regulation of Municipal Utilities* (New York, 1912); William Anderson, *The Work of Public Service Corporations. With Special Reference to the New York Commissions* (Minneapolis, 1913); and Delos F. Wilcox, *The Indeterminate Permit in Relation to Home Rule and Public Ownership: A Report Prepared for the Public Ownership League of America* (Chicago, 1926). C. Woody Thompson and Wendell R. Smith, *Public Utility Economics* (New York, 1941) is one of many general works on this subject useful for comparisons. Robert C. Post, "The Fair Fare Fight: An Episode in Los Angeles History," *Southern California Quarterly*, 52 (September, 1970), 275-298 is a unique and

thorough case study of the impact of diffused government regulation on mass transit in Los Angeles. David Nord, "The Experts Versus the Experts: Conflicting Philosophies of Municipal Utility Regulation in the Progressive Era," *Wisconsin Magazine of History*, 58 (Spring, 1975), 219-236 provides excellent background on the conflict of ideas that produced diffused regulation of the street railway industry and other public utilities. And although it does not deal with urban transport, Thomas K. McCraw, "Regulation in America: A Review Article," *Business History Review*, 44 (Summer, 1975), 159-183 sorts out many problems involved in regulation. Essential reading on another transportation industry, the railroads, that suffered from self-abuse and harmful government regulation is Albro Martin, *Enterprise Denied: Origins of the Decline of American Railroads, 1897-1917* (New York, 1917).

Much of the literature on suburbanization is by sociologists and demographers and is statistical and involves various academic theories of urbanization. Among the best works in this genre are several essays collected in Leo F. Schnore, *The Urban Scene: Human Ecology and Demography* (New York, 1965). Schnore's footnotes contain a fine guide to the literature. Peter O. Muller, "The Evolution of American Suburbs: A Geographical Interpretation," *Urbanism Past and Present*, 4 (Summer, 1977), 1-10 is a wide-ranging piece with a good basic bibliography. Several noteworthy items not included by Muller are Mark S. Foster, "The Model-T, the Hard Sell, and Los Angeles's Urban Growth: The Decentralization of Los Angeles During the 1920's," *Pacific Historical Review*, 44 (November, 1975), 459-484; Roderick D. McKenzie, *The Metropolitan Community* (New York, 1933); Richard Rhoda, "Urban Transport and the Expansion of Cincinnati, 1858 to 1920," *Cincinnati Historical Society Bulletin*, 35 (Summer, 1977), 131-143; John D. Kasarda and George V. Redfearn, "Differential Patterns of City and Suburban Growth in the United States," *Journal of Urban History*, 1 (November, 1975), 43-66; Harland Bartholomew, *Urban Land Uses: Amounts of Land Used and Needed for Various Purposes by Typical American Cities, An Aid to Scientific Zoning Practice* (Cambridge, Massachusetts, 1932); Homer Hoyt, *The Structure and Growth of Residential Neighborhoods in American Cities* (Washington, 1939); George W. Hilton, "Rail Transit and the Pattern of Modern Cities: The California Case," *Traffic Quarterly*, 21 (July, 1967) 379-393; David M. Blank, *The Volume of Residential Construction, 1889-1950* (New York, 1954); and Charles N. Glaab, "Metropolis and Suburb: The Changing American City," in John Braeman *et al.*, eds., *Change and Continuity in Twentieth-Century America: The 1920's* (Columbus, Ohio, 1968), 399-437.

17

255

TOWNSITE DEVELOPMENT ON THE WICHITA FALLS AND NORTHWESTERN RAILWAY

DONOVAN L. HOFSOMMER
Wayland College, Plainview, Texas

IN THE FIRST TWO DECADES of the twentieth century, businessmen Joseph A. Kemp and Frank Kell of Wichita Falls, Texas, fostered the construction of the Wichita Falls and Northwestern Railway (WF&NW or Northwestern). Its main line extended 305 miles from Wichita Falls into the Oklahoma Panhandle at Forgan, and boasted a fifty-seven-mile branch from Altus, Oklahoma, to Wellington, Texas. The operation of the WF&NW was an instant financial success, and Kemp and Kell discovered that the development of new townsites along those lines was remunerative.

Aside from their financial success, their railroad and townsite developments were even more important by way of corollaries. The twin undertakings served to create numerous entrepreneurial opportunities along the route, increase the value of adjacent lands, make the area more desirable for ranchers and farmers, and promote the settlement of one of the country's last frontiers.[1]

The first of the townsites along the WF&NW was located on what once was part of S. Burke Burnett's famous 6666 Ranch. Shortly after they filed articles of incorporation in 1906, Kemp and Kell purchased that part of Burnett's ranch north of Wichita Falls which fronted Red River—some 17,000 acres for which Burnett was paid $289,000.[2]

Kemp and Kell subsequently formed the Red River Land Company, cut up the former ranch land into 160-acre parcels, and sold these tracts to eager farmers. Burnett was paid about $18 per acre, but the Red River Land Company sold the same lands for $35 to $65 per acre. J. G. Donaghey of Kansas City was retained to market the lands. Later he counseled Kemp and Kell to retain mineral rights, but strangely that advice was rejected.[3]

Even before the parceling of Burke Burnett's former 17,000 acres of grassland, a small community sandwiched itself between the two principal portions of the 6666 Ranch. For nearly thirty years before the coming of the railroad, a rural store owned by J. G. Hardin stood a short distance south of

Frank Kell, *left*, and Joseph A. Kemp, businessmen and promoters who launched the Wichita Falls and Northwestern Railway, developing townsites along the way. Courtesy the author.

what later became the municipality of Burkburnett. Meanwhile, cowboys from the 6666 Ranch made fun of the farmers who lived around Hardin's store by calling them "nesters." The community was officially known as Gilbert, but was better known locally as Nesterville.[4]

Residents of the Gilbert vicinity subscribed to a subsidy to assist in the construction of the Wichita Falls and Northwestern. J. G. Hardin, also a prominent landowner, reportedly contributed $1,000 to the project. He failed miserably if his goal was to induce the construction of the railroad via Gilbert. Kemp and Kell had other plans, and they scrupulously avoided Gilbert or Nesterville which, for their purposes, was already too thickly settled. Little if any money could be made by selling townsite property in and around Gilbert.[5]

Meanwhile, the Red River Valley Townsite Company, headed by Frank Kell, was established as a separate corporation. It owned 2,000 acres of the former 6666 Ranch. On that same land Kell planned to plat a new townsite a few miles south of Red River on the newly constructed line of the WF&NW. Burnett told Kell that he wanted the town named after him, as it was to be located on what had been his land. Kell agreed and submitted the name "Burk," but the Post Office Department objected since there already was a Burke, Texas. "Burnettsville" and "Burke Burnett" also were suggested and rejected. The Post Office preferred "Gilbert." Finally, President Theodore Roosevelt, who had met Burnett in Washington in 1901 and hunted on portions of his ranch during 1904, persuaded postal authorities to designate the town Burkburnett.[6]

On June 1, 1907, the Wichita Falls *Daily Times* announced that the opening of the new town would be held in five days. The townsite company, it said, then would offer 600 lots for sale by auction. Not surprisingly, the railroad operated a free special train that day to carry interested speculators, settlers, and the curious to the new townsite for the event. On board were over 300 passengers including, a *Times* writer observed, sixteen women. For the auction of the lots, the townsite company brought in a "high-hatted stentorian from St. Louis."

Burke Burnett, as expected, purchased the first two lots at $585 each. He then announced his intention to build a two-story brick hotel on these properties, a promise he did not keep. Others quickly joined in the property-buying binge. Lots sold at the astonishing rate of one per minute. Prices varied. Downtown parcels ranged from $275 to $585, while residential offerings brought from $60 to $250.[7]

Townsite promoters provided a barrel of whiskey for the crowd. This act of civility alleviated parched throats, cultivated camaraderie and properly introduced civilization to the grassland wilderness. A bewildered rattlesnake, which interrupted the proceedings with its unwanted presence, created additional excitement. The day was successful for all parties except the snake, which was promptly killed.

Rapid growth followed. By September, F. J. Graves, editor of Burkburnett's newly founded *6666 Star*, optimistically predicted that the "queen city of the Red River Valley" soon was destined to have 1,000 people "if it kept growing at the present rate."[8]

Meanwhile, construction of the WF&NW continued. During the winter of 1906–1907, bridgemen finished the lengthy structure spanning the wide Red River valley north of Burkburnett. At the same time, graders made cuts and fills along a line of engineer's stakes which led away from the river into the Big Pasture region of what became the state of Oklahoma. The 14.09 miles between Burkburnett and Kell, a newly proclaimed station in the Big Pasture, opened to service on June 10, 1907. There was no celebration, but instead only trouble.[9]

At the turn of the century the federal government decided to allow the opening of the Big Pasture—a Kiowa-Comanche preserve—for homestead-

Steam locomotive of the Northwestern Railway. Courtesy Frank O. Kelley.

ing, but the opening was postponed until December 6, 1906. At that time the government authorized a scattering of new townsites in the area—Randlett, Quanah, Ahpeatone, and Eschiti. Absent from the listing was any named Kell.[10]

The WF&NW was given permission by the government to do no more than condemn a right-of-way, build a railroad, and install telephone and telegraph lines through the Big Pasture. A representative of the Office of Indian Affairs was surprised to discover on April 13, 1907, that a number of unauthorized buildings had been constructed in the Big Pasture on lands recently acquired by a supposed homesteader. The story of Kell, a bogus town, is a lengthy one, but on June 20, Indian Agent J. P. Blackmon of the Kiowa Agency requested that the Federal District Attorney initiate legal proceedings against the railroad, the townsite developers, and the residents of the bogus town. His request was enthusiastically endorsed, and for a while it appeared that the government might force not only the removal of the illegal townsite but also Kemp and Kell's railroad.[11]

Ultimately Frank Kell solicited the assistance of his friend Burke Burnett who once had helped Quanah Parker, the famous Comanche chief. Parker considered the Texas rancher "a great friend," so he agreed to meet with Burnett and Kell and eventually used his influence to facilitate a compromise. In 1908 the WF&NW paid $2,340.75 in damages, but retained its right-of-way and communication lines through the Indian lands. The townsite of Kell thus quickly passed to history. Heralded by bold black type in

260

the WF&NW timetables of late 1907 and 1908, Kell disappeared from station listings before Christmas 1908.[12]

Even before the fate of the Kell townsite was determined, the railroad's construction crews continued to throw up a grade which pointed toward the northwest. At this juncture, Kemp and Kell resorted to a form of blackmail that was traditional among western railroad developers. Soon after they pushed their rails into the Big Pasture, the Texas railroaders let it be known that their goal was a junction with the St. Louis and San Francisco Railroad. On May 17, 1907, they announced that they would build their line to one of two junctions with that railroad—at Frederick or five miles south of there. The latter option implied the creation of a new town which was bound to prosper as the junction of two railroads. The alert citizens of Frederick instantly recognized the necessity of courting the Wichita Falls company. Ten days later they announced that right-of-way lands, a depot site, and a $30,000 bonus would be offered as inducement to secure the WF&NW. Such generosity was not lost on Kemp and Kell who immediately ordered company surveyors to locate the line from the Kell townsite to Frederick.[13]

The businessmen no doubt demanded tribute because there was no opportunity for townsite schemes at Frederick, an established community. In any event, Frederick did not remain the WF&NW's end-of-track for long. Delegations of citizens from other established communities—Altus, Elk City, and even Woodward—sought out the Texas railroaders and urged them to lengthen the Northwestern to their respective towns. These delegations, Kell later recalled, were authorized to promise donations. As a result the

Downtown Burkburnett, Texas, during the oil boom of 1918. Courtesy the author.

WF&NW initiated its service to Altus on December 15, 1909, to Elk City on July 1, 1910, and to Woodward on May 9, 1912.[14]

Before the rails of Northwestern's main line reached just north of Altus, however, construction began in the spring of 1910 on the road's Panhandle Division, stretching westward from Altus to Wellington in the Texas Panhandle. Graders followed a survey which led toward Duke, Oklahoma, an established community. The Post Office Department authorized postal facilities at Duke as early as September 11, 1890, and in spite of the absence of rail facilities, it had seemingly prospered. In 1908, boosters of the ill-fated Altus, Roswell and El Paso Railroad boastfully predicted that the village—located on its proposed line—would eventually claim 3,000 inhabitants. Several months later, after the AR&EP passed to history, Kemp and Kell ordered construction of the WF&NW's Wellington Branch. At that time Duke claimed one bank, four mercantile stores, one drugstore, a cotton gin, two lumber yards, two hotels, a blacksmith shop, and a tailor shop. It was an impressive village.[15]

As the line neared Duke, curious observers were startled to see carpenters building a depot a mile east of town. The reason was obvious. Texas promoters had begun to develop a rival town within sight of the original community. In the end, the new townsite strangled the existing town and its residents were compelled to relocate near the railroad's station. Although the

The M–K–T succeeded the WF&NW and diesels replaced steam, but much of the land in western Oklahoma remained unchanged during that time. Courtesy the author.

venture enriched its promoters, it also triggered major antagonisms which lasted for years until an official literally conducted a hatchet-burying ceremony at the main intersection of "new" Duke. Kemp and Kell also were active in land promotions at other locations between Altus and Wellington. At Gould, for instance, they advertised the development of that town's forty-square-block plat. Nevertheless, there were no more "Dukes" on the Panhandle Division.[16]

Mangum—twenty-two miles north of Altus on the WF&NW's main line —was accorded regular service by the Wichita Falls steamcars effective January 1, 1910, but the event was mildly anticlimactic since the Chicago, Rock Island and Pacific had reached Mangum a decade earlier. Residents of Mangum wanted the benefits of railroad competition and alternate rail routes similar to those won earlier by Frederick and Altus. Since they were afforded no opportunity for townsite development there, Kemp and Kell demanded the usual financial tribute.

The Mangum citizenry responded by granting the railroad a right-of-way plus bonuses amounting to $105,000. According to one observer, "nearly every man who owned property came up and donated eight percent of the assessed value of his property to secure" the WF&NW.[17]

By early 1910, Frank Kell was fully in charge of all townsite development along the railroad. As it moved toward its eventual terminal in the Oklahoma Panhandle, Kell intensified his efforts to promote on-line townsites. One was Willow, thirteen miles north of Mangum. Kell's surveyor planned a village of forty-seven square blocks and two of the five streets were immodestly named "Kemp" and "Kell," but the project failed. Only a handful of people established residence.[18]

Before the rails of the WF&NW reached Elk City, Kell again fostered the development of a new town to starve out an old one as was done at Duke. In 1906 a religious group organized a townsite southwest of Elk City. They named it Beulah after a land of rest described in Bunyon's *Pilgrim's Progress*. Six years earlier, however, the Post Office Department had instituted mail service to a neighboring community, Carter, located about a mile south of Beulah. In 1909 Beulah and the original Carter merged whereby the Beulah townsite survived by assuming the name of Carter, presumably to satisfy the Post Office Department. Residents of Carter expected to receive full rail service when the WF&NW built through their town in the spring of 1910.[19]

The railroad failed to install sidings and station facilities there because Kell opened a new townsite a few miles north. It was proudly named Kempton after J. A. Kemp. Kell expected great things of it and his hopes were strengthened when postal service to it began on May 10, 1910, but residents of nearby Carter proved more resilient than he had expected. Much to his chagrin, residents of Kempton packed their bags to establish new homes at Carter. When the Post Office Department discontinued mail service to Kempton on January 14, 1911, the disappointed Texan admitted failure. Thus the towns fostered and named after both Kemp and Kell were equal failures.[20]

Elk City offered the WF&NW the opportunity to generate heavy traffic in freight and passengers and advanced the customary financial aids to win the line, but Kell had no chance for townsite promotion there. However, such opportunities were present just eighteen miles north and a survey was quickly effected. Just a few miles from that projected right-of-way was Hammon—the trading post for the Red Moon Indian Agency. Kell thought that Hammon should have rail service but not at its present location, so he purchased land that had been corn and alfalfa fields. The newly acquired property was quickly platted and soon "old" Hammon was removed to Kell's new townsite on the railroad's survey. It was opened in June 1910. By September it was "booming."[21]

The railroad's construction progress during 1910 and 1911 was slow. The road finally was placed in service to Hammon on January 1, 1911, and to Leedey, nineteen miles farther north, about Thanksgiving 1911. By early December, graders finished the roadbed between Leedey and the banks of the South Canadian River, another twelve miles to the north. In the process they followed a survey which had bypassed historic "old" Trail, a stopping

Building track on the WF&NW near Gate, Oklahoma, during the summer of 1912. Courtesy Earl Kerns.

Walter E. Hocker, chief of Frank Kell's townsite operations.
Courtesy Sue Hocker Word.

place on the Western or Dodge City cattle trail. The "old" town died when its few residents moved to "new" Trail, a Kell townsite located two miles northeast of the original village. Trail was handsomely designed in twenty-five square blocks with the WF&NW depot prominently situated at the west end of Broadway Avenue. Yet, Kell did little to promote it. Perhaps he felt that Trail was too close to Leedey on the south and to Camargo just over the South Canadian River to the north on the projected line of the WF&NW.[22]

In 1910, a member of the railroads board of directors, Walter L. Hocker of Elk City, was appointed by Frank Kell to take charge of his far-flung townsite operations. Hocker soon established a central office at Carmargo, next to the First State Bank, another of his financial interests. Named after a city of the same title in Illinois, Camargo had received postal service as early as September 16, 1892, but it was nothing more than a dusty postal station until the coming of the railroad in 1912.[23]

The land on which the actual town was ultimately located had been delivered to one Peter Mason under a patent dated June 15, 1901. Mason subsequently deeded all 160 acres, less one acre assigned to School District

No. 5 of Dewey County, to Frank Kell for a consideration of $6,100. Kell, in turn, issued a power of attorney to W. E. Hocker on August 17, 1911, which stipulated that the land be surveyed and subdivided into lots and blocks. Hocker was further authorized to "bargain, grant, sell, and convey these parcels of property." At the same time Kell employed J. A. Innis to survey the site which he completed by August 31, 1911. Records indicate that prices for downtown lots ranged from $150 to $750, while residential lots sold for $30 to $75. To help celebrate the arrival of steamcars in Camargo during April 1912, Kell agreed to donate one choice town lot. Interest in the drawing was remarkably keen. No fewer than 795 people registered for it.[24]

As trackmen extended the WF&NW to the north and then the northwest, Hocker vigorously promoted each of the various Kell townsite interests. Along the new route his surveyors filed a fifty-six-square block plat for Vici, and a less ambitious plan for Sharon. There were no townsite opportunities at Woodward, a long-established community, or at historic Fort Supply. Woodward, however, did pledge financial inducements to secure the WF&NW—its second railroad. Hocker authorized a twenty-two-square block plat at May, a more expansive plat at Laverne, and a smaller one at Rosston.[25]

Kell thought his townsites between May and Forgan (fifty-eight miles) were particularly successful. Rosston, Gate, Knowles, and Mocane—his new townsites on the railroad—forced the demise of older nearby communities. The case at Gate was typical. Long before the arrival of the railroad in that vicinity, a community bearing the name Gate City existed a few miles from the route eventually chosen by the WF&NW. Prior to construction of the Northwestern, a delegation from the community negotiated the ultimate location of the line. The Texas railroaders promised only that they would locate the road near Gate City and they required the usual lands for station facilities, right-of-way, and bonus money from the area's residents. They quickly struck a bargain and Gate City, the earlier community, predictably perished when Gate, Kell's new town, was born.[26]

During the summer of 1911, Kell dispatched an emissary to Beaver County where, through the assistance of J. A. Strickland, 1,600 acres of land were purchased for the Texas promoter. On January 3, 1912, Kell's townsite surveyor, J. A. Innis of Woodward, platted the village of Forgan six miles north and one mile west of long-established Beaver City. Forgan was among the most important of Kell's many townsites.[27]

In previous years, Beaver City was the acknowledged capital of "No Man's Land," and its citizenry hoped that it would become the capital city of Cimarron, a ghost state comprising all of what became the Oklahoma Panhandle. They also hoped for a railroad, but could not attract one so pinned their hopes on the WF&NW. Kemp and Kell were contacted by emissaries from Beaver City who hoped to convince them to locate the

Northwestern through their community. The Texas railroaders contended, however, that cost of construction through the broken country between Laverne and Beaver City would be prohibitive. They also said that such a line on the south side of the Beaver River would necessitate the construction of lengthy bridges spanning numerous tributaries which flowed to that stream from the south. Such bridges, of course, would require expensive maintenance. Undoubtedly these were contributing factors in their decision to bypass Beaver City.[28]

There were more plausible reasons for that rejection, however. The route north of the Beaver River was surveyed earlier by another aspirant, and the WF&NW's management considered it shrewd strategy to close off the area to competition by building its own line on that survey. Even so, the most compelling reason for choosing the northern route was that it offered greater opportunities for townsite development than its southern counterpart. So management decided that the new townsite of Forgan, a creation of Frank Kell rather than historic Beaver City, would get the railroad. The implication of this decision was that the laws of economics would deal with Beaver City.[29]

Traditionally new townsites were named for their promoters, railroad officials, or their supporters. It was simply good politics to employ such flattery. Forgan was an example. In funding the construction of the Panhandle Division between Altus and Wellington, Kemp and Kell were aided by James B. and Daniel R. Forgan, influential Chicago bankers. Kell decided there was no townsite on the Panhandle Diviison worthy of the Forgan name. That honor, he eventually announced, would be reserved for the WF&NW's terminal community in Beaver County.[30]

Forgan was incorporated on January 23, 1912, but Kell's usual lot sale was not until February 15. Sales contracts drawn on that day alone netted him $62,000. Thereafter local responsibility for the Forgan townsite was in the hands of an able area resident. The town grew rapidly, but it was not until September 7 that the rails finally reached it and several hundred people came to the countrywide celebration. Frank Kell donated a choice lot as the prize in the baby contest. Three judges evaluated 75 to 100 babies before making their decision. A banquet was held that evening followed by the traditional speeches. The railroad had finally come to Beaver County if not to Beaver City. One observer put it simply: "It was a great day in No Man's Land." A week later, on September 15, the first passenger train arrived from Woodward. By November 1, 1912, the line was fully operational.[31]

Unfortunately, extant records are not adequate to admit an accurate calculation of the assistance the promoters of the railroad and townsites received. In 1912, however, a writer for *Harlow's Weekly* contended that established towns through which the WF&NW built its lines had been "exceedingly liberal with bonuses." Frederick, for instance, obtained the

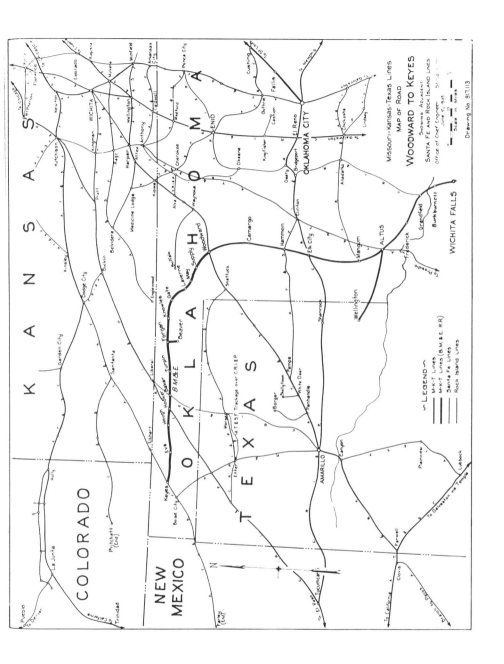

Missouri–Kansas–Texas Lines
MAP OF ROAD
WOODWARD TO KEYES
SHOWING ADJACENT
SANTA FE AND ROCK ISLAND LINES
Office of Chief Engineer   June 1, 1935
Scale in Miles

Drawing No 917.113

~LEGEND~
M·K·T Lines
M·K·T Lines (B.M.&E. RR)
Santa Fe Lines
Rock Island Lines

service only after it contributed $30,000 in cash, a free right-of-way, and a depot site. Altus and Jackson county paid $30,000 for the main line of the WF&NW and $15,000 for the Panhandle Division. Mangum attracted the railroad at the price of $105,000 plus right-of-way lands. Elk City and Woodward made similar contributions, although exact figures are unknown. Later, when the railroad approached Beaver County, WF&NW officials requested and received $3,000 from area residents.[32]

Even less is known about the profitability of individual and collective townsites. Surviving business records shed only meager light on this subject and most of Kell's papers were destroyed. As a result, one can only speculate as to the degree of profitability. Yet the costs of acquiring, platting, and dispersing these tracts of land were certainly minimal. Burkburnett and Forgan were, no doubt, the most lucrative of the townsite developments, but promotions at Hammon, Leedey, Camargo, Vici, Laverne, and Gate were likewise profitable. In the final analysis, it is safe to say that townsite development along the WF&NW was particularly remunerative for Frank Kell, but less so for J. A. Kemp who was involved with him only in the earlier ventures.

The success of townsite developments along the WF&NW was manifest in other and perhaps more important ways. The prior organization and planning of the townsites lured a wave of prospective merchants who quickly arrived to flesh out the communities. This coupled with the new availability of rail service served to increase land values and make the surrounding area considerably more attractive to cattlemen and farmers. The newly built railroads and newly created towns acted as an economic spark and was reflected in population statistics.

Six of the fourteen counties served by the WF&NW reached their maximum population coincidentally with the arrival of the railroad or shortly thereafter. Six others peaked prior to the "Dust Bowl" days of the 1930s. Of the remaining two, one leveled off in the mid-twentieth century and the other reached its maximum population in 1970. Thus, the joint legacy of the Wichita Falls and Northwestern Railway and Frank Kell's many townsite promotions along it was the rapid settlement and development of this rather obscure area of the American West.[33]

NOTES

1. V. V. Masterson, *The Katy Railroad and the Last Frontier* (Norman: University of Oklahoma Press, 1952), pp. 278–81; For a more complete history of the Wichita Falls & Northwestern Railway, see Donovan L. Hofsommer, *Katy Northwest: The Story of A Branch Line Railroad* (Boulder, Colorado: Pruett Publishing, 1976). On townsite developments elsewhere, see V. V. Masterson, *The Katy Railroad and the Last Frontier* (Norman, University of Oklahoma Press, 1952), pp. 94, 166–69,

175–82; Keith L. Bryant, Jr., *Arthur E. Stilwell: Promoter with a Hunch* (Nashville: Vanderbilt University Press, 1971), pp. 67–68, 85–113, 164–65, 192–93, 204; L. L. Waters, *Steel Trails to Santa Fe* (Lawrence: University of Kansas Press, 1950), pp. 247–48; Richard C. Overton, *Gulf to Rockies: The Heritage of the Fort Worth and Denver and Colorado and Southern Railways, 1861–1898* (Austin: University of Texas Press, 1953), pp. 98–101; Richard C. Overton, *Burlington West: A Colonization History of the Burlington Railroad* (Cambridge: Harvard University Press, 1941), pp. 182–85, 232–36, 285–89, 411–12, 473.

2. Masterson, pp. 278–81; Preston George and Sylvan R. Wood, "The Railroads of Oklahoma," *Bulletin No. 60*, Railway & Locomotive Historical Society, Boston: Harvard Business School (1943), pp. 52–53; Minnie King Benton, *Boomtown: A Portrait of Burkburnett* (Quanah, Texas: Nortex Offset, 1972), p. 38; J. W. Williams, *The Big Ranch Country* (Wichita Falls, Texas: Terry Brothers Printers, 1954), pp. 144, 198, 203.

3. Wichita Falls (Texas) *Daily Times*, June 19, 1932; Benton, pp. 6, 39.

4. Williams, p. 203; Benton, pp. 5, 30.

5. Benton, p. 31; Wichita Falls *Daily Times*, June 15, 1947.

6. Wichita Falls *Daily Times*, June 13, 1967; Louis J. Wortham, *History of Texas from Wilderness to Commonwealth*, 5 vols. (Fort Worth: Wortham-Molyneaux Company, 1924), 4:249–51.

7. Wichita Falls *Daily Times*, June 1, 7, 1907; Benton, pp. 7, 39.

8. Wichita Falls *Daily Times*, June 1, 7, 1907; Benton, pp. 7, 39; Burkburnett *6666 Star*, September 11, 1907.

9. *Poor's Manual of the Railroads of the United States; 1907* (New York: Poor's Railroad Manual Co., 1907), p. 1730; ibid., *1908*, p. 631.

10. Wortham, 4:294–96.

11. U.S., Department of the Interior, *Report of the Commissioner of Indian Affairs for the Year Ending June 30, 1903*, two parts (Washington: Government Printing Office, 1904), I:67–75; ibid., *1904*, I:91–92. Permission for the WF&NW to build across the Big Pasture was secured as a matter of course, although the railroad was restricted under the provisions of the Enid-Anadarko Act passed into law in 1902; C. P. Larabee, acting commissioner of Indian Affairs, July 24, 1907, to the secretary of the interior, Record of the Bureau of Indian Affairs, Letters Received, 1881–1907 (63336–1907), National Archives, hereafter referred to as BIA, Letters Received; John Embry, U.S. Attorney at Guthrie, Oklahoma, June 24, 1907, to Commissioner of Indian Affairs, BIA, Letters Received (59236–1907).

12. Wichita Falls *Daily Times*, August 17, 1907; Clyde L. Jackson and Grace Jackson, *Quanah Parker, Last Chief of the Comanches: A Study in Southwestern Frontier History* (New York: Exposition Press, 1963), p. 146; Benton, p. 34; J. P. Blackmon, Kiowa Agent, November 28, 1908, to Commissioner of Indian Affairs, BIA, Letters Received (83374-07-305); *The Official Guide of the Railways*, New York: National Railway Publication Company (September 1907), p. 917; ibid., (January 1908), p. 903; ibid., (December 1908), p. 860; George Shirk, *Oklahoma Place Names* (Norman: University of Oklahoma Press, 1965), p. 74.

13. Wichita Falls *Daily Times*, May 17, May 27, 1907.

14. C. P. Parks, former WF&NW conductor, Altus, Oklahoma, personal interview with the author, November 17, 1972; George and Wood, pp. 38, 52–53, 69.

15. Shirk, p. 67; *Duke, Oklahoma: The Newest City in the Newest State* (Altus, Oklahoma: n.p., n.d., 1908?), pp. 7–8; Eugene Thomas, "Cities Worth Knowing, No. Thirty One: Duke, Oklahoma," *Consolidator*, 5 (July 1950), pp. 4, 26.

16. Eugene Thomas, "Cities Worth Knowing, No. Thirty One: Duke, Oklahoma," *Consolidator*, 5 (July 1950), pp. 4, 26; AWF&H Ry., Station Map of Gould, Oklahoma (July 1910), Valuation Engineer's office, M-K-T, Denison, Texas.

17. George and Wood, p. 43; O. P. Sturm, "Review of the Past Year's Developments," *Sturm's Oklahoma Magazine*, IX (January 1910), p. 31.

18. WF&NW, Station map of Willow, Oklahoma (February 16, 1912), Valuation Engineer's office, M-K-T, Denison, Texas.

19. Shirk, pp. 22, 39, 116; Byron Clancy, "History of Carter, Oklahoma," *Prairie Lore*, Southwestern Oklahoma Historical Society, 5 (October 1968), p. 76.

20. Shirk, pp. 22, 39, 116; Byron Clancy, "History of Carter, Oklahoma," *Prairie Lore*, 5 (October 1968), p. 76.

21. Francis Elgin Herring, 1860–1938 (Obituary), *Chronicles of Oklahoma*, XVI (December 1938), p. 507; Nat M. Taylor, *A Brief History of Roger Mills County*, privately printed, n.d., pp. 7–8; Mrs. E. B. Savage, pioneer resident of Hammon, Oklahoma, personal interview with the author, November 15, 1972.

22. Shirk, p. 208; WF&NW, Station Map of Trail, Oklahoma (June 30, 1918), Valuation Engineer's office, M-K-T, Denison, Texas.

23. Sue Hocker Ward, daughter of the late Walter E. Hocker, Camargo, Oklahoma, personal interview with the author, November 14, 1972.

24. Shirk, p. 35; Camargo Townsite Ledger, First State Bank, Camargo, Oklahoma, p. 17; Abstract of Title, Dewey County, Oklahoma, number 1081, First State Bank, Camargo, Oklahoma; Camargo *Comet*, April 12, 1912.

25. WF&NW, Plat of Vici, Oklahoma (January 15, 1912), Valuation Engineer's office, M-K-T, Denison, Texas; "Woodward, Wonder City of the West," *Wide West*, 4 (November 1911), pp. 18-19; George and Wood, p. 38; WF&NW, Plat of May, Oklahoma (September 4, 1913), Valuation Engineer's office, M-K-T, Denison, Texas.

26. *A History of Beaver County*, 2 vols. (Beaver, Oklahoma: Beaver County Historical Society, 1970, 1971), I:302; ibid., II:149, 201, 238; Shirk, pp. 146, 182.

27. George and Wood, p. 15; *A History of Beaver County*, II:128.

28. *A History of Beaver County*, II;149, 201, 238; George and Wood, p. 15; Beaver *Herald-Democrat*, November 7, 1929.

29. *A History of Beaver County*, II:149, 201, 238; George and Wood, p. 15; Beaver *Herald-Democrat*, November 7, 1929.

30. *Time*, XVIII (July 13, 1931), p. 37; Francis Murray Huston, *Financing an Empire: History of Banking in Illinois*, 4 vols. (Chicago: S. J. Clarke, 1926), III:5-9, 73-74.

31. *A History of Beaver County*, II:128-30; ibid., I:166; The life of Beaver City was spared by the construction of the Beaver, Meade & Englewood Railroad. Forgan's importance peaked in 1919 and then turned downward as Beaver City (now Beaver) experienced a rejuvenation.

32. *Harlow's Weekly*, I (November 16, 1912), p. 10; O. P. Sturm, "Review of the Past Year's Developments," *Sturm's Oklahoma Magazine*, IX (January 1910), p. 31; Cecil K. Chesser, *Across the Lonely Years: A History of Jackson County* (Altus, Oklahoma: Altus Printing Company, 1971), p. 149; *A History of Beaver County*, I:116.

33. U.S., Department of Commerce, Bureau of Census, "Year of Maximum Population by Counties of the United States," Map G. E.-50, no. 37 (Washington: Government Printing Office, 1971). The WF&NW passed fully to the Missouri–Kansas–Texas Railroad in 1923. Most of it was abandoned between 1958 and 1973.

ℐhe
PACIFIC
NORTHWESTERNER

| VOLUME 17 | SPRING 1973 | NUMBER 2 |

Spokane's Interurban Era

By WILMER H. SIEGERT

Conceived in the best of business good faith 70 years ago, a major Spokane transportation enterprise flashed briefly across the city's economic sky, only to die like a spent Roman candle less than three decades later. It was Spokane's interurban railway era, laid low by technological competition from an originally ridiculed source, the horseless carriage.

The interurban's forerunner, the city streetcar, was well established in Spokane by 1903, as it was in other cities in every section of the United States. Streetcars provided a fast means of transportation where previous competition had been horse-drawn carriages and the still fairly new "safety" bicycle.

Spokane had two flourishing street railway systems, both resulting from consolidation of earlier small individual lines. One, the Spokane Traction Company, was headed by the same men who were to introduce interurban transportation to the Spokane area with the Coeur d'Alene and Spokane Railway and the Inland Empire Lines. The other system was operated by the Washington Water Power Company, to make a sideline profit using some of the electrical energy it generated, and still does. Its streetcars bore the company's name, as did the cars on its interurban lines to Medical Lake and Cheney, west of Spokane. (See map on page 27)

This story of Spokane's interurban era is not complete, because the sources of information are scanty, and usually fragmen-tary. Its research ore would have to be classified as low grade. One digs through much material, and all too often finds only small pieces which might be valuable when joined with other fragments. So, perhaps it would be better to title this paper, "Some Notes on Spokane's Electric Interurban Railway Era."

One of the earliest bits of reliable information is a news item in the July 3, 1903 issue of Railway Age magazine. It reported that the Spokane Traction Company's recently chartered Coeur d'Alene Railway proposed to build a 33-mile line from Coeur d'Alene, Idaho, to Spokane, Washington, passing through Post Falls, State Line, Trent and Green Acres, all in the then rural Spokane Valley. It was to be a combined steam and electric line, using 60-pound steel rails. Steam would be used to carry lumber and other freight, and electric power for all other traffic. F. A. Blackwell was listed as president, and J. C. White as chief engineer.

On October 30, 1903, the same magazine reported: "Tracklaying has been com-

WILMER H. SIEGERT, one of our charter members, was first exposed to Northwest history in one of Thomas Teakle's early high school classes in Spokane. He went on to earn two education degrees and will be retired next month as principal of Glover Junior High School in Spokane. After 50 years in Boy Scouts, and honored in 1945 with a Silver Beaver award, he is still an active scoutmaster. He grew up just west of Spokane near the interurban line to Medical Lake and Cheney, which may explain this paper and his model railroading hobby. For the latter his next project will be carving from basswood a scale model of one of the two-car W. W. P. interurban trains he rode as a boy

The PACIFIC NORTHWESTERNER

Vol. 17   Spring 1973   No. 2

Published quarterly by the Spokane Corral of The Westerners, P. O. Box 1717, Spokane, Wash. 99210. Subscription $3 per calendar year. Back issues available.

Articles appearing in this journal are abstracted and indexed in HISTORICAL ABSTRACTS and/or AMERICA: HISTORY AND LIFE.

OFFICERS

Sheriff .... John F. Kelley
Chief Deputy..Rev. James W. Montgomery
Program Deputy .... Ed L. Neltner
Membership Deputy ...... Paul T. DeVore
Chuck Wrangler .... Dr. Charles Anderson
Tallyman .... Ralph R. Reid
Registrar of Marks
    and Brands (Editor) ...... Cecil Hagen

pleted on this line from Spokane, Washington, to Coeur d'Alene, Idaho, 33 miles, and the first trip was made over the road on October 28. Operation will begin about November 15."

Business apparently developed satisfactorily, for the October 4, 1904, Railway Age reported: "Coeur d'Alene and Spokane is asking for bids on two three-car trains for electric service."

The big construction period seems to have started about 1905. Planning, of course, was earlier. Historian N. W. Durham in his "Spokane and The Inland Empire" Vol. I, says of 1904: "The Washington Water Power Company announced its plan of building an electric line to Medical Lake . . .

"In October, President F. A. Blackwell of the Spokane & Coeur d'Alene electric returned from the East, where he and Jay P. Graves had sought financial support for an electric road to the Palouse Country (south of Spokane). 'If the citizens of Spokane and the citizens along the line and in the towns the line will touch will subscribe for a reasonable amount of stock,' he said, 'the Spokane-Colfax electric line, running through the heart of the Palouse Country, will be built. We have practically arranged for the financing of the proposed road.'"

Durham continues: "Suiting the action to the word, Blackwell and Graves incorporated the Spokane & Interurban system in December, and announced a purpose to build to Moscow, Idaho. With them as incorporators were Alfred Coolidge, John Twohy, and F. Lewis Clark. Jay P. Graves became president, F. A. Blackwell vicepresident, H. B. Ferris treasurer, Will Davidson secretary, and A. M. Lupfer supervising engineer.

"A few weeks later, at an enthusiastic meeting of 250 businessmen and property holders, Mr. Graves outlined the company's plans, and $39,000 of the stock was taken on the spot. Graves, Twohy, Coolidge, Clark, and Thos. L. Greenough had previously put $200,000 into the enterprise. The company bought the power site and riparian rights at Nine Mile bridge 'on the Spokane River).

At that time the population of Spokane, as estimated for the U. S. Census by Mayor Boyd, was 57,249. Polk's Directory, counting Hillyard and outlying additions, said 65,267.

The year 1905 supplied an interesting number of notes. A paper of recollections by John B. Fisken, who was in the Washington Water Power Company's interurban operating department, records that in 1905 the company extended its transmission line to Medical Lake, 12 miles west of Spokane. Built on the shores of a similarly named lake, its early day residents enthusiastically touted its highly mineralized, fishless waters for their claimed medicinal qualities.

Mr. Fisken's paper also states that the same year the Washington Water Power Company started building an interurban railway line from Spokane to Medical Lake, using the abandoned roadbed of the Seattle, Lakeshore and Eastern, a railway absorbed by the Northern Pacific in the 1890's.

By 1907, a branch of this Water Power line was bringing trains to Cheney, a few miles east and a little south of Medical Lake. Cheney was the site of a state normal school, now Eastern Washington College of Education.

A comparision of the mainline trackage of the two interurban systems shows that the Washington Water Power Company never had more than about one-seventh as many miles of track as did its rival, the Spokane Traction Company.

The Spokesman-Review for July 1, 1905, carried two one-column ads, each about seven inches long, by the Coeur d'Alene and Spokane Railway Co., Ltd. Phone Main 117. The first offered:

"STEAMER 'IDAHO'

Licensed to carry 1000 Passengers
Excursions up St. Joe River
to St. Maries, Tues., July 4, '05
Ample stop at St. Maries for Dinner,
or bring your baskets and lunch
on the boat if you prefer.
Lv. Spokane 6:30 a.m. or 7:50 a.m.
via electric train from corner
Sprague Ave., and Washington St. Finest
trip in the west. Excursion tickets
good July 1 to 5 inclusive.
St. Maries $2.00 round trip.
St. Joe $3.00 round trip."

The second ad urged:

"July Fourth
Spend the 4th at
Coeur d'Alene
Boating, Bathing, Fishing.
Dancing in the Pavilion
18 trains each way.
Roundtrip $1.00"

General practice in interurban operation was to develop some sort of recreational facility at a distance from the terminal city, thus insuring much Sunday and holiday traffic. The Washington Water Power Company operated many white-flagged specials in addition to regular trains to picnics at Medical Lake. The Coeur d' Alene and Spokane developed resort facilities at Liberty and Hayden Lakes and made dockside connections with paddle-wheel steamers that plied the waters of Lake Coeur d'Alene and the shadowy St. Joe River to St. Maries, Idaho. Throughout the summer, train after train left the city loaded with white-shirted, straw-hatted passengers. They returned in the late afternoon or evening to catch a streetcar home, after enjoying a day that only the interurban made possible.

We return to the Railway Age files for developments on the new Spokane and Inland. On July 14, 1905, it said: "This Company, which is building an electric line from Spokane, Wash., south, has purchased rails for 35 miles of the line to Waverly, and it is expected to have the road in operation to Waverly by January 1."

On August 18, this report: "Spokane and Inland has purchased property for its proposed freight terminal at Spokane, Wash. This road has awarded to Porter Bros. the contract for two bridges over the Spokane River. Both will be of wood, with 150-foot spans, one bridge being four spans in length, and the other three spans . . ." The proposed freight terminal mentioned above was built just east of Division Street and north of Trent Avenue. The building is occupied today by the Riverside Warehouses, Inc.

By November 18 we can read in Railway Age: "Grading has been completed on this road to Waverly, Wash., 33 miles from Spokane, and work has been commenced on an extension from Waverly to Rosalia, 14 miles. Money has been provided for building the road from Rosalia to Garfield and Colfax.

The Spokesman-Review for July 4, 1905, reported, under this one-column headline, "Spokane and Inland Lets Contracts," that "Eleven miles of additional grading is to be done. This takes the line to Waverly. It is probable that the first section of the Spokane and Inland will be equipped and in running order before the end of the year. Yesterday the company placed an order with the St. Louis Car Company for 25 of the most modern flat cars manufactured. No order for an electric locomotive has been placed as yet."

The July 10 Spokesman-Review added, under the headline, "Activity at Medical Lake," this: "The Washington Water Power Co. yesterday let the contract for the erection of their new depot and freight house at Medical Lake. The building is to be a frame structure 40 x 80 feet, and is to be commenced at once . . . it is expected that the town will be lighted with electricity by the first of August."

The writings of 1905 presented two straw-in-the-wind stories of interest to our topic. In January, Railway Age asked: "Are interurban electric railways profitable? It is getting to be time to have figures which will throw some light on this exceedingly interesting question . . . So far, however, there are few if any examples of purely interurban roads which are paying dividends, and the number which have been unable to earn interest on their bonds is not small."

A July 1, 1905 Spokesman-Review headline shouted, "Motors Get Leeway — New Law Interpreted for Police Dept." Under it was this significant story:

"Alex M. Winston, asst. corporation counsel, rendered an opinion to Chief of Police Waller yesterday that the city could not fix the speed at which automobiles may be driven through the city streets, and that the city officials must abide by the state law regulating the speed of automobiles.

"This means a decided victory for the automobile owners. It means that an automobile may be driven eight miles an hour on the principal streets of the city and 12 miles an hour on the suburban streets. On the country roads the automobiles may speed up to 24 miles per hour. In crossing intersecting streets in the city the law merely provides that the automobile shall slow up so as to protect pedestrians, but fixes no speed at which the intersecting streets may be crossed.

"The old city ordinance provided that automobiles should not cross any intersecting street at a speed faster than four miles an hour and that the automobile must not exceed eight miles per hour speed on any city streets."

(Fifteen years later, automobiles were seriously cutting into the profits of the electric interurbans, which, in most cases, were passenger operations with freight adding only minor revenues. The electric lines were built to provide a more rapid and comfortable movement of people to and from the urban centers with a greater flexibility than the steam railroads could offer.)

By December, 1905 Railway Age was able to report: "Spokane and Inland — This electric line has been completed from Spokane south to Freeman, Wash., 18 miles, and grading is in progress from Freeman to Waverly, 16 miles. An extension, Waverly to Colfax and Rosalia, 70 miles, is proposed."

As the calendar for 1906 lost its pages, the electric line into the Palouse Country moved ahead. In January the Spokane and Inland contracted with the American Car and Foundry Co. for 30 boxcars and 20 flat cars.

In February Railway Age reported: "The Inland Empire Railway Co. has been incorporated in Washington for the purpose of consolidating the traction lines radiating from Spokane, Wash. Articles authorized a capital stock of $10,000,000, with the privilege of issuing $10,000,000 additional preferred stock for the purpose of constructing railroads connecting Spokane with other cities and towns in Washington, Idaho, Oregon, and British Columbia. Companies involved are the Coeur d'Alene and Spokane, Spokane Traction Co., the Spokane and Inland Railway, and the Spokane Terminal Railway. Officers are president, J. P. Graves of Spokane Traction and the Spokane and Inland; vice-president, F. Lewis Clark of the Spokane and Inland; secretary, W. G. Davidson of Spokane Terminal; treasurer, H. B. Ferris of Spokane Traction."

By May the line was completed to Waverly, Wash., and surveying for a line through to Lewiston, Idaho, was being reported. Equipment was being ordered. An earlier order of nine passenger cars from J. G. Brill was increased to 18, and the express car order from three to six. Six Baldwin-Westinghouse locomotives were on order, and a number of steam locomotives and steam shovels were on the job.

By June, passenger equipment ordered from St. Louis Car Company was beginning to arrive — four interurban cars with seating capacity of 50 passengers each. "It is expected that cars can be run into Waverly by early July," said news reports.

In March, 1906, the Railway Age's "New Equipment" column reported: "Washington Water Power Co. is having built by J. G. Brill Co. 20 semiconvertable cars for use at Spokane, Wash., with a seating capacity of 44 passengers, and a motor car and four trailers for the Medical Lake Suburban Line. The latter cars will seat 45 to 50 passengers each, the motor car being composed of smoking, baggage, and general passenger compartments, and the trailers having one open passenger compartment."

The June, 1906, issue of the Railway Age reported the expanding plans of the Inland Empire Lines with this item:

"F. A. Blackwell, chairman of the board, is quoted as saying that this company will have a line from Coeur d'Alene to Lake Pend Oreille, Idaho, within a year. Land has been purchased for terminals at Lake Pend Oreille, which is 24 miles from Coeur d'Alene and 56 miles from Spokane." This proposed extension of the line to Hayden Lake was never built.

By midyear, 1907, Railway Age was able to report: "The Spokane and Inland Empire Railway Co. of Spokane, Wash., operates freight and passenger service over 146 miles of track. From the first it has been operated by electric traction entirely. Its equipment includes six 50-ton locomotives and eight 72-ton machines have been ordered. "

The September 13, 1907, issue of the magazine showed pictures of the freight yard and terminal in Spokane and a grain warehouse at Crabtree, a siding near Oakesdale, Wash., as an example of the 30 new warehouses that had been built along the line. "The Company," the report continued, "expects to handle half of this year's wheat crop."

This article also lists mileages thus:

Spokane to Coeur d'Alene, 32.5; north to Hayden Lake, 8; Spokane to Spring Valley Junction (where the road branched to Colfax and Moscow), 40.2; Spring Valley to Colfax, 36.6 On the other branch: Spring Valley to Palouse, 35.8; Palouse to Moscow, Idaho (line under construction), 12.5.

The Electric Railway Journal for Oct. 10, 1908 pictures the new station at Palouse, Wash., and calls it a "handsome brick structure." The S. and I. E.'s annual report also was contained in this issue. From it we learn that "the country has gone through trying times" (what today we would call a business recession). Locally it meant a curtailment of business, and delay in constructing extensions.

However, the Palouse-Moscow line was in operation by September 15, and the double tracking of the Spokane-Coeur d' Alene division had been partly done in spring.

The annual report also stated that the hydroelectric plant of the Inland Empire system was completed at Nine Mile Bridge on the Spokane River, and began supplying power in July, 1908, for the railway and the city of Rosalia. Construction time had been two years.

The Electric Railway Journal article concludes its reporting with this: "The Spokane & Inland Empire interchanges freight with all the steam roads entering Spokane under a joint tariff arrangement exactly similar to those drawn up between two steam roads. The cars of the electric road, however, are never allowed to leave its own rails except in cases of emergency. All equipment for foreign shipments is furnished by the connecting steam roads."

The October 15 issue of this journal reported on a paper read by the advertising agent of the Inland Empire system at a convention in Atlantic City, N. J.: "This Spokane country in 1907 produced upwards of $100,000,000 worth of ores, timber, and agricultural products, and its natural resources are as yet hardly touched. People that migrate to the Spokane country seldom ever send home for money." Elsewhere the speaker emphasized the importance of setting reporters straight:

"I have in mind an accident that occurred on the Coeur d'Alene division two years ago. A man endeavored to whip up his horse and cross in front of our Shoshone Flyer, but just failed to make it, his horse running into the door of the baggage compartment of our No. 1 motor.

"It was such an unusual thing for a rig to run into a train rather than vice versa that it was 11 p. m. of the same day before I could convince the reporter of the facts of the case, and then not until he was shown the indentation of the horse's teeth in the baggage car door. As a result the story appeared in the paper correctly, laying blame where it belonged, rather than on the motorman."

Financially, the Journal report has this statement: "Revenue from operation and from other sources of the Spokane & Inland Empire Railroad, Spokane, Wash., during the year ended June 30, 1908, amounted to $1,139,186. The gross revenue from operation alone amounted to $1,118,018, and 72.2 per cent of this amount, or $807,388 was expended for operating expenses and taxes. Gross earnings from freight aggregated $291,008, or 26 per cent of the total gross revenue from operation."

In many ways 1907-1908 had the earmarks of a success year. The Palouse to Moscow extension, delayed for six months, was in operation by September 15, 1908. Good results were expected since Moscow was then the third largest town in Idaho, and the largest outside of Spokane on the railroad. The power plant had been completed at Nine Mile, so the line was operating on its own power.

The Traction division in Spokane showed a 38.3 per cent increase in gross receipts. The Coeur d'Alene division showed a 12.4 per cent increase. This was considered exceptional because the lumber and mining interests in the Coeur d'Alene country had been badly affected by the panic, many operations having closed down. Part of the Coeur d'Alene division was double tracked.

Here are additional statistics from the year's report: Number of revenue passengers carried, 1,096,817; average distance carried, 22.88 miles; total passenger

revenue, $488,605.83; average amount paid by each passenger, 44.55 cents; average rate per passenger mile, 1.95 cents; passenger earnings per train mile, 72.37 cents; revenue tons freight carried, 331,594; average distance haul per ton, 34.9 miles.

A Nov. 21, 1908 report stated briefly that the Washington Water Power Company at Spokane, Wash., was calling a special meeting of stockholders to consider authorizing a $15,000,000, 30-year 5-per cent bond issue; $7,000,000 of it to be used in betterment of its power, lighting and city street and interurban railway facilities.

The year 1909 was a momentous one for the Spokane and Inland Empire railroad. The Spokesman-Review for July 16, 1909, carried this front page tiered headline:

"LAND LOTTERY RUSH DRAWS 23,500 ON THE FIRST DAY. SPOKANE, COEUR D'ALENE, KALISPELL, AND MISSOULA ATTRACT HOME-HUNTERS."

Then followed accounts of the office difficulties in handling the rush to file for land being made available on the Coeur d'Alene Indian Reservation, and this report of rail business:

"It was reported by the traffic department of the Spokane and Inland Empire Railroad last night that the road had run trains to Coeur d'Alene on an average of every 20 minutes during the day. Daylight found several scores of people on the platform at the terminal building and on the Monroe St. bridge waiting for the first trains to the Idaho registration point.

"When the early trains were brought out the people swarmed through the windows and filled platforms in their anxiety to get to Coeur d'Alene in time to register there and catch the boat to Harrison enroute to Missoula by way of Wallace, Idaho."

Mr. Graves, speaking to the stockholders at a later date, had this to say:

"In this, my third annual report . . . I have much of good and somewhat of ill to report, and I shall put the ill report first. On July 15 of this year (1909) the registration for homesteads on account of the opening of the Coeur d'Alene Indian Reservation was commenced in Coeur d'Alene. Owing to the simultaneous opening of the Flathead and Spokane Indian Reservations, the travel over the Coeur d'Alene division of this company was phenomenal. Even with extra service . . . every train that went out of either Spokane or Coeur d' Alene was mobbed. We called the police . . . The rush was handled without a single accident until July 31.

"On that day special No. 5, westbound, collided with regular passenger train No. 20, eastbound, at Gibbs Station, about 1 1/2 miles from Coeur d'Alene. Both trains were packed, and as a result of the collision 17 persons were killed, and between 100 and 200 were injured more or less seriously.

"Both crews were experienced men. Motorman Delaney and Conductor Seymore on regular No. 20, and Motorman Campbell and Conductor Whittlesey on special No. 5. The collision was due, however, solely to their (Campbell and Whittlesey) negligence and violation of the rules.

"The Company has proceeded as rapidly as possible with the settlement of the claims against it . . . some of the claimants have shown themselves most unreasonable . . . some litigation may result. It is hoped, however, that it will be kept within the $300,000-mark, which, considering the magnitude of the disaster, should be regarded as fortunate.

"The unusually hard winter of 1908-09 and floods on the Inland division caused increased operating costs. The roadbeds have been kept in first class condition, every third tie on the Coeur d'Alene division has been renewed, and our lines are fully ballasted.

"We have built up, at Liberty Lake, a first class summer resort. Being near the city of Spokane (17 mi.) and maintaining a low rate — 75 cents for the round trip — has made it attractive for pleasure seekers . . . the rapidly growing business at Liberty Lake, and also along the line from Greenacres, where a number of new irrigated projects are under way, it will be necessary during the coming year to double track the line from Greenacres to Spokane. This will complete double tracking from Spokane to the Idaho line."

"The management expected to resume the payment of dividends on the preferred rights during the year 1909, as the earnings of the company warranted it. The deplorable accident, however, will eat up our earnings for the greater part of the year, and much as we regret it, the payment of dividends will be postponed until next year."

The Electric Railway Journal for November 27, 1909, reported: "Operations of the Spokane and Inland Empire Railroad show a revenue from the transportation of freight that is far greater in proportion to total gross earnings than most properties of this type report."

Several factors contributed to this. The Spokane freight terminal was well situated between the Great Northern and the Northern Pacific railways for easy transfer of shipments. Spokane and Coeur d'Alene line served as Great Northern's access to the lumber mills at Coeur d'Alene. The Inland Empire branches to Colfax and Moscow passed through an area of rich farm lands with rainfall sufficient for a variety of crops, though wheat was the major one, and other rail competition was small. Spokane itself was a good market for these raw materials.

Historian Durham wrote: "In October, 1909, the Great Northern Railroad acquired control of the Inland Empire properties, locally more widely known as the Graves system. By purchasing the stock held by Mr. Graves and his associates, the Hill interests took possession of the electric lines into the Palouse Country, the line to Coeur d'Alene and Hayden Lake, the Traction street railway system in Spokane, and the developed power plant at Nine Mile on the Spokane." The Electric Railway Journal for December 25 carried a paragraph of the same information.

It appears that 1909 was the last year for any major construction on the Inland Empire Lines except for this paragraph in the Electric Railway Journal:

"Work is now in progress on the driving of a tunnel which will permit all the interurban trains to pass under the business center of town, between the end of the right-of-way at the freight terminal and the passenger terminal without operating over street tracks, as now is done."

This was never completed. No doubt, the Union Station development which began about this time in the form of quiet and mysterious property purchases, as the news articles reported it, was responsible for the failure to complete the project.

The Washington Water Power Company continued its interurban development through 1910. No regular paragraphs of information for this railway appear in the two magazines frequently quoted herein. However, a couple of descriptive articles are helpful.

A 1909 report on the Washington Water Power operations in the city and county of Spokane states that the interurban lines operate on 60-pound rail over a private right-of-way, roadway fenced, and station shelters at important crossings. A terminal station was being built on Wall Street in Spokane. (It was torn down to make way for the city's river bank improvement program, which includes a parking lot in that area. The Inland Empire terminal at Lincoln and Main was razed to make way for the Sears and Roebuck building, now the main Spokane Public Library, a gift to the city by the Comstock Foundation).

The Washington Water Power Company wound up its interurban construction in 1910 with the installation of a system of automatic block signals and train stops for its single track line. The system was of the same general type as that in use on several larger Eastern lines, notably the New York Central and Hudson River, The Long Island, and the New York, New Haven and Hartford. By this system trains were kept sufficiently far apart that head-on or rear-end collisions were virtually impossible. Should a motorman ignore a signal against him, he would be automatically stopped, unable to proceed without some time-consuming labor to get his train rolling again.

The years between 1910 and 1920 were years of growing competition from improved highways, trucks, busses and especially the private automobile, which permitted its owner a greater choice of weekend pleasure sites than just those offered by the interurbans.

The Electric Railway Journal for November 27, 1920, carries an account by one of its representatives who had visited Spokane for a close range study of the Inland Empire System. He found that in 1919 the Spokane & Inland Empire Railroad had gone into receivership, out of which had come two railroads: the Spokane & Eastern to Liberty Lake and Coeur d'Alene, and the Inland Empire to Colfax and Moscow.

In other receivership changes the Hayden Lake Hotel was leased to a private operator, the developments at Liberty Lake were sold, and the company-owned baseball team, never profitable, was discontinued.

He noted that Spokane was in an area rich with timber and agricultural resources, and that during the year about 12,000 cars of lumber, 4,000 cars of grain, 1,000 cars of apples, and many of miscellaneous products had been carried by the railroad. He wrote that the promptness of service was noteworthy.

The Washington Water Power Co., over most of its operating years, scheduled eight passenger trains each way between Spokane and Medical Lake, a run of 45 minutes. Between Spokane and Cheney there were seven trains each way, and the trip took 55 minutes. On Saturday there was an extra round trip, a midnight run for each line. One freight train of two or three cars did the day's work between Spokane, Medical Lake and Cheney, carrying general freight and milk.

The Inland Empire System's time tables show the operating time from Spokane to Coeur d'Alene as 1 hour, 20 minutes, with expresses covering the distance in 1 hour. The time to Liberty Lake was 45 minutes, to Freeman, 1 hour; to Spring Valley, 1 1/2 hours; to Colfax, 3 hours; to Palouse, 3 hours; and the train was in Moscow a half hour later.

The table below showing the number of passenger train round trips per day was gleaned from four time tables of the times.

The automobile, which Spokane had tried to limit to four miles per hour at intersections in 1905, had made itself seriously felt in the interurban passenger departments by 1920. In 1922 the Washington Water Power Company ceased its interurban operation, due to the high cost of

have been viewed by this town as almost a disaster. It will be regretted now. But it is not action for which the company can be censured, nor is it a loss for which the speedy remedy is lacking.

"So rapid has been the advance in good road work and so swift has been the development of the gas car and the auto truck that neither Spokane nor its two neighbor towns should suffer even grave inconvenience for one day after the electric service ends . . .

SPOKANE & INLAND EMPIRE SYSTEM ROUND TRIPS PER DAY

Spokane	to	Hayden Lake	Coeur d'Alene	Liberty Lake	Vera, Opportunity and Greenacres
1908		3	9	2	0***
1919		2	5	1	16
1926		1	2	0	17
1939		0	2	0	0***

Spokane	to	Freeman	Spring Valley	Palouse	Colfax	Moscow
1908		3	3	1	6	2
1919		1	0*	0*	2	1
1926		1	0*	0*	2**	2**
1939		Freight only by the Great Northern				

* No turnaround. Spring Valley served by all trains to Colfax and Moscow.

** Spokane to Spring Valley as one train, divided here and one part to Moscow, one part to Colfax. Return trip the Colfax and Moscow trains became one at Spring Valley for the run to Spokane.

*** Served by trains to Coeur d'Alene.

operation and the competition of the auto-stage, according to Mr. Fisken. Relative to the closing down of the interurban railway operations by the Washington Water Power Company are the following news items from the Spokane Daily Chronicle:

March 1, 1922, editorial: "Give Cheney and Medical Lake a Straighter, Safer Highway. Six or seven years ago the dismantling of the electric lines from this city to Medical Lake and Cheney would

"The abandonment of the electric lines does suggest one step which the state and county officials should consider. This is the prompt alteration of Sunset Highway near the west city limits to get rid of the bad curves which have caused entirely too many serious accidents."

A news item on page 5 of the same issue, headlined "Cheney Commends W. W. P. Efforts," explains: "Resolutions expressing regret at the withdrawal of the Washington Water Power interurban service from

Cheney, but commending the action of the officials of the company as the only one warranted by the conditions, were adopted by the Cheney Commercial Club last night." The article indicated that service would be discontinued after March 12, 1922.

On page 8 another relative headline: "Must Pay Taxes on Rural Lines," called attention to these facts: "The abandonment by the Washington Water Power Company of its interurban lines to Cheney and Medical Lake will not affect the liability this year for taxes on the property," according to Assessor Elmer P. Bartlett. "The law provides that all property in the state at noon on March 1 is subject to taxation."

On page 12 yet another news item: "W. W. P. Shutdown to Affect NORMAL — Discontinuance of the interurban service over the Washington Water Power line, which becomes effective March 12, will mean a serious loss to the Normal School," (at Cheney) according to President N. D. Showwalter. "Students living in Spokane, as well as those between Spokane and Cheney, will be affected."

The Chronicle for March 14, 1922, carried the following headline and story:

"WANT TO BUY A STREET CAR? RIGHT THIS WAY PLEASE"

"PURCHASERS AWAITED FOR 13 CARS, OTHER EQUIPMENT FROM ABANDONED LINES"

"Three of the 16 passenger cars used by the Washington Water Power Company on its interurban lines between Spokane and Cheney and Medical Lake have been sold to a traction company in Wichita, Kansas. The others are standing idle in the street car company's car barns here. They are being offered for sale to interurban lines over the west, but no sale has been found for them yet . . . Work of dismantling the lines into Cheney and Medical Lake is being continued this week. The substations have been closed and the transformers brought to Spokane, and today work of dismantling the passenger stations on the two lines was started. Some rails are being taken up."

At the time of this rewriting (1973) the Spokane passenger station has been gone from the scene since 1965, making way for river front improvement and parking, as mentioned earlier. The station in Medical Lake was destroyed by fire a few years after closing of operations. In Cheney the old station lives on as a nursing home. The brick substation at Jamieson Station still stands, and for many years housed a tannery. In more recent years it has been a "Spook House" at Halloween, operated by the present owner of the property. It is located near the south edge of the Spokane Airport.

Passenger traffic on the Inland Empire lines was also dwindling in this period as the table of round trips elsewhere shows, but the Inland Empire Lines continued to operate a gradually curtailed service. In 1922 the passenger trains began to operate out of the Great Northern passenger station, and the Inland Terminal Building became a source of office space for rent until it was razed in 1927 to make way for the construction of the Sears, Roebuck and Company building, now the Spokane Public Library.

Inland Empire (at that time marked Great Northern) passenger trains operated out of the Great Northern depot until 1940. A Spokane Daily Chronicle story for March 20, 1939, announces that the last passenger service to Moscow, Idaho, would be April 1, with Conductor T. R. Smith, who took the first passenger train into Moscow on Sept. 15, 1908, making this last run. He would then take over the run to Coeur d'Alene from Spokane. Great Northern officials said that daily revenues for the past year had averaged about $2.

The Spokesman-Review for July 14, 1940, carried a picture of motor car 20 with Conductor Tom R. Smith and Engineer C. L. Jones, and the story: "The Spokane, Coeur d'Alene and Palouse Electric Line made its last passenger run yesterday, arriving from Coeur d'Alene at the Great Northern station at 5:40 p.m. The line opened in 1903 . . . The last passenger off the train was T. D. Miller, E. 511 Mission, who was a passenger on the line's first run."

The rails to Coeur d'Alene, Colfax, and Moscow continue in service as a Great Northern freight carrier. The electric interurban had run its course in the Spokane area. Its passenger train average of 28 to 38 miles per hour had not been fast enough for a whole new way of life.

With the arrival of 1973, the Great North ern passenger depot was one of the build ings on Havermale Island that was raze to make way for Spokane's Expo '74 Only its tall clock tower remains, sol downtown reminder of the electric trai days — for those who remember.

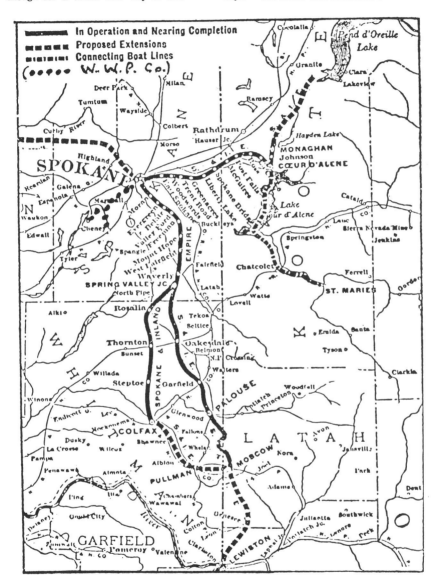

This map of the Spokane and Inland's system was copied from a 1900 issue of the Electric Railway Journal. The round-dotted W. W. P. system was drawn upon it by the author. None of the four extensions optimistically shown on this map was ever built.

Above: This three-car summer Sunday or holiday W. W. P. Co. special was passing present Finch Arboretum about 1910 bound for Medical Lake. There steam and gasoline motor launches transported its passengers to former Camp Comfort's picnic and camping facilities on the west side of the lake. Below, left: Spokane and Inland Empire interurban train loading at the old Great Northern depot in 1940 for a run to Coeur d'Alene. Three, four and five cars were once common for these trains, and also those that ran to Mosoow and Colfax. Below, right: Spokane and Inland Empire two-engine freight train at Mt. Hope in 1940. These locomotives, built by Westinghouse Electric, had far more power than speed. They hauled regular freight cars, same as the transcontinental steam lines. All that remains of Spokane's interurban era are the lines to Coeur d'Alene, Moscow and Colfax, now diesel branches of the Burlington Northern.

By Paul Barrett

TEACHING ASSISTANT IN HISTORY
UNIVERSITY OF ILLINOIS, CHICAGO CIRCLE

Public Policy and Private Choice: Mass Transit and the Automobile in Chicago between the Wars

❧ *Public policy, grounded in the conception of urban transit as a private business and of the automobile as a public good, played a crucial role in the decline of public transportation and the triumph of the automobile in Chicago.*

In the early years of this century, all across the United States, journals were speculating that soon "the automobile is going to solve the 'traction question' by putting the (street) railroad car out of business." George Hooker of Hull House disagreed. In a long letter to the Chicago *Record-Herald* in 1904 he went on for thirteen column-inches to show that "the tremendous and steadily increasing importance which the streetcar system is destined to have in any great city enforces the lesson that streetcar development should steadfastly be kept within the power of public authority to direct its course." [1]

Until recently, Hooker's letter would have seemed an amusing if understandable piece of naiveté. His thinking bespoke the Hull House environment — the dense slums of a nineteenth-century city, a reformist circle that sought to make urban life more bearable by altering the structure of political power and modifying social values. By the 1950s it seemed that the dilemmas against which these reformers threw lifetimes of futile labor had technological, not sociopolitical, solutions. Technology could be changed. Social values and institutions evolved at a glacial pace. So it appeared at the height of the automobile age; social problems might indeed have technological solutions.

The coming of the automobile did free man, not only from rigid patterns of urban development, but from some social and political obligations as well. And it seems to have been inevitable. With

Business History Review, Vol. XLIX, No. 4 (Winter, 1975). Copyright © The President and Fellows of Harvard College.
[1] George E. Hooker to Chicago *Record-Herald*, March 5, 1904.

approval or dismay, scholars of the motor age have described the rise of individual transportation as a result of several factors, most of which were beyond anyone's control. To some, like John Bell Rae, the irreversible social and economic impetus to the dispersion of population and employment, augmented by motorized transportation, created a world in which mass transit was not only inconvenient but also less economical than the automobile. And Brian J. L. Berry, among others, found patterns of change in land and transit use so uniform within groupings of cities as to make the effects of varied public policies upon the dispersion process seem nil.[2] Other scholars, including Martin Wohl and Wilfred Owen, point to the popular emotional identification with the automobile as an additional factor in its rise.[3] Working from a different perspective, Glenn Holt discovered a partial cause for the demise of mass transit in man's aversion for close contact with his fellow man, class prejudice, and the backward and self-seeking policies of the people who controlled urban transportation.[4] These factors, too, while essentially the products of human choice, could hardly have been altered in the context of the American system as it existed.

But is it all so? Was the automobile's victory, in the form and degree it assumed, really inevitable? More recent perspectives again make George Hooker's argument worth hearing. Indeed, the evidence shows that the usual explanations do not account for all that has happened to transportation in Hooker's city of Chicago since 1904. Some factors that we might be inclined to take as givens (in particular, technical progress and "middle class aspirations") turn out to be in part the results of public decision making. Furthermore, the determining factor in Chicago's decisions about its trans-

[2] The dispersion emphasis is well illustrated by John Bell Rae, *The American Automobile: A Brief History* (Chicago, 1965), 224–225. Wilfred Owen, *The Accessible City* (Washington, D.C., 1972), 11, 15, sees the desire to disperse as "sustained by automobiles." See also J. R. Meyer, J. F. Kain, and M. Wohl, *The Urban Transportation Problem* (Cambridge, Mass., 1966), 17, 18, 23–24, 44–55, 74–82; Brian J. L. Berry and Elaine Neils, "Location, Size and Shape of Cities as Influenced by Environmental Factors: The Urban Environment Writ Large," in Harvey S. Perloff, ed., *The Quality of the Urban Environment: Essays on "New Resources" in an Urban Age* (Baltimore, 1969), 278–79, 289–95.

[3] E.g., Martin Wohl, "Congestion Toll Pricing for Public Transportation Facilities," in Selma Mushkin, ed., *Public Prices for Public Products* (Washington, D.C., 1972), 256; Alfred Politz Research, Inc., "The Automobile in the Daily Life of the American Population" (unpublished, April 1951), cited in Wilfred Owen, *The Metropolitan Transportation Problem*, rev. ed. (Washington, D.C., 1966); Bureau of Labor Statistics, *Family Expenditures in Selected Cities*, 1935–1936, vol. 6, Travel and Transportation, U.S. BLS Bulletin #648, p. 3.

[4] Glenn Holt, "The Changing Perception of Urban Pathology: An Essay on the Development of Mass Transportation in the United States," in Kenneth T. Jackson and Stanley K. Schultz, eds., *Cities in American History* (New York, 1972), 337, 338. On the automobile as providing privacy and a sense of control, compare Lawrence J. White, *The Automobile Industry Since 1945* (Cambridge, Mass., 1971), 236–237.

portation dilemma has been not political influence or corruption but the interplay between crisis and intellectual inertia, the two real tyrants of urban public policy.

Chicago is historically a model metropolis, a ready subject for sociologists to ring and slice and grid with classic sectors. The traffic flowing through its archetypal crosshatch streets has obligingly approximated the national pattern over the years, with the appropriate rises in automobile registration and accompanying decline in transit ridership. Yet when Chicago's ridership record over the years between the wars is compared with those of other cities, Chicago shows differences that size and terrain alone may not explain. Only New York, with its vast and continuing subway program, clearly outstripped Chicago in the growth of its mass transit ridership before the Great Depression. Between 1918 and 1927, Chicago mass transit ridership per capita increased by 10 per cent while that of most other cities fell. The few other cities with increased per capita ridership all experienced significant increases in service during the 1920s. Even on the clutch-destroying hills of San Francisco and the winding, ancient streets of Boston, mass transit was only able to hold its own in the postwar decade. But Chicago, with its flat ground, stagnant transit system, and thinning population, produced per capita ridership increases through most of the 1920s, and increased usage by weekday commuters through at least 1939.[5]

Why was Chicago different? What role must be assigned to technology, to class-linked values, and to policy decisions in the changing travel patterns of this one city's citizens? Commonly cited factors — dispersion, the real and apparent intrinsic advantages of the automobile, popular prejudice, and transit management policies — go far toward accounting for the decline of transit ridership dur-

[5] "Survey Shows Transit (Grows Despite Auto Boom)," *Electric Railway Journal*, [hereafter *ERJ*], 74 (November, 1929), 1030–1033. By 1927 only five cities (New York, San Francisco, Chicago, St. Louis, and Boston) averaged 300 or more revenue rides per capita.

The same survey's data indicate that these changes in mass transit rides per capita are not *simply* the result of increased automobile ownership. The cities with the most rapidly increasing auto registration per 1,000 were St. Louis (3.1 per cent gain in transit rides per capita), Memphis (12.2 per cent loss), New York City (37 per cent gain), Cincinnati (5.4 per cent loss), Kansas City (15.8 per cent loss), and Chicago (10.1 per cent gain in mass transit rides per capita). Thus, a simple "more autos equal less use of transit" equation does not seem to apply. On weekday vs. weekend ridership, see Chicago Surface Lines, "Traffic and Scheduling Reports," 1939, p. 3 (notebooks in possession of Mr. Harold Hirsch, Department of Scheduling, Chicago Transit Authority).

Chicago data are from "Riding Habit in Chicago" (graph prepared by CSL, undated [1942]), in McIlraith Collection (documents collected by E. J. McIlraith now in possession of the author), vol. I, "Traffic," Section C, Item 16; national data is from U.S. Department of Labor, *How American Buying Habits Change* (Washington, D.C., 1956), 186.

ing the first half of this century. Yet they cannot account for Chicago's peculiarities, nor can they do much to help us learn how workers, given a choice, decided to drive alone to work or ride the streetcar with their fellow citizens. Most important, they cannot tell us how historians, urbanologists, and political leaders arrived at an essentially *business* definition of what constituted efficiency in mass transit, a definition that has done much to shape our idea of the relationship between mass transportation and suburban development.

THE RISE OF THE AUTOMOBILE

The dispersion of work and housing preceded the automobile, and in the period between the two world wars it prevented the rise in trade and population once thought to be the inner city's natural patrimony. Between 1920 and 1940, the population of Chicago spread out at a rapid rate within the city. Business dispersed even more rapidly, with the central business district's share of sales transactions in the city dropping from two-thirds in 1910 to one-third by 1935.

While this redistribution helped to even out the city's population, it did not produce a major thinning out of the most crowded districts. But a move to the newer areas of Chicago (really suburbs within the city limits) almost always meant a move away from good mass transportation.[6] Data from the 1938 W.P.A. Land Use Survey and from local government reports show that only traffic-hobbled surface vehicles followed new work and housing sites to the outer edges of the city. It is virtually impossible to find an area of single-family dwellings in "good condition" (as defined by the Land Use Survey) that are in reasonable proximity to a rapid transit line. On the other hand, while many new jobs appeared in the area away from rapid transit lines after 1930, many more were created within the optimum area of transit service.[7] Thus, while

[6] The absolute number of persons living within a half-mile of rapid transit lines increased slightly between 1930 and 1940 (from 1,627,857 to 1,633,665). The point is that the sort of housing to which most Chicagoans presumably aspired was not available in these areas; data in Gerald William Breese, *The Daytime Population of the Central Business District of Chicago, with Particular Reference to the Factor of Transportation* (Chicago, 1949), 79; Malcolm J. Proudfoot, *The Major Outlying Business Centers of Chicago* (University of Chicago Library, 1935), *passim*; Harold M. Mayer, "Patterns and Recent Trends in Chicago's Outlying Business Centers," *Journal of Land Use and Public Utility Economics*, 8 (1942), 8–16; Brian J. L. Berry, *Commercial Structure and Commercial Blight: Retail Patterns and Process in the City of Chicago*, Chicago University Department of Geography Research Paper #85 (Chicago, 1963), 8–11.
[7] "Map of Chicago Surface Lines [with] Population Change in Chicago Between 1920 and 1930," in McIlraith Collection, unlabeled binder; "Map of Chicago Surface Lines

industrial employers did not quickly move away from the traditional lines of transportation, thousands of families were moving into areas where mass transportation was inconvenient, if it existed at all. As good housing became synonymous with poor mass transit even within the city, people appear to have become willing to abandon public transportation for the sake of the real advantages of modern, single-family dwellings.

But habits of thought and action do not change so quickly. Complaints about the lack of service to new middle class residential areas fill the records of the Committee on Local Transportation of the Chicago City Council between 1910 and 1940. They came from individuals as well as groups and developers. One area even began its own (illegal) bus line on a middle-income residential street during the 1930s.[8]

The meaning of this evidence is clouded by the fact that surface extensions into new areas had a remarkable rate of failure.[9] But the important fact is that these middle class areas that persistently begged for good mass transit service never received it. Extensions to factories and residential areas alike were always held to the maximum level of service that would pay for itself according to the rules of business accounting. It is not correct, therefore, to say that mass transit was irrelevant to the dream of the dispersing middle class. The sort of mass transportation created by the business definition of efficiency could not prosper save in the world of the multiple-unit dwelling, but this reality was in part a function of public policy, which helped to define the implications of dispersion.

Dispersion itself was not a result of public policy.[10] The automobile, however, the second factor associated with the national decline of mass transit, depended heavily on the support of public agencies.

[with] Population Change in Chicago Between 1930 and 1940," in *ibid.*, both maps undated; Chicago Plan Commission, *Residential Chicago*, vol. I of *Report of the Chicago Land Use Survey* (Chicago, 1942), Map I; Board of Supervising Engineers, Chicago Traction (BOSE), *36th Annual Report 1944* (Chicago, 1946), Map XII; Chicago Association of Commerce and Industry, *Directories* (Chicago, 1936 and 1940).

[8] The author has thus far seen over 400 requests for service from middle class areas between 1907 and 1944 in the *Proceedings* of the Chicago City Council Committee on Local Transportation and other sources. (A request is here classified as a separate issue, no matter how many groups of petitioners are involved.) While the number diminishes with time, many requests continued to flow to the Council into the late 1930s.

[9] F. A. Forty (Superintendent of Schedules) to E. J. McIlraith, September 3, 1938, bound with car allocation data, 1915–1953, in unmarked and coverless set, CTA Office of Traffic and Scheduling (pages not numbered but arranged chronologically); interview with LeRoy Dutton, November 14, 1973, Tape 1, Side 2.

[10] But on Federal dispersion policy in World War II, see John F. Kain, "The Distribution and Movement of Jobs and Industry" in James Q. Wilson, ed., *The Metropolitan Enigma* (Garden City, N.Y., 1970), 1–43, especially 6–10, where Kain finds a major role for Federal policy.

Without good roads and other facilities, the chief intrinsic value of the automobile would have been symbolic.

The status value of the automobile has been widely discussed. A glance at automobile industry publications from the early 1920s shows that the automobile was seen by the industry as an artifact of upper middle class and upper class culture. Through the early twenties, the average automobile owner was conceived by the industry as being a man of means, and the industry long continued to emphasize an association of its product with wealth and leisure in its national advertisements.[11]

Ownership data for Chicago are scanty. The 1926 Chicago traffic study published a map that showed an association of automobile registration with higher status neighborhoods, but many cars were not registered and, of course, high status residence and automobile ownership are both functions of wealth. The correlation of automobile use with central business district (CBD) employment and high status residence is further illustrated by Miller McClintock's 1931 traffic survey for the City Council Committee on Traffic and Public Safety. Through boom and depression, the volume of automobile traffic entering the CBD between 1926 and 1941 remained equal to about half the total number of automobiles registered in the city.[12]

But did the desire for higher social status help to cause the increasing use of the automobile for the journey to work, or is the correlation illusory, merely the result of the fact that the well-off could better afford a convenience that was really equally attractive to all classes?

If Chicagoans resented mixing with their social inferiors on the way to work, they rarely said so in print. It is particularly dangerous to infer from complaints of crowding that ordinary people harbored a strong distaste for their fellow men. As *News* reporter George Stone explained in 1923, crowding meant more than an occasional brush against a foul-smelling stranger: [13]

[11] *Literary Digest*, 100 (January through March, 1929), advertisements. See also, for example, National Automobile Chamber of Commerce (NACC), *Facts and Figures 1921*, p. 16; "Who Owns a Motor Car?" *Literary Digest*, November 11, 1922, 78; Eugene Taylor, "Progress on the Chicago Plan," *Enginering News-Record*, March 22, 1923, 528; Samuel Shelton, "The Greatness of Automotive Transportation," *Motor Age*, August 5, 1926, 12; "King Automobile," *Outlook*, November 23, 1927, 355–356; Clyde Jennings, "Why and How Automobile Dealers Can Boost the Tourist Business," *Motor Age*, October 25, 1923, 9–11; "Good Weather Will Help the Middle West," *Automobile Topics: The Trade Authority*, June 16, 1923, 437; Lee J. Eastman, "The Closed Car: Its Development and Its Great Future," *Automobile Trade Journal*, vol. 28 (December, 1923), 22–23.

[12] Chart from McIlraith Collection, vol. I, "Traffic." All the automobiles entering the central business district need not, of course, have been Chicago-registered.

[13] Chicago *News*, February 2, 1923. Few of the hundreds of rider complaints reviewed

[At] 5:30 p.m. yesterday thirty and more people were waiting for a car at Madison and Paulina [two miles directly west of the center of the CBD]. Two cars went by, already overloaded . . . . A through car finally arrived. All hands stormed the rear platform. Women took their chances with the men. They weren't savage about it: just fiercely persistent, . . . crowding wherever the opportunity offered.

The car started off before all had climbed on. It was full anyhow. As it swung away three women, one white haired and frail, were standing on the step, kept off the platform by the crowd inside, yet held to the step by the men clinging behind them . . . .

[At the next two streets] each corner had another small crowd of which only the strongest found clinging room on the car. The weaker ones waited as they had been waiting no doubt for many minutes. It was cold outside too, and damp.

This was what letter writers meant by "indecent crowding," crowding that altered the dispositions of the riders and endangered their health. Contemporary historians may be failing to distinguish between class prejudice and the simple desire not to have one's ribs cracked when falling off a streetcar if they try to infer much about alienation in urban life from complaints of the type seen by this writer.

But here is another reason why the status connotations of mass transit *per se* should not be overemphasized. Chicago's mass transit system had long provided ample opportunity for skittish riders to choose the character of their fellow travelers. As early as the 1880s one South Side woman, complaining of the lack of "heating" straw on the floors of streetcars, observed to the *Tribune* that "the rich have their [Illinois Central commuter] trains to ride." And early streetcar routings took class into account, as Northwest Side community leader Tomaz Deuther discovered when he asked Chicago Railways president John Roach to send cars directly down State Street from Deuther's working class neighborhood. "You can't mix silk stockings with picks and shovels," Roach replied. Deuther was satisfied and marked Roach down as an honest man.[14] As late as 1947 patrons in many districts could choose among streetcar, elevated, interurban, boulevard bus, and commuter railroad service for a trip to the CBD. Each line had its own fare structure and

---

by this author reflected a clear distaste for contact *per se* with other citizens. Most crowding complaints centered on the fact that passengers could not board the cars (selections from 215 citations, about half press, half Committee on Local Transportation, most before 1926).

[14] Chicago *Tribune*, January 10, 1886. Tomaz Deuther, *Civic Questions Concerning Chicago Traction* (Chicago, 1924), 8–9; the quote is from 1911. "The "silk stockings" Roach was worrying about were almost certainly not middle class riders but "Gold Coast" businessmen going to their downtown offices in an era where the streetcar could still compete with the horsedrawn "cab" for the patronage of the wealthy.

routing and, we may assume, its distinctive clientele. In short, the argument that aversion to class mixing helped to kill mass transportation must be considered in the context of the unique transit system each city developed for itself by means of local policy decisions.

The lot of the Chicago transit rider fluctuated between 1907 and 1947, but always the experience was basically the same. One waited, struggled aboard, paid the 5¢ to 7¢, and shared a jerky ride at 8 to 11 miles per hour through crowded streets in greater or less discomfort. The route, the roadbed, and the vehicle were likely to be the same in the last year as in the first. Against this static experience, the automobile held the attractions of privacy and status discussed above, but it was also an increasingly practical way to get to work for those who could afford it. To this practical value public policy made no small contribution. But the reasons are often misunderstood. Chicago's civic leaders seem to have favored automobiles chiefly because those free-wheeling vehicles presented them with crises that demanded immediate solutions and that seemed to threaten the very existence of the CBD. Such crises led to positive, creative policies of a sort mass transit could never elicit.

Like the citizens of most other cities, Chicagoans entered the twentieth century with a set of ideas that helped to pave the way for the automobile. Certainly they felt that it was the city's duty to provide free access on public ways for all who wished to use them. This view gained impetus before 1910 from farmers who had to bring their goods to market and from an organized effort to clear CBD streets of wagons and to ease the passage of streetcars.[15] The Burnham Plan, 1906–1908, summed up the early twentieth-century attitude toward public ways in the credo that better streets and roads were always a benefit to the whole community.[16] Between 1910 and 1930, as automobiles appeared in substantial numbers, the early twentieth-century point of view respecting public roads was applied to them in two ways: in traffic regulation and in the provision of facilities for the automobile.

Three sorts of individuals helped to assure that traffic control in Chicago would encourage the increased use of automobiles. First,

[15] For example, Interurban Roadways Committee of the Commercial Club of Chicago, *Inter Urban Roadways about Chicago* (Chicago, 1908), *passim*; J. P. Altgeld in *Good Roads*, II (1892), 299–304.

[16] Daniel H. Burnham and Edward H. Bennett, *Plan of Chicago Prepared under the Direction of the Commercial Club During the Years MCMVI, MCMVII and MCMVIII* (Chicago, 1909), 34–36, 41–42, 75, 88–89; E. S. Taylor (Manager, Chicago Plan Commission), "The Plan of Chicago in 1924, With Special Reference to Traffic Problems and How They Are Being Met," *Annals of the American Academy of Political Scientists*, vol. 116 (Philadelphia, 1924), 230.

engineers of all sorts proved to be the most credible and informed spokesmen for equal access to the street for all vehicles. Bion J. Arnold, a city consulting engineer for twenty-two years, deeply involved in the development of transportation plans in Chicago and other cities before World War I, opposed any attempt to ban automobiles from parts of the Loop on the ground of fairness to all vehicle owners.[17] R. F. Kelker, longtime engineer for the City Council Committee on Local Transportation and a good political bellwether, opposed any traffic law that gave the appearance of restriction, including most stop signs.[18] Miller McClintock, an early auto traffic expert of national reputation, concurred that the task of regulation was to facilitate increased traffic.[19] Similarly, Evan McIlraith, Chief Engineer of the street railway company and a major figure in the development of traffic regulation in Chicago, held that maximum efficiency in street usage meant the maximum number of vehicles as well as the maximum speed and efficiency for streetcars; McIlraith thought that the two were quite compatible.[20] Though they often wrote in different terms, traffic engineers and legal experts around the country held similar views.[21]

Second, and perhaps more predictably, automobile dealers and owners associations lobbied vigorously for pro-automobile regulation. The Chicago Motor Club and the National Automobile Chamber of Commerce attacked Chicago for restrictive regulation and warned it of stagnation and decline if it did not accommodate the needs of the automobile.[22] The drivers' reaction against restrictive

[17] Bion J. Arnold, "City Transportation," *Journal of the Western Society of Engineers*, vol. 19 (April, 1914), 32–33.

[18] CLT, *Proceedings*, November 13, 1923, pp. 751–758 of volume.

[19] Miller McClintock, *Report and Recommendations of the Metropolitan Street Traffic Survey*, prepared under the direction of the Street Traffic Commission of the Chicago Chamber of Commerce (Chicago, 1926), 2, 3, 16, 27.

[20] McIlraith Interviews, December 11, 1973, Tape 10, side 2 near end (not yet transcribed); Evan J. McIlraith, *Address Delivered by E. J. McIlraith, Staff Engineer, Chicago Surface Lines, before the Canadian Electric Railway Association* (Toronto, June 7, 1928), 12–14.

[21] For example, Howard Parsons, "Traffic: A Major Engineering Problem," *Engineering News-Record*, January 4, 1923, 30–31; Alvin McCauley (President, Packard Motors), "Adapting the City to the Automobile," *Public Works*, vol. 55 (September, 1924), 285–286; Samuel Shelton, "Automobile Makers and City Officials Discuss Traffic Problems," *Motor Age*, (April 21, 1927), 9, 20; John C. Long, in *Transactions of the American Society of Civil Engineers* (1927), 930–931; editorial, *Motor Age*, November 15, 1923, 31.

[22] Testimony, expert and otherwise, to the indispensability of good roads to the survival of a city includes: "Road Space for Cars in Cities' Saturation Problem," *Motor Age*, December 13, 1923, 36; Samuel Shelton, "Greatness of Automotive Transportation," 12; McClintock, *Report and Recommendations*, 2, 70–71; Hugo E. Young (Engineer, Chicago Plan Commission), "Day and Night Storage and Parking of Motor Vehicles," *American City*, vol. 29 (July, 1923), 44–46; "Traffic," *Motor Age*, December, 1923, 28, and November 15, 1923, 31, *Motor Age* remarked, with seeming horror, that "The Mayor's Traffic Commissioner thinks traffic regulation is just a matter of making rules for the motorist." After 1923, the industry's attitude became congratulatory, as noted below. See also "Chicago's New Traffic System," *Automotive Industries*, May 1, 1926, 775, which draws together developments to that date.

regulation, which may have been in part a middle class reaction to their first contact with unfriendly police officers, seems to have been a national phenomenon.[23]

Finally, business interests in the central business district were particularly suspicious of any regulation that might make drivers reluctant to patronize their district. Despite their concern over the growing congestion of the early 1920s, they opposed the prohibition of parking on Loop streets until they were assured that it would not drive away patronage and that off-street parking facilities could be provided.[24]

These three influences (together with an increase in congestion during the early 1920s, which led many civic leaders to speculate about the impending death of the CBD) helped lead Chicago to a series of innovations in traffic control that earned the enthusiastic praise of the automobile industry.[25] Streetcars were rerouted, left turns were forbidden in the CBD, parking was restricted, off-street parking facilities increased dramatically, penalties for violations were reduced so that they might more easily be imposed, and Chicago began the development of coordinated signal control of automobile traffic in 1925 with the installation of a system designed by the Chicago Surface Lines' Chief Engineer to restructure the movement of traffic into evenly spaced waves.[26] The result of these improvements was a dramatic increase in the number of vehicles passing through the CBD, with an equally dramatic drop in con-

[23] Hon. Joseph Sabbath, "What a Traffic Court Can Do," *American City*, vol. 17 (August, 1917), 126–127. In the pre-war era at least, drivers often simply ignored policemen's orders (BOSE, *9th Annual Report 1916*, 226). As late as 1931, "ordinary good citizens" would not obey traffic lights unless a police officer were present (Citizens Police Committee, *Chicago Police Problem* [Chicago, 1931], 22, 152–166). Indicative of the auto lobby's position on restrictive legislation are calls to arms in automobile trade journals, for example, "Legislation" (an editorial), *Automotive Trade Journal*, July 1, 1923, 55 (which praises the "tireless efforts of a few" lobbyists); "An AAA Warning," *Motor Age*, January 15, 1925, 4; "Car Owners Are Voters," *Motor Age*, February 15, 1925, 31; "'Killing Bills Is the Year's Big Problem,' Says AAA Chief," *Motor Age*, February 26, 1925, 36.

[24] Business opposition to parking restrictions was reflected in repeated attempts to prevent their implementation; see Young, "Day and Night Storage," 44–45; *ERJ*, November 5, 1927, 876; "No-Parking Foes Defeated in Chicago," *American City*, vol. 47 (September, 1932), 80. By 1937 off-street garages provided space for 86,300 cars a day at a daily cost to motorists of $21,810 (Philip K. Harrington, Rudolph F. Kelker, Jr., and Charles De Leuw, *A Comprehensive Local Transportation Plan for the City of Chicago* [Chicago, 1937], 71). Even existing parking laws were really not enforceable through 1926 since officers had to serve violators with a personal summons if they were to be fined (McClintock, *Report and Recommendations*, 176).

[25] E.g., J. Bibbins, "The Growing Transportation Problem of the Masses: Rails or Rubber or Both?" *National Municipal Review*, vol. 18 (August, 1929), 520; John C. Long, "'Go' Is Present Keynote in Traffic Regulation," *Automotive Industries*, January 21, 1928, 78.

[26] George Barton, *Street Transportation in Chicago: An Analysis of Administrative Organization and Procedure* (Evanston, Ill., 1948), 21–26; McIlraith Interviews, Tape 6, Side 1 and Tape 8, Side 1; Edward J. Kelly, *Annual Report to the Judges of the Circuit Court of Cook County*, submitted by Edward J. Kelly, South Park Commissioner, December 31, 1924 (Chicago, 1925), 19 (plans for a $10,000,000 garage in a public park).

gestion. While enforcement of all traffic laws, even under the Uniform Code of 1925, proved difficult due to the multiplicity of police departments within the city, the remarkable success of this early burst of facilitative regulation probably helped to affirm the belief that all traffic problems could be solved by competent engineering and that the old-style CBD was indeed compatible with the triumph of the automobile.[27]

Good roads, even more than friendly regulation, were a necessity for the automobile. Pavements designed for horse traffic either wreaked havoc on the auto or crumbled beneath it. However, the Burnham Plan called for boulevards in most sections of the city, and a good number of these were built by 1926. In that year it was possible to travel through the city and around the central business district on broad, well-surfaced promenades free of most public vehicles and with right of way at all intersections.[28]

This dealt with only half the problem. The state's road system grew ahead of city streets and boulevards, and poured new waves of traffic in upon them. While the boulevards had been built in line with a plan to beautify and order the city, another sort of street widening and bypass construction was already under way. This work was designed chiefly to alleviate automobile congestion. Between 1907 and 1937, over $340,000,000 had been appropriated for street improvements in Chicago — enough to build a comprehensive subway system twice at 1923 prices.[29]

It is difficult to tell how these roads were paid for. Many were built by Park District and city bond issues in an era when a good deal of bond issue money was diverted to other city "expenses."

[27] Robert Naw, "No Parking: A Year and More of It," *American City*, vol. 40 (March, 1929), 85–88. See also: City Council, *Journal 1920-21*, June 29, 1920, 654; *Journal 1923-24*, April 16, 1923, 104; *Journal 1924-25*, October 31, 1924, 3887; CLT, *Proceedings*, vol. 80, November 18, 1920, pp. 23–24 of meeting, plus vol. 90, November 13, 1923, pp. 751–758 of volume, and vol. 92, January 25, 1924, p. 321 of volume. See also West Chicago Park District, *52nd Annual Report for the Year Ending February 28, 1921* (Chicago, 1921), 18, 85, *53rd Annual Report 1921-22*, 13.

[28] Burnham and Bennett, *Plan of Chicago*, 40; Chicago Plan Commission, *The Plan of Chicago in 1925, A Report to the City of Chicago Setting Forth What Has Been Accomplished by a United Civic Effort During the Past Fifteen Years* (Chicago, 1925), 5, 12, 16–22.

[29] Eugene S. Taylor, "Chicago's Superhighway Plan," *National Municipal Review*, vol. 18 (June, 1929), 371–377, 393; McClintock, *Report and Recommendations*, 2, 3, 16, 27, and "Vehicle Traffic Pressures in Greater Chicago," *American City*, vol. 47 (October, 1932), 89; City Council, *Journal 1925-26*, December 9, 1925, 1792, 1796; Chicago Park District, *1st Annual Report 1935*, 105, 125, 253, *3rd Annual Report 1937*, 106–107, *4th Annual Report 1938*, 113, 114, 116–117; Philip Harrington, Acting Commissioner, Department of Superhighways, City of Chicago, *A Comprehensive Superhighway Plan for the City of Chicago* (Chicago, 1939), 7, 8; Virgil Gunlock, Commissioner of Subways and Superhighways, in Chicago *Tribune*, April 18, 1946; Harrington, Kelker, and De Leuw, *A Comprehensive Local Transportation Plan*, 68–69; CLT, *Proceedings*, vol. 90, Special Committee to Formulate Traction Plans, November 23, 1922, p. 34 of meeting, vol. 93, October 27, 1924, p. 325 of volume.

We do know that these bonds were redeemed from real estate taxes on adjacent property, so that homeowners and renters often paid the cost of better roads. By 1926, however, Chicago had fully accepted the idea that good roads benefited the whole city, and vehicle taxes from all portions of the city were used for lesser street improvements. Altogether, bond issues totaled at least $164,000,000, vehicle taxes totaled nearly $70,000,000, and rebates on state taxes during the 1930s amounted to over $63,000,000.[30] It may be useful to compare these figures with the combined valuations of the city's surface and rapid transit lines at their highest point during our period — $273,000,000. The entire city budget for 1926, the last year before the usual chaos of Chicago's finances was exacerbated by economic depression and a court decision requiring reassessment of all real estate valuations, was slightly more than $132,000,000.[31]

Thus Chicago drove itself into debt with street improvements quite early in the game, though much of the problem lay with excess condemnation costs and the city's low debt ceiling. The Illinois roads that helped to funnel suburban automobiles into Chicago were built almost entirely from general funds until 1930.[32]

Throughout the 1920s and 1930s, then, the automobile and its facilities steadily improved, while the cost of the car, its fuel, and tires dropped considerably.[33] The public image of the road, as well as the willingness of automobile owners to insist on and in part to pay for the improvements, were major factors in the steady up-grading of automobile facilities.

THE DETERIORATION OF MASS TRANSIT

While the automobile improved, mass transit deteriorated, loading the traveler's "choice" more heavily in favor of the automobile. This widely known fact is easily misunderstood. The decline of

[30] City Council, *Journal 1907–08*, February 3, 1908, 3829; CLT, *Proceedings*, vol. 80, November 18, 1920, p. 11 of meeting; City Council, *Journal 1907–08* through *Journal 1939–40*, Appropriations Ordinances. Park District and federal government data are drawn from the *Annual Reports* of the South Park Commission, West Chicago Park Commissioners, and Chicago Park District; state information is from *Laws of Illinois* for the appropriate years. The total includes $17,269,445 of general city funds and $11,799,653 of federal money. All figures are probably underestimates.

[31] Werner Schroeder, *Metropolitan Transit Research Study* (Chicago, 1955), 1–6, 7; City Council, *Journal 1925–26*, 2368.

[32] Chicago Bureau of Public Efficiency, *Chicago's Financial Dilemma* (Chicago, December 17, 1917); Chicago Bureau of Public Efficiency, *Excess Condemnation* (Chicago, September, 1918); National Industrial Conference Board, *Taxation of Motor Vehicle Transportation* (New York, 1932), 67, 129, 136, 156–157 (Illinois motor vehicle taxes per car were consistently among the lowest in the nation). See also Thomas Agg and John F. Brundley, *Highway Administration and Finance* (New York, 1927), 178–180.

[33] U.S. Department of Labor, *American Buying Habits*, 186.

transit service was traceable to a paralysis of all the public and private agencies involved, not simply to political corruption and corporate greed. Hallowed concepts about mass transit, shared by public and reformers, politicians and bankers, were a major cause of that paralysis.

In 1937 an engineering report commissioned by Mayor Edward Kelly described the failures of mass transit in Chicago in terms that reflected the complaints of generations of political writers and newspaper correspondents. The equipment was old, inadequate service produced dangerous overcrowding, and routes were obsolete, with the result that 83 per cent of the passengers had to transfer on each trip.[34]

The first point is indisputably true. By 1937 most of the city's streetcars and all of its rapid transit equipment was more than ten years old, the noisy, sluggish product of an earlier era of mass transit technology. Street railway and bus technology improved dramatically around 1934, but Chicago's rolling stock did not. The slower acceleration and longer loading time required for older streetcars made service less attractive, and the fixed-rail vehicles that dominated Chicago's streets until after 1947 were more subject to the vicissitudes of traffic than were the relatively few buses on the street.[35] Rapid transit equipment was even more inadequate. Most of it was built of wood and was of archaic design. Lack of modern signalling combined with largely wooden equipment produced a deadly crash in the fall of 1936, and car fires were a frequent occurrence.[36]

Rapid transit scheduling, and therefore overcrowding, is hard to evaluate. The company did not alter schedules to meet expected loads, but simply added cars to trains if conductors reported an overload.[37] More precise information is available for the Chicago Surface Lines. In the 1930s, Traffic and Scheduling Reports show that the company cut runs in periods of ridership decline. This step, inevitable in a system operated for profit, increased the "distance" between cars and, thus, the waiting time. As volume went down, therefore, service must have become less attractive in this important respect. But crowding and consequent tension among

[34] Harrington, Kelker, and De Leuw, *A Comprehensive Local Transportation Plan*, 20, 66.

[35] On the disadvantages of older cars, see CSL, "Loading and Acceleration Data" (July, 1937), in McIlraith Collection, vol. I, "Traffic," Equipment Section.

[36] CLT, *Proceedings*, vol. 133, November 28, 1936, pp. 15–34 of meeting; Central Electric Railfans Association, Bulletin #113, *Chicago's Rapid Transit*, vol. 1, pp. 19, 34, 75.

[37] Dutton Interview, Tape 2, Side 1.

riders probably did not increase, as schedules were cut *after* fall-offs in riding, not in anticipation of them.[38]

An investigation of the final charge — that the high volume of transfers showed that transit routes did not go where people needed to go — leads to an interesting revelation. Frequent Chicago Surface Lines transfer checks illustrate that Chicagoans traveled in such diffuse patterns that no economical through-routes could serve them all.[39] After 1914, in the context of universal free streetcar transfers, Chicago had developed travel patterns that required transfers — or automobiles. For a long time, many people seem to have chosen mass transit for journeys to rapidly decentralizing places of work and business.

The case of rapid transit is different. The quick rise in Chicago Rapid Transit patronage after the institution of free transfers between it and the Chicago Surface Lines in 1936 suggests that a good deal of potential rapid transit patronage was lost until that time because of the Rapid Transit's limited routes.[40] The Chicago Motor Coach Company, serving 12 per cent of the city's transit passengers at its peak, provided a specialized service to a special clientele; it cannot therefore be criticized for its limited coverage.

Service, clearly, was increasingly inadequate. The role of management in this deterioration may not have been as important as contemporaries assumed. Management's role may be divided roughly into what it put into the operation, in the form of adjustments to changing ridership volumes, and what it extracted, in the form of profits and interest.

Management financial policy — overcapitalization — had a major effect on the quality of transit service in Chicago, though not in the way usually supposed. Beyond this, conservative, "backward" management policies were not of great importance in Chicago after 1923. One has rather to speak of exploitative and erratic policies on the part of the rapid transit ownership, and a Chicago Surface Lines management limited as much by its own vision of progress as by the cautious hand of the company's banker-owners.

[38] BOSE, *39th Annual Report 1947*, Exhibit 9, and Harrington, Kelker, and De Leuw, *A Comprehensive Local Transportation Plan*, Exhibit 2; CSL, "Traffic and Scheduling Reports," 1924, 4; CSL, "Summer Schedules, June, 1938," p. 9 of group bound in volume erroneously labeled "Par Operating Speeds" in CTA Office of Traffic and Scheduling. The CSL blamed lax enforcement of traffic laws for a decrease in speeds after 1938 (CSL, "Traffic and Scheduling Reports," 1941, 4.). In the absence of corroborating evidence, this effect of city policy on transit service, which could be of considerable significance, cannot be evaluated.

[39] E.g., checks of four North Side lines in August, 1928, May, 1936, and May, 1942, in McIlraith Collection, vol. II (unlabeled), Section B.

[40] McIlraith Collection, vol. I, "Traffic," 1941 study just ahead of Section D marker.

Overcapitalization usually occurred in one of two ways: a company issued more securities than necessary to cover the cost of an improvement and someone pocketed the difference; or it paid dividends with money that should have been used to retire the securities held against equipment that would eventually be scrapped. By the end of the nineteenth century all of Chicago's transportation companies were seriously overcapitalized.[41] A new franchise issued to the various surface companies in 1907 eliminated $59,-000,000 of "water," but in the process it accepted overcapitalization on principle as a part of the traction business. In order to induce the companies to undertake a far-reaching modernization program, the city agreed that unexpired franchises, contractors' fees, and bank charges and brokerage fees on bonds would be enshrined as part of the companies' capital value. That valuation was fixed in law as the price the city must pay should it ever attempt municipal ownership; no provision was made for amortization, and every subsequent track extension or renewal was added to the total.[42] In order to obtain what it hoped would be good service under private ownership, the city had had to accept terms that encouraged the companies to keep and extend obsolete facilities, held back the merger of competing companies, and prevented municipal ownership for forty years. Yet the surface companies paid their bond interest during most of the depression years and managed to spend over $179,000,000 on renewals by 1943.[43]

The Chicago Rapid Transit was also overcapitalized. From an alleged value of $91,000,000 in 1944, the SEC cut it back to $11,-

[41] CLT, *Report of the Special Committee on Local Transportation of the City Council of the City of Chicago* (Chicago, 1901), 3, 6–10; Milo Ray Maltbie, *The Street Railways of Chicago: Report of the Civic Federation* (Chicago, 1901), 8–9 (reprinted from *Municipal Affairs*); *Report of the Special Commission of the City Council of Chicago on the Street Railway Franchises and Operations of [11 companies]*, March 28, 1898 (Chicago, 1898) ("The Harlan Report"), *passim*. Robert David Weber "Rationalization and Reformers: Chicago Local Transportation in the Nineteenth Century" (Ph.D. dissertation, University of Wisconsin, 1971), 282, argues that much of the overcapitalization was a necessary result of rapid technological change.

[42] Chicago Traction Valuation Commission, *Report on the Valuations of the Chicago City Railway Company* (Chicago, 1906), *passim*; Walter L. Fisher, *The Traction Ordinances: An Address to the Chicago City Club*, November 23, 1907 (Chicago, 1907), 9; Bion J. Arnold, "The Transportation Problem in Major Cities," *California Progressive* (April 1, 1911), 13–14.

[43] BOSE, *39th Annual Report 1947*, Exhibit 8; Schroeder, *Metropolitan Transit Research Study*, 1–6 (in 1944 the SEC ruled that $110,000,000 of the Chicago Surface Lines' $172,000,000 valuation represented no useful property); Walter L. Fisher, *Analysis of the Traction Ordinance: A Report to the Hon. James H. Wilkinson, Judge of the United States District Court* (Chicago, 1930), 10–14. For the latest strong argument that the companies were overcapitalized, see Paul H. Douglas, "Chicago's Persistent Traction Problem," *National Municipal Review*, Vol. 18 (November, 1929), 669–675. Because Chicago Surface Lines counted day-to-day maintenance in its maintenance and renewals figures, its totals were considerably higher than those of the Board of Supervising Engineers (BOSE, *39th Annual Report 1947*, Exhibit 8).

000,000.[44] This overcapitalization probably did serious damage to the Chicago Rapid Transit's service. By 1923 Samuel Insull, who controlled the elevated lines, declared that he could not get bank credit to finance extensions or new equipment, and he turned to selling bonds to employees and patrons to foot the bill. With the collapse of the Insull empire, the Chicago Rapid Transit quickly became unable to meet the obligations on any of its securities.[45] But even the fate of CRT securities should be viewed in context: no other rapid transit system in the nation was unsubsidized after World War I. In part, the rapid transit's problems were due to the willingness of Chicago and of Insull to place upon private enterprise a burden that Insull knew it could not sustain.[46]

The inflated official "purchase prices" of both companies probably had two major effects on Chicago. The desire to keep the cost of tracks and other streetcar equipment in the companies' purchase price gave Surface Lines management a vested interest in obsolescence and kept the Chicago Surface Lines far behind most of the nation's surface systems in the use of motorbuses. The issue of bus substitution was the subject of strong disagreement between the engineering staff and the bankers who controlled the companies.[47]

Even more important was the disruptive influence of the companies' high valuation on every plan of unification or municipal ownership presented between 1907 and 1944. The 1905 plan of Mayor Edward Dunne for municipal ownership of transportation facilities was defeated in part because state law forbade the city from selling enough bonds to purchase the companies' securities. Valuation was a major factor in the determination of the companies' rates of fare by the State Commerce Commission after 1914, and plans for consolidation under private ownership put forth in 1912 and 1918 both met defeat at referendum largely on the issue of valuation and potential fare increases. Mayor William Dever's 1925 plan for modified municipal ownership was also defeated in part

[44] Schroeder, *Metropolitan Transit Research Study*, 7.

[45] *ERJ*, January 2, 1924, 84, January 26, 1924, 156; Forrest McDonald, *Insull* (Chicago, 1962), 92, 153–158, 205; *ERJ*, June 21, 1924, 998, August 9, 1924, 203; In the District Court of the United States for the Northern District of Illinois, Eastern Division, *In the Matter of Chicago Rapid Transit Company, Debtor, and Union Consolidated Elevated Railway Company, Subsidiary Debtor, Proceedings for the Reorganization of a Company*, vol. I, January 22–September 20, 1937, p. 21.

[46] Walter A. Shaw, *Report* (to the Receivership Court, June, 1938), bound with In the District Court of the United States for the Northern District of Illinois, Eastern Division, *Harris Trust and Savings Bank, Trustee, Plaintiff v. Chicago Railway Company et al., Defendant in Equity*, Consolidated Case 6839 (CSL, *Receivership Record*), vol. VI, 31–48; *ERJ*, January 2, 1909, 12–14, and June 26, 1909, 176.

[47] McIlraith Interviews, Tape 7, Side 1. The city, for its part, tried to prevent the companies from purchasing new streetcars (CSL, *Receivership Record*, vol. V, 4164).

because of a provision for possible fare increases to meet payments on outstanding securities of the old companies. A consolidation plan worked out between 1927 and 1930 with the blessing of Samuel Insull survived a referendum but proved impossible to finance after Insull's collapse, partly because of its high capital valuation and the endless quarrels among junior holders who wanted a more secure position in the new financial structure. Finally, two reorganizations, in 1943 and 1944, were ruled untenable by the Illinois Commerce Commission, which considered the valuations of the proposed new companies to be too high.[48]

Poor management in the form of overcapitalization was, then, a major factor in retarding improvements and preventing unification, but it was not a major cause of deterioration of service on lines that might otherwise have been profitable. However, the quality of management could also affect the level of day-to-day service. In the case of rapid transit management, that effect was negative. Samuel Insull, who took control of the lines in 1911, was fond of talking of the benefits of rapid transit, but the only major alterations he made in rapid transit service were three long suburban extensions, which Insull himself declared were designed to encourage the development of communities to which he hoped to sell power.[49]

[48] BOSE, 7th Annual Report 1914, 12–13; ERJ, November 20, 1915, 1050; Harry Weber, Outline History of Chicago Traction (Chicago, 1936), 121–128, 215–229; Paul R. Leech, Chicago's Traction Problem (Chicago, 1925, Chicago Daily News Reprint #17); ERJ, January 10, 1925, 75, February 14, 1925, 271, March 14, 1925, 426, April 11, 1925, 591; An Ordinance Providing for a Comprehensive Municipal Local Transportation System, Passed by the City Council of the City of Chicago, February 27, 1925 (Chicago, 1925), 3, 5, and 8. See also ERJ, December 28, 1918, 1149; William Hale Thompson, The Thompson Plan for People's Ownership and Operation of the Street Railways at a 5¢ Fare, submitted to the City Council of the City of Chicago, September 9, 1919 (Chicago, 1919), 10–18; Chicago Municipal Ownership League, Public Ownership: The Solution to Chicago's Transit Problem (Chicago, April, 1919), 4; George C. Sikes, An Argument Against the Traction Ordinances, Presented Before the City Club of Chicago, September 20, 1918 (Chicago, 1918), 1–3; Transit Journal, vol. 78 (November, 1934), 449; Philip Harrington, "Report and Recommendations for Expediting Settlement of the Chicago Transportation Problem" (typescript, Chicago Commission on Subways and Superhighways, August 7, 1944, available in the Chicago Historical Society), 17–19; CSL, Receivership Record, vol. XI, Document 9610; "Chicago Needs 10¢ Fare to Sell Bonds," Mass Transportation, vol. 43 (May, 1947), 188–189.

[49] Journal of Commerce, November 15, 1923; CSL, Receivership Record, and First National Bank of Chicago v. Chicago City Railway Company, Equity Consolidated Case 9915, "Brief in Support of the Abbott Plan" (Chicago, 1937), 17–20, in Gottlieb Schwartz, Inc., "Records" (Chicago Historical Society, Manuscript Division, Box 483AA); Illinois Commerce Commission, Cases, #19127 — People v. Chicago Rapid Transit Company, 349 Ill. 309, 313 (1932); Samuel Insull, Chicago's Future, An Address Delivered at the 42nd Dinner of the Chicago Real Estate Board, (Chicago, n.d.), 5–7; Insull, Rapid Transit Development for Chicago, Delivered before the Chicago and Cook County Bankers Association at the Union League Club, February 21, 1924 (Chicago, 1924), 3; Insull, "Memoirs" (typescript in the Insull Papers, Loyola University Library, Chicago), Box 17, 191–193; Committee on Public Affairs of the Liberal Club of Chicago, Illinois, The Chicago Traction Situation (pamphlet, June, 1930), 26; Douglas, "Chicago's Persistent Traction Problem," 675; Barton Aschman Associates, Inc., Needs and Opportunities for Coordination and Transportation Improvements (Chicago, 1963), 10, 14; C. K. Mohler, Report on the Union Elevated Railroads of Chicago, submitted to the Loop Protective Association, Inc. (Chicago, 1908), 13–15.

CHICAGO MASS TRANSIT   489

Surface Lines management presents a different picture. After a period of decline and confusion under the direct control of Chicago bankers and old-style "railroad-oriented" engineers between 1914 and 1923, with franchises soon to expire and public opinion hostile, the bankers turned the operation of the companies over to professional transit managers and engineers. These men and their subordinates altered the image and the daily practices of the companies, placing the Chicago Surface Lines in the forefront of the development of modern transit equipment (of which it bought little), traffic control, scientific scheduling, and public relations.[50] But in the long run, the progressive attitudes of CSL management probably contributed more to the development of the city as a whole than to the survival of mass transit. Like Chicago engineers before them, the Surface Lines' experts treated mass transit traffic as a natural outgrowth of the nineteenth-century urban way of life. Ridership, they thought, could not be substantially increased or decreased by city or company policies. Only the efficiency of the company could be altered. Much of their real sense of civic responsibility was thus channeled into improvements that benefited the automobile. This perspective, rather than simple inertia, conditioned the Chicago Surface Lines management's attitude toward the automobile's competition.[51]

ATTITUDES AND POLICY

So far a number of facts of the sort customarily used to account for the decline of mass transit have been examined. Chicago Surface Lines management policies may help to explain some of the early resistance of Chicagoans to change in their travel habits. So may the fact that riders soon became accustomed to long, indirect journeys to work via public transportation (the average Chicago Surface Lines ride in 1934 was 4.5 miles, the longest in the coun-

[50] *ERJ*, November 1933, 403; BOSE, *20th Annual Report 1928*, x, 96; McIlraith Interviews, Tape 5, Side 1; *Transit Journal*, vol. 28 (May, 1932), 197. Much of the Chicago Surface Lines' activity fostered the use of automobiles. The Surface Lines led the movement for city-wide parking restriction and off-street facilities. Company engineers also participated actively in the writing of Commerce Secretary Hoover's model traffic code (CSL, "Traffic and Scheduling Reports," 1929, pp. 5, 8, 1930, p. 7, 1932, p. 9).

[51] E. J. McIlraith, "Modern Urban Transportation," *Mass Transportation*, vol. 34 (November, 1938), 365–369; Guy Richardson (CSL president), "Looking Ahead in Urban Transportation," *ERJ*, September 15, 1931, 503–506 — both on the "natural" limits of automobile usage. Cf. also McIlraith Interviews, Tape 9, Side 2. The official attitude changed only with the creation of the CTA (CTA, *Annual Report 1946*, 9–10). For an early statement of the problem, see Charles T. Yerkes, *Investments in Street Railways: How Can They Be Made Secure and Remunerative?*, reprinted from the *Report of the 18th Annual Meeting of the American Street Railway Association*, Chicago, October 17–20, 1899 (Chicago, 1899), 3–4.

try).[52] But while financial manipulation, population dispersion, and the automobile performed the roles usually assigned them with differences described above, it is apparent that these factors were themselves in part functions of popular ideas and local government decisions.

How then did the attitudes and policies of Chicagoans toward their mass transit system come to be what they were? Why was Chicago the last city in the nation to consider an operating subsidy for rapid transit? Why did the city develop an adversary relationship with its transportation companies when dealing with issues of fares and service, while cooperating heartily with them in traffic control? Why above all did everyone involved with mass transportation come to feel that it could justify itself only insofar as it could be financially self-supporting?

Five characteristics of public attitudes and policy affected the quality of Chicago's mass transit service: regulation by public bodies was largely negative in character; mass transit was taxed to pay for other city services; whenever possible, the companies were kept on short-term franchises; and proposals for municipal ownership seemed always to run aground on the public's suspicion of elected officials. Underlying all these facts was the local tradition that trapped the system and its planners in a nineteenth-century framework — the belief that public transportation must pay for itself, "from the fare box." The bare facts can be briefly outlined.

State regulation of Chicago transportation was first proposed by streetcar executives as a method of escape from city control.[53] But the first successful step toward regulation came on the city level, as part of an attempt to protect mass transit from political manipulation. As a part of the watershed Ordinances of 1907, a Board of Supervising Engineers, members of which were appointed by the city and the companies, was to see that the surface companies carried out other provisions of the Ordinances, such as equipment and service improvements. The Board also fixed the companies' valuation, but it could not order improvements or extensions not called for in the Ordinance itself. That power lay with the City Council, whose decisions could be appealed to the Board of Supervising Engineers or to the courts. Control of mass transit was thus

[52] Leslie Vickers, "Fare Structures in the Transit Industry" (Ph.D. dissertation, Columbia University, 1934), 197–198.

[53] Yerkes, *Investments in Street Railways*, 4–5; Chicago *Tribune*, February 19, 1897, March 11, 1897, March 21, 1897; Carter Harrison, *The Stormy Years* (Chicago, 1935), 138; Forrest McDonald, "Samuel Insull and the Movement for State Regulatory Commissions," *Business History Review*, XXXII (Autumn, 1958), 248–253.

divided between city agencies and companies in a way that could serve only to diffuse responsibility and put the "traction problem" out of focus.[54]

Even this limited power was removed from the Council and the Board of Supervising Engineers with the creation of a state Public Utilities Commission in 1913. Thenceforth, rates, service complaints, and routes for all Chicago transportation were under the control of this commission of gubernatorial appointees. The major effects of thirty-three years of regulation by this commission (and its successor, the Illinois Commerce Commission) were that: (1) it kept fares low while the cost of service rose, (2) it prevented the abandonment of little-used services, (3) it created a third competing service, the Chicago Motor Coach Company, and (4) after 1936 it ordered transfers among all the competing companies. In none of these cases did the commission *initiate* action.[55]

In the long run, regulation served neither the companies nor the city. Another consequence of private ownership, the taxation of mass transit property, was likewise of questionable value. Between 1907 and 1947 the Chicago Surface Lines alone paid over $163,-000,000 in taxes, an amount little short of its total valuation at its highest point. Of this amount, $47,500,000 was paid to the city's "Traction Fund" under the 1907 agreement, by which 55 per cent of the surface companies' net after interest became the property of the city. In addition, the Chicago Surface Lines was obliged to sweep, plough snow, and lay pavement in the area between its tracks. Other companies paid smaller amounts of "compensation." [56]

The franchise system also failed in its presumed aim of forcing the companies to behave "responsibly." The Chicago Rapid Transit had several 50-year franchise ordinances, which expired during the

[54] CLT, *Report of the Committee on Local Transportation to the City Council of Chicago, With Ordinances to the Chicago City Railway and Chicago Railway Company*, January 15, 1907 (Chicago, 1907), 1–18; BOSE, *1st Annual Report 1908* (Chicago, 1908), 52, 182, 220ff, and *13th Annual Report 1920*, ix; "Duties and Organization of the Board of Supervising Engineers," *ERJ*, October 5, 1912, 578–582.

[55] Edward F. Dunne, "Message to the 48th Assembly" (1913), reprinted in William L. Sullivan, ed., *Dunne: Judge, Mayor, Governor* (Chicago, 1916), 397–400; Dunne, "Statement of Governor Dunne Regarding Public Utilities" (House Bill 907), Springfield, Ill., June 30, 1913, in *Dunne*, 439; *Chicago Tribune*, June 16, 1913; *ERJ*, March 7, 1914, p. 555, August, 1914, 714, October 10, 1914, 789; and February 24, 1912, 318. See also State Public Utilities Commission of Illinois, *Operations and Orders for the Year Ending November 30, 1914*, vol. 1 (Springfield, 1915), Order #2586, pp. 1137–1138.

[56] BOSE, *39th Annual Report 1947*, Exhibit 8. The best available source (Harrington, Kelker, and De Leuw, *A Comprehensive Local Transportation Plan*, Figure 2) shows $1,140,000 for the CMC to 1936, and $1,270,000 for the CRT. Neither figure includes federal or state taxes. In 1937 one-third of the Traction Fund was in tax anticipation warrants used by the city to pay debts, including interest on road-building bonds (CLT, *Proceedings*, vol. 134, January 19, 1937, 56–59).

mid-1940s. The Chicago Motor Coach Company operated on a certificate of convenience and necessity from the Illinois Commerce Commission, while the Chicago Surface Lines operated on a franchise granted in 1907, which expired in 1927. After this point, franchises offered it by the city were "terminable permits," which could be abrogated by city purchase at the agreed valuation. But in no case did the companies have to grant significant concessions to gain new franchises.[57]

City purchase was always unlikely. Chicagoans voted for municipal ownership and operation in two early twentieth century referenda, but the voting majorities steadily decreased. In 1907, the voters agreed to grant the private companies a 20-year franchise under a plan which seemed to promise eventual city purchase with money from the Traction Fund.[58] But opinion on municipal ownership remained polarized. Influential Chicagoans attacked municipal ownership on the ground that politicians were too corrupt to be trusted with such a mine of jobs and plunder, while after 1907, municipal ownership advocates rejected any compromises that seemed to repeat the errors of 1907.[59] In the end, the modified form of municipal ownership represented by the Chicago Transit Authority was arrived at without much enthusiasm on any side. The legislators creating the Chicago Transit Authority took care to see that it would be operated essentially as a private business and run by a non-partisan board.[60]

Finally, the assumption that mass transit could and must pay for itself set the parameters for all mass transit planning through the

[57] *ERJ*, March 23, 1918, 98, 369; franchise information is summarized in Harrington, "Report and Recommendations," Introduction.

[58] Advisory Referenda — 1902: 142,000 for, 27,000 opposed; 1904: 152,000 for, 59,000 against; 1906, 240,000 for, 130,000 opposed. Publicity for the 1907 Ordinances suggested that they were the fastest way to achieve municipal ownership (e.g., Chicago *Tribune*, March 10, 1907, March 18, 1907; Chicago *InterOcean*, April 3, 1907; Citizens Non-Partisan Traction Settlement Association, leaflet dated March 9, 1907 [in Hooker Collection, University of Chicago, vol. 3]).

[59] Street Railway Commission of the City Council of the City of Chicago *Report, December 1900* (Chicago, 1900). Prominent public figures were sent a questionnaire asking their views on the several topics taken up by the Commission; George Hooker preserved one (in Hooker Collection, University of Chicago, vol. 1). Lawrence Laughlin of the University of Chicago made the point concerning the morality of government and municipal ownership specifically in his contribution to the *Report* (98); Milo R. Maltbie concurred (74). Among many other examples are John M. Harlan and E. F. Dunne, *The Facts on Municipal Ownership* (Chicago, March, 1905), *passim*; Chicago *Record-Herald*, March 18, 1907; *Chicago Inter Ocean*, April 3, 1907. A later example of this attempt to link traction with politics and business corruption may be found in Tomaz Deuther, *Civic Questions*. See also John C. Kennedy, *Speech Delivered by Alderman John C. Kennedy to the City Council of the City of Chicago* (n.p., n.d. [August 9, 1917]), 5, 30; Sikes, *An Argument*, 1-2; Henry P. Bremer, "How Chicago Is Attempting to Solve Its Traction Problem," *Harvard Business Review* (July, 1931), 459.

[60] Schroeder, *Metropolitan Transit Research Study*, 137–144; "Proposed Consolidation in Chicago," *Mass Transportation*, vol. 41 (January, 1945), 28, vol. 43 (October, 1947), 372–374; Chicago *Tribune*, April 13, 1945, May 28, 1945, June 3, 1945, June 4, 1945.

CHICAGO MASS TRANSIT   493

mid-1960s. Bion Arnold told his fellow engineers in 1914 that the city would build subways only to the extent that it could rent them to the Chicago Surface Lines at a price that would soon cover their full cost. The most comprehensive survey ever made in the period of private ownership, the report of the Traction and Subway Commission of 1916, also declared that any system of subways must be self-supporting. The Western Society of Engineers observed in 1924 that a comprehensive subway plan was impossible for Chicago since it could not pay for itself. Both the ordinance of 1925 and that of 1930 contained provisions for fare increases to make sure the system would not lose money. The fact that extensive subways could not pay for themselves was also the justification used by Mayor Edward Kelly for the abandonment of broad transit planning after 1937. The *Chicago Area Transportation Study*, published in 1960 contended that, while transit in the central business district might reasonably be subsidized as a public benefit, improvements in outlying areas would have to be self-supporting. By 1927, reform Mayor William Dever, whose roots — like Hooker's — went back to the settlement movement, was declaring that any honest transportation system must be a paying proposition. Only Big Bill Thompson, whose reputation for corruption and bungling was already well established, stood forth on behalf of a possible subsidy for rapid transit.[61]

The crisis created by the automobile in the early 1920s was acute. Mass transit's crisis was perennial and, despite occasional alarms by the press, it could always be temporarily alleviated by halfway measures. The very resilience of public transportation made it possible for Chicagoans to avoid major changes in their conception of its role.

CONCLUSION

When George Hooker came to telling the *Record-Herald*'s readers what must be done with mass transit, he chose a phrase that

[61] ERJ, September 14, 1912, 416; Arnold, "City Transportation," 12, 20–21; William Barclay Parsons, Bion J. Arnold, and Robert Ridgway, *Report of the Chicago Traction and Subway Commission, 1916* (Chicago, 1916), 44; "Engineers Would Better Chicago Transport," *Engineering News-Record*, February 28, 1924, 377; *An Ordinance . . . 1925*, 8; *An Ordinance . . . 1930*, 10–12 and especially 12–13 (Section 19), 13–15 (Section 21); Harrington, Kelker, and De Leuw, *A Comprehensive Local Transportation Plan*, 15; Mayor Edward Kelly, "Message to the City Council," in CSL, *Receivership Record*, Documents 9754, 9756; Chicago Area Transportation Study, *Final Report*, vol. III, *Transportation Plan* (Chicago 1960), 82–88. One explanation of William Hale Thompson's plan for a metropolitan transit district with taxing power is in Chicago *Journal*, January 7, 1921; see also Chicago *News*, December 22, 1921. Dever held that "no . . . mandate can compel a private enterprise to give something for nothing, and the same goes for municipal ownership" (Chicago *Journal*, August 23, 1923; Chicago *Tribune*, February 15, 1924; see also Chicago *Post*, January 16, 1921).

embodied the nineteenth-century attitude toward transportation. We must, he said, "control" its "expansion." This idea — that mass transportation was a dynamic, almost natural creature whose power must be checked for the common good — is at the root of much public thought about mass transit during and long after the period when it was a daily necessity for the majority of citizens. Belief in the power and profitability of mass transit, along with a feeling that each individual transit rider must pay his own way, still make good mass transit an impossibility in many suburban areas despite regional transit authorities and long-term fuel shortages.[62] Yet the image of mass transit that steers public opinion is rooted in the facts of a world long vanished.

The supposed profitability of mass transit was fixed quickly in the public mind. Several well-documented reports issued at the turn of the century illustrated the profits to be made from mass transportation in the late nineteenth century. Later authors down to Paul Douglas in 1929 reiterated the idea that mass transit was potentially a highly profitable affair.[63]

Transit's reputation for omnipotence and unscrupulousness is also easily accounted for. Mass transit service improved remarkably between 1880 and 1900. Speeds doubled, and electric lights, heat, and power replaced smoking lamps, straw-covered floors, and odorous horses. Yet people traveled further than ever before and so encountered new levels of inconvenience.[64]

The transit companies, the first large corporations with which most people had direct daily contact, contributed much to their own bad reputation. Charles T. Yerkes in particular must take a large share of the blame. He divided the lines under his control in order to keep the length of ride short. He scoffed at the pleas of little children, and scorned reporters. He frightened Marshall Field, and he drove the president of the South Side company to near-

[62] E.g., Chicago *Tribune*, March 10, 1974. Opposition to use of public funds for mass transit from those who did not use its services is, of course, longstanding. A few examples are: Chicago Motor Club, *Factors Suggested for Investigation Relative to Superhighway Construction* (Chicago, 1939), 33–37; "Protect Your Highways," *Motor News*, vol. 43 (May, 1957) — a reaction to the CTA's first subsidy plea; and Rep. Giddy Dyer to Chicago *Daily News*, November 28, 1973.

[63] Special Committee, *Report . . . on Street Railway Franchises* (1898), 18, 44–69; George Schilling, *The Street Railways of Chicago and Other Cities* (Chicago, 1899), 31, 43, 54, 61, 67; Street Railway Commission, *Report 1900*, 38–41; Delos F. Wilcox, "Public Regulation of Motor Bus Service," *Annals of the American Academy of Political Scientists*, vol. 116 (Philadelphia, 1924), 109; The Liberal Club of Chicago, Illinois, *Chicago's Traction Situation: A Factual Analysis* (Chicago, 1930), 6; Douglas, "Chicago's Persistent Traction Problem," 671–675.

[64] On length of rides, 1886–1897, see Chicago City Railway, *The Humphrey Bill and Comparisons of American Street Railways* (Chicago, 1897); on speed, see Yerkes, *Investments in Street Railways*, 3.

distraction.[65] Yet before 1901, Yerkes was Chicago's chief spokesman for the idea that public transportation is a community benefit, a destroyer of slums and an opener-up of parklands.

Like Insull after him, Yerkes had his own reasons for stressing the social role of transit. Like Insull, he ultimately became a discredited man, and the ideas he advocated found few supporters.[66] Meanwhile, transit service did improve when the companies were under heavy public pressure after 1907 and again in 1924–1928. While other factors were involved in each case, the image of an omnipotent and malevolent transit management can only have been sharpened by such crises.

The policies that grew out of this distorted and unrealistic vision of mass transit had a few positive effects. The Illinois Commerce Commission kept transit fares at an attractive level, required a special renewals fund under its own supervision, helped to provide a higher class of service (the Chicago Motor Coach Company), which may have retained some riders for mass transportation, and, rather late, brought about a truly universal system of transfers. Short-term franchises may have spurred the Chicago Surface Lines to make the improvements discussed in the mid-1920s. But overall, the effects of public policy were not good. Rapid transit service with many obsolete stations was unnecessarily slow and unattractive. Such factors as competition between surface companies, taxes, low fixed fares, and short franchises can only have held back investment in these properties, especially before the mid-1920s when there was little sign that the old fare and franchise policies would be abandoned. Improvements by private capital were discouraged, but municipal ownership was impossible. Finally, the idea that transit can and should be paid for by those who use it has survived the reformers and the traction barons to haunt another generation of Chicagoans.[67]

[65] Chicago *Tribune*, January 12, 1893, January 21, 1894, February 7, 1894; Chicago *Post*, December 11, 1893, December 22, 1893; Chicago *News*, December 20, 1894; Chicago *Tribune*, December 21, 1894; Chicago *Chronicle*, October 24, 1895; H. Wayne Andrews, *The Battle for Chicago* (New York, 1946), 196; letters between Yerkes and President Holmes of Chicago City Railway, October 18 through December 23, 1889 (in CSL, *Archives*, Box 8, Folder 53 of the North Chicago Street Railway files, at the Chicago Historical Society).

[66] Chicago *Tribune*, January 11, 1899; Yerkes, *Investments in Street Railways*, 6. Insull made frequent speeches in which he portrayed himself as a disinterested promoter of the city's good. His image of himself as a modern Medici is well illustrated in Samuel Insull, *Why I Am in the Public Utilities Business, An Address before the Midday Luncheon Club of Springfield, January 25, 1922* (Chicago, 1922).

[67] CSL service was the cheapest per mile of any unsubsidized system in the nation in 1939 ("Mass Transportation in Chicago Moves Forward," *Mass Transportation*, vol. 35 [January, 1939], 6). This fact may have made Chicagoans all the more unwilling to pay increased fares after World War II. For the effect of regulation on rapid transit service, see CTA, *4th Annual Report 1948* (Chicago, 1949), 12–14; CTA, *Station Location on*

The precise reasons for the fluctuations in the ridership habits of Chicagoans remain unclear. Some early journey-to-work studies and ridership statistics recently uncovered by the author suggest that distance from central place and the presence of good single-family housing are better correlated with declining transit ridership than is the quality of service. But ridership does seem to respond to innovation in mass transit facilities, and a fair number of areas do not conform to the expected pattern.[68]

Study of these materials is just beginning. But the role of public policy, grounded in the popular conception of transit as a private business and of the automobile as a public good, is already clear. Technology, as it turns out, did not set all the terms for the solution of social problems. Private aspirations and the assumptions on which public policy was based helped to determine the meaning of technological change. Thus, Chicagoans changed their mode of transportation for reasons which often had little to do with either transportation or technology.

Rapid Transit Lines (Research and Planning Report RPX 70175, March 11, 1970), 3. Other conclusions are based on evidence cited above.

[68] These generalizations are based on detailed ridership data in the possession of the Chicago Transit Authority (car assignment sheets and schedule revision records, 1914–1947, traffic and scheduling reports of the Chicago Surface Lines, 1923–1947) and the synthesized results of two journey-to-work studies undertaken by the Surface Lines ("Residence and Means of Transportation of Carnegie-Illinois Steel Employees" [November, 1941] and "Residence of Armour & Company Workers" [June, 1942], in McIlraith Papers, vol. I, "Traffic"), along with housing data in Chicago Plan Commission, *Residential Chicago*, 73, 93, 113. The statements in the text are based on preliminary calculations made by the author from this data.

A Sociologist Looks at America's City Traffic Problem

John E. Owen

America's transportation system is moving rapidly to a crisis, and is rapidly worsening with increasing population, urban crowding, and an excess of automobiles in city streets. The USA is a nation dependent on the automobile—100,000,000 today and a possible 350,000,000 by the year 2000. Cars are multiplying faster than the population and one business in six is already dependent on the auto industry. But America has never had a national transport policy, though public awareness of the problem is rising.

AMERICA'S transportation system is moving rapidly to a crisis. It is ironic that the nation which sent a man to the moon cannot find a way to move its city workers to and from their work conveniently, quickly, and safely.

Despite advances in town planning and automobile technology, the problem shows all signs of becoming more intense rather than easing, as population increases, urban congestion accelerates, and Americans insist on driving their own vehicles. The nation today has over 100,000,000 automobiles, one for every two persons. Official estimates are that by the year 2000, there will be 200,000,000 cars, trucks, and busses on the road. Some reports claim that the figure may be as high as 350,000,000.

Four out of every five US families own one car, and over 25 percent of all families own two or more cars. Nine out of every 10 American men have a driver's license, as do two thirds of the women over 15. Eighty percent of all US workers go to work by car. The automobile has in fact become the prime symbol of US society. In 1969

Americans spent only a little less on their cars than they spent on national defense. But the impact of the automobile on American cities since World War II has affected the quality of life for literally millions, in overcrowded highways, air pollution, nervous tension, and accidents.

Surveys have shown that the average American worker spends an hour in his car going to and from work for every eight hours that are actually spent on the job. From 7:15 to 8:15 a.m. on weekdays, 40,000,000 Americans are on the highways, trying to get to work. The journey to work is the hardest part of the day for many, involving bumper-to-bumper driving on the city highways that are becoming more traffic-strangled every year. Before 1910, horse-drawn traffic in New York averaged 11 miles an hour. Today, automobiles there average only eight miles an hour. The annual cost of traffic delays in midtown Manhattan has been estimated at $150,000,000.

The problem is partly a result of America's urban concentration of her people, in which 75 percent of the population is jammed into 1 percent of the total land area. The problem is particularly acute in New York and the

Dr. John E. Owen was born in England, educated at Manchester Grammar School, and has been a US citizen for many years. A graduate of Duke University, he received his M.A. and Ph.D. in sociology from the University of Southern California. He has taught in several US and Canadian universities, the University of Helsinki, Finland, and Dacca University, Bangladesh. He has contributed chapters to four sociology texts and authored over 200 articles published in 16 countries. Currently, he is a professor of sociology, Arizona State University.

311

North-Eastern corridor from Boston to Washington, where over one fifth of the US population is crowded into less than 2 percent of the land. If present trends persist, in less than 30 years almost two thirds of all Americans will be concentrated in three corridors, a continuous strip of cities from Boston to Washington, in a strip around the southern rim of the Great Lakes, and between San Francisco and San Diego, California. By that time cars and trucks (also trains and planes) carrying twice as much cargo will be competing with each other to reach destinations that are hard to reach even today.

But cars are increasing in America much more rapidly than people. In the last 15 years, the population has gone up by 20 percent but the number of cars by 50 percent. America now produces 10,000,000 cars annually. Automobiles are in fact not merely a foundation of the nation's transportation system, but of the US economy. One business in six is dependent on the manufacturing, distribution, servicing, and use of motor vehicles. Almost 14,000,000 workers (one in every six) are employed in the industry, which accounts for more than half of all rubber consumed in America and one fifth of all steel. Hence, auto sales are a prime indicator of the condition of the economy. Among key factors in the demand for autos are expansion of personal incomes, a high rate of abandoning old cars (6,500,000 annually) under America's system of "planned obsolescence," a booming teen-age population, and rising ratio of young married people. In consequence, auto sales alone account for fully one tenth of the gross national product. The total business of moving people and goods accounts for at least 20 percent of the gross national product, not including military movements. It is doubtful if the cost of transport is

nearly as high in any other country. As one instance, of every dollar that Americans spend for food at a grocery, 32 cents pays for getting the food to the shop, one third of the cost. Since everything that Americans do requires transportation, it is not surprising that the total bill is not much below that of the entire federal budget.[1]

In light of these facts, it is paradoxical that America has never had, and still does not have, a national transport policy. But in the next two decades, she will have to double the carrying capacity of a transport system that it has required the lifetime of the nation to achieve. But just as US cities have been rapidly built on an unplanned basis, so transportation has been allowed to grow in haphazard fashion with little consideration for the social consequences of the new technology on the roads, rails, or in the air.

The system contains some odd contradictions. For example, US cars consume twice the fuel of the average imported car, but nearly 90 percent of all automobile trips in America are journeys of less than 10 miles. But there is no public policy to restrict the size and horsepower of cars, and it is only recently that regulations have been imposed for safety equipment and antipollution devices. The political power of the auto industry, in which three corporations produce 90 percent of all US cars, and the force of public taste have to be contended with. Similarly, the nation has never had a long-term consistent policy on land use, though the time is approaching when the auto will take up all the available city space.

A master plan for transport and its human effects is desperately needed.[2] The efficient movement of traffic cannot be achieved by a patchwork of uncoordinated regulations in which cars, trucks, trains, and planes are, in effect, competing for the consumer's dollar, for tax subsidies, and for space to move. The entire situation is complicated by the relationship in every transport facility between industry and government. A city mayor, state governor, and a cabinet officer may agree on a new transport idea, but without the cooperation of the public facility's private users the system will not work. Highway commissioners in the different states plan where to build roads, but private cars and trucks use them. Similarly with airports and canals. There is no consensus as to transport goals at any level. The $50,000,000,000 a year trucking industry is opposed to subsidizing railways at the expense of new highways. The powerful "highway lobby," made up of the auto and gasoline industries, tire makers, and haulage contractors, has great influence on federal legislation affecting plans and funds for road-building.

Other interest groups compete for their particular source of livelihood, and the result has been an unplanned development of transport facilities arising from competing interests and resistance of powerful groups to any change affecting them adversely, together with lack of national consensus on aims and goals, and inadequate planning and analysis. There are no congressional committees empowered to look at all aspects of the transport problem, and legislators are too frequently pressured by special interest groups in their home districts. But public awareness of the problem is on the rise, mainly because Americans are increasingly finding out what it means in physical discomfort and lost time to move from place to place.

Similarly, awareness of the tragedy of auto accidents is greater today than it has ever been in America's history, partly because it is reaching cri-

sis proportions, and the increase in highway accidents has been well publicized in the press and on TV.[3] About 56,000 persons are killed on the roads annually and almost 4,000,000 are injured. Road accidents are the top killers of US youths, more than all diseases combined. In 1969, almost one third of motoring deaths were in the 15 to 24 age group. During the 1960's, highway accidents made 1,500,000 Americans permanently disabled.

Significantly, the ratio of highway deaths diminishes in the higher age levels, older drivers having the best safety records. Growing numbers of youth, increasing affluence that permits more of them to own cars, their tendency to drive delapidated cars and motorcycles, and rising use of drugs and alcohol by US youth are possible factors. One government report describes the young motorist as "an inexperienced driver and an inexperienced drinker." But some states still license 14-year-olds to drive.

The belief that secondary-school courses in "driver education" produce safer drivers has been challenged by psychologists. These programmes reach 2,500,000 students annually, but there is little actual evidence that they reduce accidents.

Alcohol is involved in approximately half of all highway deaths, and official estimates are that one motorist in 20 on US roads is an alcoholic. Recent studies have shown that some alcoholic drivers also display symptoms of mental illness, and that other personality types utilize their cars for the release of aggressive impulses.

Many experts have seen the solution to accidents and crowded highways in a greater use of mass transit systems—underground and elevated railways, trams, and busses.[4] But America's excess dependence upon the auto has resulted in very little being spent on systems of public transportation. In fact, America is the only major developed nation in the world to allow its public transport system to decay. The number of electric railway cars in service today is only a quarter of the number in 1940. The number of bus passengers carried today is little more than half the 1940 figure. And since 1950 almost 200 transit companies have gone out of business, most of them in towns of less than 50,000 people. In 1964 Congress did pass a mass transportation bill to furnish research, equipment, and construction funds. But the funds have been totally inadequate to meet the demand. In 1968, only $177,000,000 was spent.

In many a US city, including some of the largest metropolises, public transport systems are quite unequal to meet the needs of the tens of thousands of commuters who require this service to and from work.[5] This lack particularly hurts the old and the poor. It prevents many people on public assistance from being able to take jobs for which they would be qualified. The worker who has to depend on public transport in America is severely handicapped. And most urban transport systems have a record of failure, because in city after city workers prefer to use their cars. In one ill-fated experiment in Flint, Michigan, comfortable air-conditioned busses were installed to pick workers up at their doors. But fewer than 300 chose to ride the busses each day. They preferred the privacy and freedom of using their own cars.

There is a deep psychological import in the individual's preserving control over his freedom of movement. Medical studies have shown that the citizens of New York are under severe pressure. New York is a metropolis where congestion has forced 85 percent of all commuters to use a system in which there are no alternatives for mobility, an unhealthy situation. Chicago, where the figure is 90 percent, is comparable.

Many psychologists have also observed that the auto has become an extension of the owner's ego, that he is psychologically attached to his car and the greater liberty of action (despite crowded highways) that it offers him compared with crowded rush-hour underground subways or bus-riding. Many of the public transit systems are obsolete, creaking with age, and losing money year after year, even though fares have more than tripled since 1945.

For the last 10 years American cities have been asking Washington for fiscal aid with their transit problems. The government's response was inadequate for metropolitan needs, and 16-fold was allotted to highways as to mass transit systems. Highways were politically popular and the auto-makers' lobby in Washington used its influence against mass transit schemes. But rapid transit has not been dropped as an answer to US urban commuters' needs. Its exponents point out that it uses a separate right-of-way, it does not produce lead or smog, and it can be made financially feasible. Its problems are not primarily technological but political and financial.

Only five rapid transit systems operate in America today—in New York, Boston, Cleveland, Chicago, and Philadelphia. The newest of these began half a century ago. But current plans will add to these systems, on the premise that a well-planned scheme can inject new economic vitality into deteriorating inner cities. The five existing rail-transit systems are being extended and new ones are being built or planned for nine other metropolises.

One argument of the American Transit Association is that one rail track can move 70,000 persons an hour. One exclusive bus line can transport 40,000 riders per hour, but only 4,500 motorists can travel on one lane of a motorway in an hour. In Chicago 138,000 persons leave or enter the downtown area daily by underground, elevated trains, and commuter railroads. If all these people were to travel by private cars, the city would have to build (after finding room) 140 additional motorway lanes, 70 in each direction.

The first completely new rail-transit system since 1905 was planned for the spring of 1972 in San Francisco, a 75-mile network to tie together three counties (with 2,500,000 people), using trains moving at 80 miles per hour. If successful, this could stimulate underground transit in many other communities. The total cost of $1,300,000,000 is financed through a bond issue, the federal government, state legislature, and passenger fares ranging from $.25 to $1. Known as BART (Bay Area Rapid Transit), it cuts travel time between San Francisco and Oakland from 45 to nine minutes, by carrying 200,000 passengers per day and reducing peak-hour commuter congestion through major intercity corridors by as much as 50 percent. Eventually serving nine counties, whose population is expected to reach over 7,000,000 within 20 years, it will be the world's first completely automated transit system.

Work has also begun on a rapid-transit system for the nation's capital. Washington and Peking are the last of the world's major capitals to lack such a system. London in 1863 was the first. The first US underground came in Boston in 1898. Ground was broken at the end of 1969 for the first Washington station, with the entire plan to be completed by 1980, and thus serve Washington's 3,000,000 people. Other cities now planning new rapid-transit lines are Atlanta, Detroit, Kansas City, and Baltimore. Under an Urban Mass Transportation Assistance Act passed in 1970, over $3,000,000,000 is authorized for mass-transit aid by 1975, with the federal government providing up to two thirds of the funds.[6]

It is hoped that these new schemes will prove attractive alternatives to private car use, but some critics claim that proposed funding is inadequate, in view of the fact that for every baby born today, two new cars are produced. The nation's transit systems have already filed with the government requests for grants, but it is highly doubtful that they can be met. The new Department of Transportation's powers are limited by Congress. Despite the excitement generated by monorails and vehicles floating on cushions of air, the future probably rests with four-wheel cars moving on steel rails. Unless the government provides sufficient financing, the future of mass transit is bleak.

Considerable sums have been spent in America on building new major roads. Once a motorist leaves the larger metropolis, he leaves behind problems of parking, congestion, and traffic delays. In 1956 an Interstate Highway System was started with federal funds. It envisioned a nationwide network of modern high-speed motorways, 40,000 miles of roads to cost $27,500,000,000 over 10 years. The system has now been expanded to 42,500 miles and the cost will be $77,000,000,000 on completion, scheduled for 1978.[7] It already offers vast stretches of wide highway, free of intersections or traffic lights, the goal having been for a motorist to drive coast to coast without a red light. Washington pays 90 percent of the cost of this new road system, plus half the cost of primary roads that need modernizing.

In the cities there is sharp resistance to building more motorways through established neighborhoods, and much controversy looms ahead over funding distribution for roads vis-à-vis mass-transit systems. Recent estimates are that US metropolitan centers will require $123,000,000,000 over the next two decades to solve their transport problems— $90,000,000,000 for streets and motorways and $33,000,000,000 for mass-transit systems.[8]

Toll charges, plus federal taxes on gasoline, oil, and tires defray the cost of motorway-building. The nationwide motorway system reduces accidents and cuts down on long-distance driving time, but many engineers claim that it is the city roads that are in most need of improvement and hence more funds should be allocated to them. The immediate crisis is, in fact, localized in 10 cities where the need is for a bold plan to integrate rapid transit systems, air lines, and highways. At present, this does not exist, but planners are aware of the urgent necessity for an integrated approach to national transportation problems as a unified whole.[9]

It is this intensified awareness of the need for planning and action that, as with so many US problems, affords grounds for hope. The official recogni-tion of the situation and the need to cope with it, and to bring to its alleviation imaginative, innovative results of research and technological development, augurs well for the future. If adequate funding can be combined with far-sighted planning, the movement to and from US cities will be made safer and less stressful for their inhabitants.

Notes

[1] John Burby, *The Great American Motion Sickness*, Boston: Little, Brown, 1971, Chapter 1.
[2] "Transportation Needs a Drastic Overhaul," *Business Week*, November 14, 1970, pp. 68-69.
[3] E. Langer, "Auto Safety: New Study Criticizes Manufacturers and Universities," *Science*, January 21, 1966, pp. 277-279.
[4] N. Owings, "Mass Transit and the Cities: Mobility and Place in America's Future," *Current History*, August 1970, pp. 95-99.
[5] "Mass Transit: What Works and Why," *American City*, December 1969, p. 8; "Transportation and Cities," *Science News*, January 25, 1969, p. 91.
[6] Grant Davis, *The Department of Transportation*, Lexington, Massachusetts: Heath, 1970.
[7] "Traffic Jam: Arthur D. Little, Inc. Questions Goals of Interstate Highway System," *Reporter*, April 7, 1966, p. 14.
[8] L. Schneider, "Fallacy of Free Transportation," *Harvard Business Review*, January 1969, pp. 83-87; M. Schmertz, "Strong Voice Is Raised for Integrated Transportation Planning: International Conference on Urban Transportation," *Architectural Record*, April 1966, p. 125.
[9] W. Wagner, Jr., "Are Highways Seven Times More Important Than Cities?" *ibid.*, May 1969, p. 9.

Transit Strategies
for Suburban Communities

Robert R. Piper

Transit is asked to cure a variety of ills in modern, low-density cities: isolation of those who cannot drive, traffic congestion, air pollution, energy waste, and inner-city decay. Transit planning is difficult because of the wide dispersion of origins and destinations and the temporal rhythms of travel. The options in deploying buses are reviewed relative to those attributes that attract passengers: accessible service that goes directly from here to there without taking too long. The planner who understands these fundamental requirements can both evaluate transit proposals and modify land use plans to facilitate transit service.

American urban areas built up since World War II exhibit a dispersion of land uses that has made vehicular travel almost indispensable in reaching both the necessities and pleasures of life. Low-density, scattered cities have become the norm.

Urban sprawl developed simultaneously with the growth in ownership of automobiles that made it possible. The proliferation of automobiles brought other consequences: roadway congestion, petroleum consumption, and several varieties of environmental degradation. Sprawl also had unhappy consequences for public transportation and for those unable to travel by automobile: the young, the old, the poor, members of one-car families when the car is elsewhere, and others. Public transit deteriorated. With the decline of public transit, those without access to automobiles became effectively isolated from many social activities.

Planners are under increasing pressure to belatedly fit these automobile oriented communities with transit, both to provide mobility to the isolated and to reduce the unwanted effects of the automobile by converting motorists into transit riders. The purpose of this article is to assist planners in responding to these pressures.

Transit productivity
and the energy crisis

This article originated as a study of energy conservation. The oil embargo of 1973–74 showed that there are few alternatives in conserving transportation energy:

1. Make fewer trips; this was the solution widely adopted during the embargo
2. Adopt vehicles that consume less fuel; rising fuel prices and government pressure are pushing the vehicle market in this direction
3. Car pool; car pooling has yet to become popular,

Dr. Piper received his M.B.A. and Ph.D. degrees from the Stanford University Graduate School of Business where he specialized in transportation, performing research on transportation, land use, and energy conservation. He participated in designing dial-a-ride transit systems in California. Previously employed in marketing and administration with the British Columbia Bureau of Transit Services he recently became director of transportation for the City of Berkeley with responsibility over parking, traffic engineering, and transit.

largely because it negates attractive features of automobile travel, such as independence of movement and privacy

4. Expand transit service; for this conservation strategy to succeed, the additional energy expended by transit must be less than what the passengers who switch to transit from automobiles would otherwise have consumed.

Barring major technological advances, the diesel bus is the primary contender for transit duties in the scattered city. The only proven electrically powered vehicles on the market require expensive (capital intensive) overhead wires (trolley buses) or guideways (rails, etc.) or both, such that they are suitable only in heavily traveled corridors.

Analysis of bus fuel consumption is complex. For example, a bus deadheading or on a lightly loaded backhaul burns fuel just as irretrievably as when rolling full. In addition, an analyst must distinguish *switch riders*, those who switch from a less efficient mode, from *captive riders*, those who would not have traveled at all in the absence of transit. No direct energy savings can be ascribed to captive riders; however, they may be the raison d'être of the transit system. Marginal switch riders conserve almost all their driving fuel if they occupy excess capacity in a bus that is running anyway to carry captive riders. Conversely, when the bus is full, marginal captive riders cause a corresponding fuel *penalty* if they displace a switch rider. If another bus is put in service to handle an overload, the fuel that it consumes must be considered.

Happily, these niceties have little relevance in planning transit that is intended both to serve the autoless, captive riders and to attract switch riders, whether to conserve fuel or for other reasons. People in both groups react similarly to service characteristics. Both have alternatives to riding transit. The motorist can drive. The nonmotorist can forego making the trip, beg a ride from a friend, hitchhike, take a taxi, or resolve travel desires in other ways. If all the autoless in society were truly captive, there would be far fewer empty seats on transit buses than one sees today.

Filling bus seats with riders attracted from other alternatives is not easy in the scattered city (Ward and Paulhus 1974, Regional Plan Association 1976). Transit must cope with a spatial pattern of travel demand that is from everywhere to everywhere and a temporal pattern that makes capacity inadequate at one moment and surplus the next. The bus itself can move along no faster than other traffic—except in a few instances where special lanes or shortcuts have been provided.

There are only a few alternatives in deploying buses. The most common is the scheduled, fixed route bus. Its conventional variants include local, express,

and skip stop service. A specialized form is the commute bus or bus pool; examples include school buses, commuter club buses, and suburban transit routes scheduled only at the peak times. A deployment not tied to fixed routes is dial-a-ride, a concept that attracted attention during the late 1960s and still survives in a few locations.

Conventional, fixed route service

Conventional, fixed route services are those that follow fixed routes on regular and reasonably frequent schedules. With fixed route service, supply (buses on the street) is offered whether demand materializes or not. It is offered because the existence of demand (passengers waiting at stops) cannot be predicted in advance of any run and the service must be reliable in order to attract patrons. Success of fixed route service is measured in terms of the demand (number of passengers) that springs forth in response to the supply.

Volumes have been written to the effect that demand for transit service does not materialize because people "don't know any better." Neanderthal transit managements, the story goes, spend far too little on publicity with the result that people are not even aware of the travel opportunities that are available. While this line of reasoning has some merit, word-of-mouth advertising counts for something in American society. It has had plenty of time to draw people to transit. A more likely explanation why suburbanites prefer the automobile to transit is that they do know what transit offers and it does not offer enough. A look at suburban transit systems shows that this perception is accurate more often than not.

Typical fixed route services

The fixed routes of primary interest are those whose buses run back and forth along arterial streets, serving all local stops. In old town centers, the arterial tends to be the original main street, sometimes with another at right angles. The residences are distributed over a predominantly rectilinear grid of streets with frequent connections to the arterial(s). As main streets became congested in the 1950s, a new concept of arterials emerged, one permitting smooth flow of through traffic. In some variants, the arterial was lined with commercial establishments (more precisely, their parking lots) which formed a buffer between the street and residential areas. In more recent developments, the arterial is a limited access street fronted directly by neither commercial nor residential development.

The residential subdivisions that have sprouted around the new arterials are different from the old ones. Access to the arterial is by collector streets, from which branches of feeder streets spring, providing the links to residential culs-de-sac. Initially, homes and

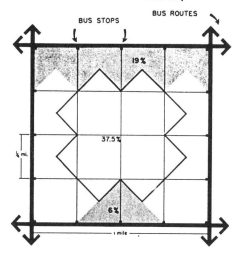

BUS STOPS
BUS ROUTES

19%

37.5%

¼ mi.

6%

1 mile

driveways would front on all street categories, but the trend has been increasingly to sterilize the collector streets; in new subdivisions, homes front only on feeders and cul-de-sac streets.

When land is developed for apartments or condominiums, a similar, though less pronounced trend to shun the street prevails. Modern, garden apartment and condominium complexes are focused inward on courtyards (with swimming pools, playgrounds, etc.); parking lots and the back sides of buildings face the street. Those complexes that border arterials often include units a quarter mile away, and complexes are increasingly being developed off collector rather than arterial streets.

Whether the arterials are new or old, they are frequently separated by a mile or more, when parallel, and lead to different activity centers. Dire consequences for fixed route bus service result. Assume that patrons will walk ¼ mile to a bus stop, and imagine a one-mile square, largely residential, bounded by arterials along which fixed route buses serve stops every ¼ mile (Figure 1). In an old neighborhood with a rectilinear street pattern, only ¹/₁₆ (6 percent) of the area is within walking distance of each bus stop. Over 37 percent of the area is beyond walking distance of any stop. If the arterials are bordered by commercial development, then the fraction of the interior residential area accessible to bus stops becomes smaller still. Less than 19 percent of the area is within walking distance of stops served by each bus route. Each route serves directly only the activity centers that it passes.

In brief, each route serves less than 20 percent of the area and competes for only a fraction of the trips that originate in that 20 percent. The situation is even worse in more modern developments, where passengers must walk along a maze of curvilinear feeder and collector streets (without sidewalks in many cases) to reach the arterial. In some locations, walking distances to the arterials could be reduced by providing walkways, but many homes and activity centers would still remain far from bus stops.

It is partly because of stop accessibility problems that some transit operators have resorted to routing buses off the arterial, through the residential area, and then back onto the arterial. There results a meandering route pattern, the most common being a loop, along which there are sometimes journeys that can be effected as rapidly on foot as by bus. Exasperation of through passengers, from a point on the arterial before the diversion to another further along, is inevitable. These meandering routes rarely can be considered serious competition to the automobile.

Underlying these observations are two assumptions about people: (1) the well-documented characteristic that they do not walk far to and from transit stops (Piper, Chan, and Glover 1976) and (2) the reasonable presumption that they compare a trip by transit with the best alternative, the same trip by automobile. In other words, a person is unlikely to patronize transit if the stops are too far away from his origin or destination or if the routing is so circuitous as to be markedly longer and more time consuming than a direct trip by automobile. Other service attributes such as predictability, safety, and vehicle condition play important roles but usually lie beyond the scope of a planner's authority.

In transit planning, the requirement for direct routing frequently conflicts with the requirement that the stop be close to where the action is. To serve a large shopping center, for example, a bus may have to make a two to five minute detour from a direct route, to the detriment of through service. The situation is worse still if a senior citizen center is located, as they often are, well away from the beaten track where the rest of society travels. When confronted with land use patterns that present such conflict, the transit planner is stymied. He cannot serve both the activity and the through passenger well.

Express buses

Local bus service is frequently overlaid with express bus routes that link major activity centers. When a motorist has a long distance such as ten miles to travel, the driver will seek to minimize travel time by using high speed roads. The same strategy is followed by transit agencies desiring to compete for these long trips. Express buses do so by running along high speed roads of one sort or another for as much of the

382

line haul as possible and by making few stops en route. The success of express buses is determined by factors similar to those influencing local service: (1) the time dissipated between the line haul segment of the trip and the origin or destination stops and (2) the ease of passenger access to the stops.

Just as subdivisions and commercial developments are commonly laid out in ways that frustrate local bus service, highways and their interchanges or intersections are seldom constructed in ways that facilitate express bus service. Interchange designs incorporating bus pull-offs and space for stations exist on paper but few have been converted into concrete (Wilbur Smith 1966). To function properly, such pull-offs require access by pedestrians from the surrounding area, from feeder buses, and from park-and-ride or auto drop-off facilities.

Shuttles and bus pools

Two special cases of fixed route buses are shuttles and peak period only commute services. Both follow fixed routes on published schedules but they are tailored to serve specific market segments. The shuttle bus travels back and forth along a direct route linking two or more major activity centers. Shuttles are viable where a substantial travel demand between or among these centers persists through the day. Though shuttles are usually associated with central cities, suburban applications exist. Examples include links between different shopping centers or between residential clusters and activity centers.

Commute services normally offer only a few trips, inbound from residential areas to employment centers in the morning and outbound in the evening. In some systems, passengers are picked up at stops along residential streets, a sort of bus pool. In other cases, the buses rendezvous with commuters at park-and-ride lots. The commute bus suffers a high cost per passenger when, as is frequently the case in transit operations, both bus and driver are dedicated solely to transporting one coachload of regular riders. Even if the bus and driver are idle during the midday, the driver, under conventional labor agreements, is paid as though he had worked a full shift; he may even receive a split shift premium if his workday extends beyond the contractual maximum.

Few operators offer shuttle or commute service unless sufficient demand materializes to fill all or most of the seats, at least in the predominant direction. Both attract passengers because the buses lose little time at stops and follow direct routes, thus duplicating automobile travel closely.

Valuable though shuttle and commute services are, they compete for only a tiny fraction of the travel demand in the scattered city. For the vast majority of suburbanites, fixed route service, whether local or express, suffers one or more of the following drawbacks:

1. People cannot get to a bus stop
2. The bus that people can get to does not take them where they want to go
3. The bus takes too long to go anywhere.

These drawbacks long have been recognized and provided the rationale behind dial-a-ride, which was an attempt to solve the problems of walking distance and dispersed destinations.

Dial-a-ride

Dial-a-ride was intended to overcome people's reluctance to walk far to and wait at transit stops. It was billed as a door-to-door service that would enable transit to compete with the automobile. The passenger would request transportation by telephone and transit would respond by sending a bus to the door. Passengers would share the ride with others over varying portions of one another's journey.

The concept has borne a variety of names: dial-a-ride, dial-a-bus, demand activated transit, telebus, and others. It has been offered over a wide range of sophistication. In the simplest, the driver tours a limited, geographic zone, returning to a central point at regular intervals for a new dispatch sheet that lists the pick up and drop off addresses of the next tour. Addition of radio communication enables the dispatcher to insert another passenger pick up address while the bus is under way. In the most sophisticated systems, the bus is continually on tour and entirely radio directed; at each stop the driver is informed of the next stops (either pick up or drop off) on the itinerary. This last version can offer a broader selection of origins and destinations and a quicker response than the others.

The initial enthusiasm for dial-a-ride soon faded, for reasons that become evident from analysis of how it works. Dial-a-ride can eliminate the walk to and from the bus stop only by sacrificing direct delivery of passengers to their destinations. Because passenger pickup and drop off points are randomly distributed geographically, the bus must follow an erratic path. It proceeds only part of the time in the direction desired by each passenger. To make matters worse, bus dwell times in picking up passengers at random locations are much longer than those of a fixed route bus. Dial-a-ride envisages front door pickup, with the passenger waiting inside in comfort and security. Not all passengers are ready to board when the bus arrives. Even if they were, the walk from door to bus takes time; for example, one cannot expect elderly patrons to charge out to the bus with the verve of O. J. Simpson heading for a first down.

Largely as a result of circuitous routings and the slow average speeds imposed by long dwell times, dial-a-ride attracted fewer passengers than originally hoped. Some would-be patrons also disliked the delay and uncertainty in bus arrival time after telephoning for service. Patronage was low and costs per passenger were high. The reasons for poor performance are explored in the appendix. Not only is dial-a-ride bus productivity low but labor productivity suffers relative to fixed route service because an additional employee is required in the control room for every five to ten vehicles on the street. From an energy standpoint, productivity is so low that a dial-a-ride bus system would consume up to twice as much fuel as would be used by the switch passengers in automobiles.

While dial-a-ride is an unproductive deployment of buses and personnel, the concept still has two valid roles. One is to improve the productivity of taxicabs in communities where ride sharing is legal. The other is for transportation of people with physical handicaps which prevent them from traveling to and from bus stops or boarding and leaving regular buses. Door-to-door service is appropriate and low productivity is inevitable in transporting the handicapped.

Transfers

The walk to the bus stop and wait there that dial-a-ride promoters seek to eliminate are only two drawbacks of fixed route service. An even greater disadvantage is: the bus that people can walk to does not go where they want to go. The route may intersect another that does lead to their destination, but transfers from one to the other are haphazard. Many transit operators optimize route schedules individually rather than as components of an integrated network. The transit route map may look as though the community is well served, but in practice, very few of the from-everywhere-to-everywhere movements typical of suburban travel can be made reliably or conveniently.

In a few transit systems, a different type of network has been tried featuring timed transfer focal points (TTFP). *Timed* indicates that vehicles (buses or other transit modes) are scheduled to meet at a focal point where sufficient space and facilities exist. During the brief period that all the vehicles are congregated, all feasible transfers can be made with very little walking or waiting. The focal point is, ideally, at a prominent center of activity. A sort of cobweb route structure results; routes from different directions come together at the focal points or nodes. Local, fixed route buses feed all nodes and link the minor ones. Express buses link the major nodes.

In timed transfer systems, the interline transfer is the keystone around which schedules are built rather than being a fortuitous by-product. The range of destinations conveniently accessible to transit passengers is vastly expanded because they can easily and reliably transfer from one route to another. The competitive posture of transit vis-a-vis the automobile is greatly strengthened.

TTFP systems require more planning skill and effort than do routes that are optimized individually. Delays at the transfer point must be kept to a minimum; otherwise, the journey by transit will take too long relative to automobile travel. TTFP also requires stringent schedule adherence; driver supervision must be tight. For these reasons, old line transit managers may balk at the prospect of operating TTFP; however, the rewards of service integration are substantial if the concept is properly implemented. Ridership on systems offering TTFP has been significantly greater than on alternative systems in similar communities (Boleen 1976).

Conclusion

Suburban transit systems vary greatly in attractiveness as links between residential neighborhoods and activity centers. These variations are reflected in system productivity, such as passengers or passenger miles per employee year. Fuel efficiency in passengers or passenger miles per gallon approximately parallels productivity. Productivity and fuel efficiency suffer from deadheading, low-demand backhauls, and empty seats at the outer ends of residential routes.

In the conventional scattered city, where activity centers are widely separated, a hierarchy of services must be offered to meet the demand for everywhere-to-everywhere movements. Stops must be located within walking distance of activities. To be perceived as competitive with automobile travel, each route must be direct, with a minimum of detours. If the services are well integrated, the range of alternatives open to transit patrons is greatly expanded. The key to service integration is the transfer. On a well-integrated system, both productivity and fuel efficiency will be greater than the same personnel and equipment deployed on independently optimized routes and services.

Dial-a-ride cannot generally be justified as a transit strategy on either productivity or fuel efficiency grounds. It is appropriate in transporting handicapped persons and can be used to improve the productivity of shared ride taxis.

Appendix.
Dial-a-ride productivity analysis

The productivity of scheduled, fixed route bus service is limited by demand: how many people are willing to walk to the stop, wait, and ride. The productivity of dial-a-ride, by contrast, is limited by constraints imposed on supply. The dial-a-ride bus, in order to attract patronage, is dispatched so as to deliver passengers to their destinations within an elapsed time chosen to be reasonably competitive with automobile travel. This quality of service constraint can be met only by limiting the number of passengers served.

In practice, the bus usually is dispatched so the trip will last no longer than some factor (such as two) times the automobile travel time from origin to destination. Even if the bus speed is assumed as fast as an automobile in traffic, this task is complicated because the bus travels further than the automobile as it makes detours to serve other passengers. It also loses dwell time at stops where other passengers board or alight. Mathematically, the sum of the bus time in transit and the dwell times must be less than some multiple of the automobile travel time.

$$\left(\frac{rL}{v}\right) + 2n\left(\frac{t}{60}\right) \leq q\left(\frac{L}{v}\right)$$

where:

L is the average passenger trip length in miles by the most direct route;

n is the number of passengers served as the bus traverses L;

q is the quality of service factor: the ratio of elapsed time on the bus to automobile driving time between the journey end points;

r is the deviation factor: rL is the distance traveled by the bus in traversing L;

t is the average dwell time in minutes per passenger per stop; and

v is the average speed (m.p.h.) in traffic.

The relation assumes an equilibrium passenger load such that one passenger boards and another alights at succeeding stops. The maximum number of passengers served in traversing L is then:

$$n \leq \left(\frac{30L}{tv}\right)(q - r).$$

The productivity, P, or number of people carried per hour is:

$$P \leq \left[\frac{30}{t}\right]\left[1 - \left(\frac{r}{q}\right)\right].$$

The operator has little control over the deviation factor, r. For a rectilinear street network, the random walk factor $r = \sqrt{2} = 1.41$ typically prevails. In modern subdivisions with few access points to

the arterial street system, r can be substantially greater. On a cul-de-sac, for example, $r = 2$.

The quality of service factor, q, is under the operator's control but demand slackens as q grows much greater than 2. How many customers will the service attract at $q = 3$, which means that a ten-minute trip by car takes half an hour on the bus?

Dial-a-ride operators have tended to focus on dwell time, t. In many-to-many service, a separate stop is made to pick up and to let off each passenger. If they are picked up separately but all go to one destination (*gather*), only one stop is necessary for all to debark. If they all leave from one origin (*scatter*) the average boarding dwell time is greatly reduced. Dwell times for pickup are the most critical as the bus waits for the passenger to take notice, walk out, and board. The time savings are greatest, therefore, in scatter. In computing systemwide averages, however, round trip travel by each passenger must be assumed; as a result, every scatter assignment is matched by one or more gather assignments.

Based on conversations with dial-a-ride operators, good average values of t are 30 seconds in many-to-many service and 20 seconds for a combination of scatter and gather. For a bus serving regular subscribers, t may diminish further; however, subscription service, like the fixed route commute bus, is usually viable only at peak hours and does not greatly improve systemwide performance. When dial-a-ride serves handicapped persons, dwell times grow significantly. With wheelchair lifts a t of 2 minutes is not unusual.

If $r = 1.41$ and the operator sets $q = 2$, the best productivity achievable is about eighteen passengers per hour in many-to-many service and twenty-six in combined gather and scatter. A service for handicapped persons would move fewer than eight persons per hour in many-to-many service. These numbers represent a sort of theoretical maximum in dial-a-ride performance. To the author's knowledge, productivities of actual dial-a-ride systems have rarely averaged as much as 50 percent of these levels. The analysis is optimistic.

The same analysis extended to fuel efficiency yields theoretical maximums of fifteen passenger miles per gallon in many-to-many service and twenty-five to thirty in combined gather and scatter. These computations assumed an average length of four miles for suburban trips (Wilbur Smith 1966).

Actual performance would be about half as good: eight to fifteen passenger miles per gallon. This range is typical of that achieved by automobiles on short trips (Hirst 1974). Given that not all dial-a-ride passengers are displaced from automobiles, one must conclude from a narrow, fuel conservation standpoint that dial-a-ride buses are best left parked in the garage.

References

Boleen, G. 1976. *The timed-transfer focal point concept.* Unpublished. Faculty of Commerce and Business Administration, University of British Columbia.

Hirst, E. 1974. *Direct and indirect energy requirements of automobiles.* Report ORNL-NSF-EP-64. Oak Ridge, Tenn.: Oak Ridge National Laboratory.

Piper, R. R.; Chan, E. K. Y.; and Glover, R. S. 1976. Walking distances to bus stops. *Proceedings, seventeenth annual meeting, Transportation Research Forum* 17, 1: 307–312.

Regional Plan Association. 1976. Where transit works: urban densities for public transportation. *Regional Plan News* 99, August: entire issue.

Ward, J. D., and Paulhus, N. G., Jr. 1974. *Suburbanization and its implications for urban transportation systems.* Report DOT-TST-74-8 prepared for U.S. Department of Transportation, Assistant Secretary for Systems Development and Technology. Washington, D.C.: U.S. DOT.

Wilbur Smith and Associates. 1966. *Transportation and parking for tomorrow's cities.* New Haven, Conn.: Wilbur Smith and Associates.

Highways as a Barrier to Equal Access

By YALE RABIN

ABSTRACT: There is a widening gap between growing concentrations of blacks and other minorities in the central cities, and whites and the expanding supply of employment opportunities in the suburbs. While exclusionary zoning controls have been seen by many as the most immediate barrier to suburban opportunities, transportation facilities and the lack of them play an important role. The federal highway program in particular, while a powerful stimulus to dispersed development, has, in its implementation, failed to protect equal access to the benefits of development such as housing and employment—benefits often made possible entirely by the provision of highway access where none existed before. As a result the comprehensive planning of metropolitan areas is seriously undermined, and new barriers are erected which threaten to perpetuate the burdens and disadvantages which a long history of racial discrimination has produced. Regional planning agencies are needed with adequate authority to make and implement integrated land-use and transportation decisions based on a clearly expressed metropolitan development policy that includes the goal of eliminating all barriers to equal access. In the interim the discriminatory aspects of current transportation policies and projects should be challenged.

Yale Rabin, member of the American Institute of Planners, has been a consultant to the NAACP Legal Defense Fund and other community legal services agencies, as well as to the U.S. Department of Housing and Urban Development. He recently served as consultant to the U.S. Commission on Civil Rights in a study of the impact of the Federal-Aid Highway System on low-income and minority groups.

THE sprawling decentralization of this nation's metropolitan areas in the period since World War II has opened new opportunities for improved housing, employment, and schools for millions of Americans, predominantly whites, while restricting both the range of opportunities and the quality of life for blacks and other minority groups who are being relegated in steadily increasing numbers to a growing dependence on the diminishing resources of the central cities. The result is a growing polarization—racially and economically—which is persistently self-reinforcing, and which threatens to perpetuate the burdens and disadvantages which a long history of racial discrimination has produced. Many, if not most, of the great concerns generally characterized as urban problems are probably attributable to the nature of these metropolitan changes.

This polarization is a pervasive aspect of the continuing process of metropolitan decentralization and is a major condition resulting from that process. It therefore appears reasonable to assume that an effective strategy for altering this condition must deal directly with the process by which it continues to be generated. While it is recognized that a great many diverse forces contribute to the process of decentralization, there is much evidence to suggest that transportation policies, programs, and facilities play an important role.

Based on the judgment that the effects of racial and economic polarization are grave enough to require effective remedial action, this article examines the decentralization-polarization process and those aspects of the process that are transportation related, and discusses the implications of that relationship for present and future strategies for change. The elements of the discussion are:

1. The extent of black [1] concentration in central cities, and the dispersal of whites and jobs to the suburbs, are increasing. This process has resulted, and continues to result, in reduced access to growing employment opportunities in the suburbs for inner-city blacks.

2. This reduced access is caused by both the dispersed locations of jobs in relation to black central city concentrations and the dependence for access to those jobs on automobile ownership. These in turn derive from the transportation policies and programs of the federal government, and in particular the authorization and funding of the multi-billion dollar system of metropolitan area highways, which have exerted a major influence on metropolitan dispersal.

3. State highway departments, in designing and constructing the highway systems financed by federal and state funds, have played a major role in determining the spatial distribution of suburban development. Decisions by government agencies at the federal and state levels which determine and approve the locations of highways and their points of access have been and continue to be made without regard to their impact on the redistribution of housing and employment opportunities or the comprehensively planned development of metropolitan areas.

4. In implementing these highway programs, agencies of government have, by providing new or improved access, created billions of dollars in new land values, enriching land owners and developers, and adding substantially to the tax revenues, and consequently the amenities, of countless suburban municipalities; but have failed to take any steps to protect equal access to benefits

1. In this article the term "black" is used to describe blacks, Puerto Ricans, Mexican-Americans, and American Indians.

such as housing and employment. Pending projects should be reexamined in this light.

5. While restrictive land-use controls are the most apparent obstacle to low- and moderate-income housing in the suburbs, their removal will not of itself produce housing accessible to low-income blacks, or halt the decline of central cities, or alter the pace and pattern of employment and population dispersal, or create access between existing housing and existing jobs. These changes will require a metropolitan mechanism capable of: making and implementing land-use and transportation decisions in order to locate and provide new low- and moderate-income housing in relation to and in proportion to job opportunities; channeling the growth of centers of employment and residence; and improving transportation access between those centers.

While these circumstances require changes in plans and planning, there are immediate problems that must be the concern of lawyers and lawmakers. Vigorous attempts need to be made to protect the rights of low-income and minority groups under existing transportation programs; and new laws must be enacted that enable the necessary changes in planning to take place.

It is recognized that transportation facilities and policies are not an isolated force and that other public programs have facilitated dispersed development.[2] The focus is on transportation because it is a major factor in dispersal and because transportation facilities are shaped by government policies and paid for by public funds.

2. Probably the most significant of these have been the Federal Housing Administration (FHA) mortgage insurance program and the water and sewer grants program of the Department of Housing and Urban Development.

It is not intended here to make a qualitative judgment about decentralization per se. The ecological and other considerations necessary for such an evaluation go far beyond the scope of this article. Nor is it intended to suggest that racial and economic polarization are inevitable consequences of decentralization. Decentralization, assuming the ability to alter some of the underlying forces, could conceivably result in patterns of spatial, social, and economic distribution substantially different from those that characterize the changes now taking place.

METROPOLITAN DECENTRALIZATION
AND POLARIZATION

Racial concentration

The most persistent aspect of metropolitan change is the growing concentration of blacks in the central cities and the continuing exodus of whites and jobs to the surrounding suburbs. The percentage of central city population that is black has doubled since 1950. Between 1950 and 1960, the percentage of blacks living in the central cities of Standard Metropolitan Statistical Areas (SMSAs) increased from 12 percent to 18 percent. In the decade between 1960 and 1970, in the sixty-six SMSAs having populations of 500 thousand or more, the central cities lost 1.92 million whites, while black population there increased by 2.811 million to 10.82 million. In the suburban rings, white population grew by 12.468 million to 54 million, while black population increased by 762 thousand to 2.577 million. The percentage of black population in suburban rings increased from 4.2 percent in 1960 to 4.5 percent in 1970, while the percentage of blacks in the central cities increased from 18 percent in 1960 to 24 percent in 1970. Of the 2.5 million blacks who lived in the suburbs, almost a quarter live in cities of 50 thousand or more which are located within suburban rings.[3]

In the Philadelphia SMSA, for example, the number of blacks in the suburban ring rose by 48 thousand, increasing the percentage from 6.1 percent in 1960 to 6.6 percent in 1970. However, the suburban ring in the Philadelphia SMSA includes the older cities of Burlington, Camden, and Gloucester in New Jersey; and Bristol, Chester, Coatesville, Conshohocken, and Norristown in Pennsylvania. These cities, some of which have higher rates of unemployment than does Philadelphia, accounted for almost two-thirds of the increase in black population in the seven suburban counties in the SMSA. Consequently, the black population that gained access to the newer developing communities of the suburban counties comprised less than 3 percent of the over half a million increase in those places.[4]

In the thirty-one SMSAs with a population of 1 million or more in 1970, 85.5 percent of all black households earning under $4 thousand per annum lived in the central cities. For whites, 53.6 percent of households earning under $4 thousand a year lived in the suburban ring. In the category of $10 thousand or more annually, 30.9 percent of whites and 76.8 percent of blacks remain in the central cities.[5] The high percentage of low-income white households in suburban areas is not necessarily an indication that barriers of cost do not deter whites. It is more likely the case that a substantial portion of this low-income white group are long-time residents of the older cities located within the

suburban ring and not recent arrivals to the newer expanding suburban communities.

Between 1960 and 1970, the increase in the proportion of city population that is black was almost five times as great as the increase in the proportion of suburban population that is black. And, much of the small gain in the suburbs is offset by the many blacks whose new "suburban" housing is in places like Camden, New Jersey, or East St. Louis, Missouri.

Dispersal of jobs

The continuing departure of whites for the growing opportunities of the suburbs has been accompanied by a steady flow of industrial and commercial employment. In a study published in 1968, John Kain of Harvard University concluded:

By any measure metropolitan growth, since World War II, has been rapid but unevenly distributed. Outlying portions of metropolitan areas have been growing quickly while the central areas have been growing very little and, in an increasing number of instances have actually declined. During this period, what began as a *relative* decline became an *absolute* decline for a lengthening list of central cities. Losses in retail trade and property values, declining profits for central city merchants, and falling tax bases have usually followed from these employment and population declines. Moreover, depopulation was selective; the young, employed, well-to-do, and white moved to suburban areas leaving behind the aged, the unemployed, the poor, and the Negro.[6]

Kain found that between 1954 and 1963 the central cities of the forty largest metropolitan areas lost an average of 25,798 manufacturing jobs and

3. U.S. Bureau of the Census, Statement of Dr. George H. Brown, Director, before the U.S. Commission on Civil Rights Hearings on Barriers to Minority Suburban Access, Washington, D.C., June 14, 1971.
4. U.S. Census of Population, 1970.
5. Statement of George Brown.

6. John F. Kain, "The Distribution and Movement of Jobs and Industry," in *The Metropolitan Enigma*, James Q. Wilson, ed. (Cambridge, Mass.: Harvard University Press, 1968), p. 1.

that this loss was almost exactly offset by an average growth of manufacturing jobs in the suburbs of 25,948 jobs.[7] Significant declines were also found in retailing and wholesaling employment with the greatest percentage of decentralization occurring in wholesale employment.[8] Kain found that the data ". . . suggest an *acceleration* of postwar trends toward metropolitan dispersal."[9]

This forecast has certainly been borne out by more recent data. According to an analysis of 1970 census data published by the *New York Times*, the fifteen largest SMSAs in the country provided about 19 million jobs in 1960. Of these approximately 12 million were located in the central cities and 7 million in the suburbs. During the decade between 1960 and 1970, employment in the suburban areas of those SMSAs rose by over 3 million for a gain of 44 percent. During the same period, employment in the central cities declined by 836 thousand or 7 percent, and the central cities' share of total SMSA employment fell from 63 percent to 52.4 percent. By 1970, in nine of these fifteen SMSAs the number of suburban jobs exceeded the number of jobs in central cities; and for all fifteen SMSAs, 72 percent of all workers who lived in the suburbs were also working in the suburbs.[10]

When examined in greater detail, these changes are even more dramatic. In New York City between 1960 and 1970, the number of jobs dropped by almost 10 percent, a loss of almost 340 thousand, while employment in the suburbs increased by 353 thousand. In six

other cities employment fell by over 10 percent. In Detroit employment within the city fell by 23 percent, or 156 thousand jobs. In the Philadelphia SMSA the number of jobs in the city fell by 98 thousand or 11.3 percent between 1960 and 1970. By contrast, jobs in the suburbs increased during the same period by 61.4 percent, a gain of 314 thousand jobs.[11]

Adjusting for the one-year difference in time, omitting the disproportionate impact of New York, and assuming conservatively that only 75 percent of the loss was in manufacturing, the average job loss in the remaining fourteen cities for the 1960–70 period was over 20 percent greater than the average job loss that Kain found for the 1954–63 period.

More recent data covering the period from January 1970 to December 1972 indicate that the trend continues unabated. In each case there has been a decline in the city's share of SMSA employment, and in each case the rate of decline has been equal to or greater than that of the 1960–70 period.[12]

Inaccessibility

The nature and extent of racial polarization and employment dispersal are widely recognized, and disparities between the unemployment rates of blacks and whites are well known. Unemployment rates among blacks have for many years been approximately double the unemployment rates for whites. There are strong indications that, while a number of other factors including poor education, lack of skills, and racial discrimination contribute to higher unemployment rates among blacks, the inaccessibility resulting from polarization and dispersal plays an important role.

7. Ibid., pp. 19, 20.
8. Ibid., p. 16.
9. Ibid., p. 22.
10. Jack Rosenthal, "Large Suburbs Overtaking Cities in the Number of Jobs Supplied," *New York Times*, October 15, 1972. These figures appear to exclude employment in state and local governments.

11. Ibid.
12. See U.S. Department of Labor, Bureau of Labor Statistics, *Employment and Earnings* 17, no. 9 (March 1971); ibid., vol. 19, no. 8 (February 1973).

Several examinations of this relationship provide compelling evidence that this is so.

Kain, in a study published in early 1968, examined these relationships for the cities of Chicago and Detroit and found that distance imposed unreasonable cost burdens on centrally located blacks, and that public transit, because it focused on the central business district, was badly oriented for traveling from the ghetto to outlying centers of employment.[13] He examined the effects of residential segregation on black unemployment and the effects of dispersal on black unemployment and concluded:

While the estimates presented in this paper of Negro job loss due to housing market segregation are highly tentative, they nonetheless suggest that housing market segregation may reduce the level of Negro employment and thereby contribute to the high unemployment rate of metropolitan Negroes.[14] [Furthermore] . . . the empirical findings do suggest that postwar suburbanization of metropolitan employment may be further undermining the position of the Negro, and that the continued high levels of Negro unemployment in a full employment economy may be partially attributable to the rapid and adverse (for the Negro) shifts in the location of jobs.[15]

Subsequent conclusions by others have not been couched in such cautious terms.

A paper presented to the Conference on Poverty and Transportation in 1968 also acknowledged that existing public transit facilities were designed primarily to bring residents of outlying areas to the employment concentrations to be found in the central business district

and then noted:

In contrast suburban employment concentrations are being developed during an era of widespread private ownership of automobiles. They rely on their employees to commute to work by private automobile and generally are poorly serviced by transit, if at all. Central city residents, particularly low income residents and Negroes, may find suburban employment centers difficult or expensive to reach, because their incidence of private automobile ownership is relatively low. A normal expedient would be to relocate one's residence near the area of current or prospective employment. However, non-whites may be deterred from doing so by residential segregation. These factors interact so as to limit the number of job opportunities available to lower income Negroes.[16]

The author also found that distance reduced the level of information about job availability and that the scattered nature of suburban employment was a deterrent to job seeking.[17]

The National Commission on Urban Problems appointed by President Lyndon B. Johnson in 1967 examined five metropolitan areas: Baltimore, New York, Philadelphia, St. Louis, and San Francisco; and found that "commuting from the central cities of these five metropolitan areas to suburban jobs is both time consuming and costly."[18] They also found that "existing in-and-out commuter public transit systems are generally not suited to 'reverse commuting.'"[19]

A study of the Philadelphia area done at Villanova University in 1971 to ex-

13. John F. Kain, "Housing Segregation, Negro Employment, and Metropolitan Decentralization," *Quarterly Journal of Economics* 82 (May 1968), pp. 180, 181.
14. Ibid., p. 196.
15. Ibid., p. 197.

16. Edward Kalachek, "Ghetto Dwellers, Transportation, and Employment" (presented at the Conference on Poverty and Transportation, Brookline, Massachusetts, June 7, 1968), pp. 3–5.
17. Ibid., pp. 10, 13.
18. "Building the American City," Report of the National Commission on Urban Problems to the Congress and to the President of the United States, December 12, 1968, p. 48.
19. Ibid.

amine accessibility of suburban employment by public transportation found that only 11 percent of the trips from low-income, inner-city residential zones to suburban industrial parks can be made in less than forty-nine minutes and that 42 percent would take between seventy and eighty-nine minutes. In addition, 63 percent of all trips had weekly costs between $3.50 and $8.37. The remaining 37 percent ranged in cost from $10.20 to $16.00 per week.[20] Because the rate of automobile ownership in inner-city residential zones was 11.4 persons per automobile, most residents were dependent on other means of transportation.[21] It was concluded that ". . . people who reside in low-income residential areas of Philadelphia are virtually trapped because of lack of mobility in reaching regional industrial parks."[22]

Data on distribution of employment by place of residence in the fifteen largest SMSAs for 1970 also suggest that central cities are more accessible from the suburbs than the suburbs are from the central cities. Workers who live in cities hold 70.0 percent of all city jobs and 14.4 percent of all suburban jobs, while workers who live in suburbs hold 85.6 percent of all suburban jobs and 30.0 percent of all city jobs. Suburban residents hold more than twice the proportion of city jobs that city residents hold of suburban jobs; and 70 percent of all employment-related commuting between city and suburbs is by workers living in the suburbs. In addition, the share of city jobs held by suburban residents has increased at a much greater rate

20. John Collura and James J. Schuster, "Accessibility of Low-Income Residential Areas in Philadelphia to Regional Industrial Parks," Institute for Transportation Studies, Villanova University, 1971, pp. 6, 7, 28, 29.
21. Ibid., p. 41.
22. Ibid., Abstract.

than the share of suburban jobs held by city residents.

While it is not possible here to assign specific dimensions to the role played by inaccessibility in increasing the employment disadvantages of black city residents, it appears clear that access to suburban employment opportunities is substantially reduced.

Some related effects

The consequences of this seemingly inexorable polarization process produces secondary effects which tend to compound and reinforce disparities in opportunity and tendencies toward dispersal. Thus the departure of commerce, industry, and middle-class residents to the suburbs may reduce the city's tax revenues. To offset this loss, new taxes are levied or rates raised, adding further to the impetus to leave. Such cities, faced with the need to spread declining tax revenues over an expanding demand for services and facilities, are unlikely to be able to compete with the more affluent suburbs for competent personnel and may be forced to hire less-qualified applicants, or reduce the number of their employees. These cutbacks result in reductions in the level and quality of services—for example, in increased classroom sizes, less frequent street maintenance, or restricted library hours.

These effects simply add to the pressure for leaving and at the same time reduce the quality of life for those who must remain. The exodus of middle-class households then hastens the further departure of retail and consumer services establishments, once again reinforcing the cycle.

The situation in the city is further compounded by the scattered successes achieved in attempts to lower suburban cost barriers. Since these have the effect of slightly lowering the income thresholds necessary for access to the suburbs, they enable some households at the upper level of the economically restricted population in the city to leave, thereby increasing the proportion of low-income households which remain.

HIGHWAYS AND DECENTRALIZATION

Highways and suburban development

The forces at work in the process of decentralization, while diverse, appear to fall for the most part into two general categories: (1) those that tend to create pressures to disperse and (2) those that enable or limit decentralization and influence its characteristics. The first may include, among others, lack of space for expansion, high land costs, obsolete facilities, changing transportation needs, new transportation options, high taxes, poor services, inadequate housing, poor schools, fear of violence, and racial discrimination.

The second category includes suitable and affordable land to which necessary services are available, or can be reasonably made available, and which is accessible to necessary support services such as labor, materials, markets, housing, schools, shopping, and employment.

Transportation facilities are significant factors in both categories. However, it is in the second category that they assume a dominant role; for no matter how inexpensive or suitable land may be, and no matter how adequate the supporting services and facilities may be, to make development feasible, the land must be accessible. Adequate access, or the reasonable expectation that it will be provided, is essential if development is to take place.

The relationship between land use and transportation has long been understood, and the principle that both should evolve from a planning process that fits transportation facilities to travel needs generated by planned patterns of land use is a basic element of planning theory. However, as is so often the case, practice is another matter. In metropolitan areas it is the highway system which stimulates development. The National Commission on Urban Problems was emphatic on this point:

Probably there is no more important single determinant of the timing and location of urban development than highways. Highways in effect "create" urban land where none existed before by extending the commuting distance from existing cities. The low-density pattern found in most of the Nation's suburban areas would never have been possible without the effect of high-speed highways in reducing the importance of compact urban development. As highways stretch out from existing urban areas, development quickly follows, with even the most carefully considered plans and zoning ordinances rarely providing a match for the development pressures generated. The phenomenon of strip commercial development along non-limited access roads is one example of the irresistibility of such pressures.[23]

An understanding of the development potential of highways has not been limited to planners. Highway departments across the country have, over the past twenty years, produced dozens of studies that demonstrate the "beneficial" role of highways in stimulating development and increasing land values.[24]

A striking example of highway impact is to be found in Parsippany–Troy Hills, N.J., where five highways, three of them new interstates, will intersect. Population since 1950 has increased from 15,290 to 55,112 in 1970. The value of all property in the town has risen from $107 million to $483 million in the past ten years.[25]

23. "Building the American City," p. 231.
24. U.S. Department of Commerce, Bureau of Public Roads, *Highways and Economic and Social Change* (Washington, D.C.: U.S. Government Printing Office, 1964), pp. 204–21.
25. David K. Shipler, "New Highways

The principal incentive behind this growing system of metropolitan highways is the U.S. Highway Trust Fund [26] which provides about $5.5 billion annually in funds whose use is restricted by law to highway construction. These funds pay 90 percent of the costs for highways that are elements of the interstate system and 50 percent of the cost of U.S. primary highways, state highways, and urban highways.

Between 1956 and the end of 1972, over $55 billion in Highway Trust funds had been spent on the interstate system alone. By comparison, the 1970 Urban Mass Transit Assistance Act, with its authorization of $3.1 billion over a five-year period, appears as little more than a token gesture to placate the critics of poor transportation planning.

These restricted Trust Fund subsidies, and principally the 90 percent interstate contributions, are powerful incentives to the continuing construction of highways; and in the absence of adequate funding for mass transit or control by regional plans, they are also powerful incentives for continuing uncontrolled dispersal.

Highways and public transit

It has already been noted that existing public transit systems are generally not capable of meeting the journey-to-work needs of those who live in the inner city who might work or seek work in the suburbs. In addition, the great volume of auto commuting generated by highway-oriented patterns of dispersal has resulted in substantial reductions in patronage on existing public transit facilities.

Public transit reached its peak patronage of 23.3 billion passengers per year in 1945. By 1970 this figure had declined to 7.3 billion, 40 percent less than the 12.1 billion passengers per year recorded in 1912.[27]

Faced with greatly reduced revenues, some transit companies have ceased to operate, thus eliminating service entirely. In most cases the result has been severe cutbacks in service and substantial increases in fares—often by as much as 400 percent since 1950—burdens that are most heavily borne by low-income inner-city residents who are most dependent on public transit for access between home and workplace.

Not only does widespread affluence intensify the traffic problem by substituting automobiles for buses but this very reaction ricochets back and intensifies the poverty problem. With the majority of the urban area residents, and practially all of the suburbanites, commuting by car, too few mass transit users remain to support good service. Thus an increase in per capita income has improved the transportation position of the majority, except perhaps in the peaks of traffic congestion but has left the poor worse off. Elderly residents, physically unable to drive, are also left worse off. . . . Simultaneously with the decline of mass transit, manufacturing, retailing, and other activities have been suburbanizing. With suburban densities far too low to support the extension of the lines of even a healthy mass transit system, the elderly, those financially unable to own a car, those unable to drive and others find that dependence on the central city mass transit system has narrowed their employment opportunities very appreciably.[28]

Shaping Future of City's Suburbs," *New York Times,* August 19, 1971.
26. For more detailed discussion of the Trust Fund see Alan Mowbray, *Road to Ruin* (Philadelphia: Lippincott, 1969) ; Helen Leavitt, *Superhighway, Superhoax* (Garden City, N.Y.: Doubleday, 1970) ; and Ben Kelley, *The Pavers and the Paved* (New York: Donald W. Brown, Inc., 1971).

27. Wilfred Owen, *The Accessible City* (Washington, D.C.: Brookings Institution, 1972), p. 27.
28. Wilbur R. Thompson, *A Preface to Urban Economics* (Baltimore: Johns Hopkins Press, 1965), p. 375.

Highways and planning

The planning of highways is carried out, not by metropolitan planning agencies as might be reasonably expected, but by state highway departments whose officials have repeatedly contended that they have neither the responsibility nor the authority to deal with land-use matters. Thus highway officials do not include among their considerations the developmental impacts of the routes, access points, and intersections of the systems they design.

This attitude has persisted, with the tacit approval of the Bureau of Public Roads which must approve all federally funded projects, in spite of a provision in the Federal-Aid Highway Act of 1962 that:

after July 1, 1965, the Secretary shall not approve . . . any program for projects in urban areas of more than fifty thousand population unless he finds that such projects are based on a continuing comprehensive transportation planning process carried on cooperatively by States and local communities in conformity with the objectives stated in this section.[29]

Procedures for compliance with this provision were developed by the Bureau of Public Roads, and their effect was to preserve, virtually unchanged, the prerogatives of state highway departments.

As transportation planning processes were organized in metropolitan areas, planning agencies at both local and metropolitan levels were bypassed by state highway departments which deal directly with local governing bodies and elected officials.[30]

The Bureau of Public Roads made no requirement that highway plans be consistent with land-use plans, even where such plans had been adopted by local authorities. Even the refusal of a local government to participate in cooperative planning, or to approve of highway elements within its jurisdiction, was not considered evidence that the process was ineffective, or grounds for disapproving a project. All that was required was that the state highway department make "scrupulous efforts" to obtain a local government's cooperation.[31]

Recognition of the need for comprehensive planning was not limited to the 1962 Highway Act. Subsequent highway and mass transit legislation in 1964, 1966, 1968, and 1970 has repeated and expanded the concerns of Congress and instructed highway officials to coordinate roads with other forms of urban transportation, give due consideration to the impact of highways on the future development of metropolitan areas, and, in consultation with the Department of Housing and Urban Development (HUD), "assure that urban transportation systems most effectively serve both national transportation needs and the comprehensively planned development of urban areas."[32]

In compliance with the 1962 Highway Act and subsequent legislation, many regional transportation agencies were established, but these were limited to advisory roles, thus leaving the authority of the state highway departments intact. All of the initiatives in the "continuing comprehensive transportation planning process" were maintained by those who allocate and spend the resources of the Highway Trust Fund. Having observed this process for several years, the National Commission on Urban Problems also found that state

29. Federal-Aid Highway Act of 1962, Public Law 87-866, section 9(a).

30. Thomas A. Morehouse, "The 1962 Highway Act: A Study in Artful Interpretation," *AIP Journal* 35, no. 3 (May 1969), p. 164.

31. Ibid., p. 163.

32. Department of Transportation Act, Public Law 89-670, 1966, section 4(g).

highway departments

. . . are able to ignore totally the desires of local officials; and no State agencies— outside the legislatures and Governors—are established to reconcile differences. The result is that State highway departments to a considerable extent go their own way, leaving local officials "holding the bag" after a poorly planned and designed highway has damaged sections of built-up areas (and left the job of relocating displacees to already hardpressed local housing officials) or completely ignored and effectively destroyed development plans. . . . In practice, then, highways are seldom used as affirmative tools for development guidance.[33]

While the Congress wisely urged comprehensive planning, the states took care to see that the light shed by planning studies should not adversely affect existing commitments. In California, for example, the legislature in compliance with the 1962 Highway Act established the Bay Area Transportation Study Commission (BATSC), but

. . . wisely provided that the Commission's existence should not interfere with or in any way impede "execution by Federal, State, or local public agencies of any projects in the Bay Area which have already been planned by such agencies, *or which might be planned during the course of the study.*"[34]

The BATSC report, issued in 1969, was frank in conceding that the highway-dominant system proposed failed to deal with problems such as the isolation of racial minorities from employment owing to lack of public transportation and low rates of automobile ownership.[35] The report explained that, "BATSC analysis and projections are based largely on extension of current

policies and planning as it exists in the Bay Area."[36] The major reason for this is that "no *enforceable* general regional plan to which transportation might be fitted exists."[37]

Meanwhile, new regulations issued by the Department of Transportation in 1968 required that public hearings on highway proposals include consideration of twenty-three possible effects of highways including social, economic, and environmental impacts.[38] However, since the new regulations contained no criteria or guide lines by which these impacts were to be judged, no enforceable requirements were produced other than that these factors be "considered." During the four years that the new public hearing regulations have been in effect there is no record of a single instance in which the U.S. Department of Transportation, *on its own initiative,* rejected a highway project proposed by a state highway department because of its adverse social, economic, or environmental impact.

Approved highway plans are by definition enforceable, and their developmental consequences, in the absence of meaningful standards and enforceable regional plans, have been and continue to be determined by private developers and sanctioned by the zoning powers of local municipal governments.

Congressional concern for the comprehensively planned development of urban areas and the coordination of all forms of urban transportation is unmistakeable. It seems inconceivable in these circumstances that the implementation of the highway program to date can

33. "Building the American City," p. 231.
34. Bay Area Transportation Study Commission, "Bay Area Transportation Report," May 1969, pp. 3, 4, emphasis added.
35. Ibid., p. 11.

36. Ibid., p. 74.
37. Ibid., p. 74, emphasis added.
38. Policy and Procedure Memorandum 20-8 BPR, January 17, 1969. Original hearing requirements relating to the interstate system required only that a hearing be held on the "economic effects" of the proposed highway.

be construed as complying with their wishes. Development is guided not by plans, comprehensive or otherwise, but by private investment decisions; and a major result of ..the process has been the erection of massive new economic barriers to overcoming the disadvantages produced by a long history of racial discrimination.

IMPLICATIONS FOR CHANGE

Edward Banfield, writing recently in this journal, skillfully disposed of "the several factors widely held to be the principal causes of the 'urban crisis,'"[39] and concluded with a characteristic flourish:

If these are not the principal causes of the "urban crisis," what are? My answer is that they are mainly changes in the way things are perceived, judged, and valued, and in the expectations that are formed accordingly—in a phrase, changes in the state of the public mind.[40]

While this point of view serves as a convenient rationale for the further conclusion that ". . . the 'urban crisis' is not to be solved or alleviated by government programs, however massive,"[41] it ignores both the increasing polarization which is taking place and the role played by "massive government programs" in bringing it about.

Efforts to date

Urban economic problems, real not imagined, underlie every aspect of the "urban crisis" and are largely a consequence of the inequitable distribution of metropolitan resources and opportuni-

ties. Efforts, to date, to alleviate racial and economic polarization have been directed primarily at the exclusionary barriers to the construction of low- and moderate-income housing in suburban communities. These efforts have taken several forms. There have been numerous law suits attacking the restrictive provisions of local zoning ordinances; a few states have enacted legislation that establishes procedures for overriding local zoning restrictions; a few regional planning agencies have prepared "fair share" plans for allocating low- and moderate-income housing among suburban communities; and the U.S. Department of Housing and Urban Development has revised its site location criteria in order to promote racial integration. The immediate goal of most of these efforts has been the production of moderate- rather than low-income housing.

These measures, while producing scattered instances of success, have, nevertheless, had no material effect on the gap between the growing concentrations of minority groups in central cities and the steady departure of employment opportunities to the suburban hinterland.

This is true because zoning suits generally affect only one municipality. State "anti-snob" zoning statutes depend on the initiative of developers who have been denied permission to build, and their reluctance has rendered these statutes ineffective. In New York State, a state housing agency has the authority to override local zoning and build low- and moderate-income housing; but there has been a marked reluctance to do so in wealthy suburbs of New York City such as Westchester County. Fair share plans, while they serve a useful purpose in quantifying disparities and developing allocation models, lack any authority to require compliance. New HUD site location

39. Edward C. Banfield, "A Critical View of the Urban Crisis," THE ANNALS 405 (January 1973), pp. 8–14. The factors that Banfield considers and dismisses are: overcrowding, white flight to the suburbs, the physical environment, white racism, and the multiplicity of metropolitan governments.
40. Ibid., p. 13.
41. Ibid., p. 14.

criteria have resulted in a virtual halt to the construction of inner-city sub-sidized housing [42] and have resulted in the construction of some units in the suburbs, the great majority of which have been occupied by whites.

Highway projects have also been challenged in the courts. And, a few of these challenges have succeeded in halting individual projects, either on the grounds that relocation resources were inadequate or that public parkland was being improperly taken. No attempt, however, has yet been made to invoke the wishes of Congress and challenge highway proposals because the auto-dominant pattern of dispersal which they generate reduces both the mobility and the basic opportunities of low-income inner-city residents. Highway proposals should be vulnerable simply on the grounds that they subvert "the comprehensively planned development of urban areas."

In the light of growing national support for mass transit, it is also important to understand that new rail transit systems such as Bay Area Rapid Transit (BART) in California and the Linden-wold Line in New Jersey are equally ineffective for improving access for inner-city residents to opportunities in the suburbs. Both systems are designed to bring suburban residents to their places of employment in the offices and stores of the central business districts. The same is also true of the rail transit systems approved for Washington and Atlanta. Both focus on the central business districts and government centers.

Metropolitan development policy, equal opportunity, and regional planning

While it is important that immediate attempts be made to insure that the benefits of present programs are more equitably accessible, long-range improvements will require more fundamental changes. If disparities in opportunity are to be effectively reduced, and hopefully eliminated, there must be intervention in the process itself to alter the forces that generate, shape, and sanction racial and economic polarization. Such intervention must be regional in character and must be based on a functional rather than a political definition of the region. That is, it must not be arbitrarily restricted by state boundaries.

Regional planning agencies could effectively meet this need if granted additional authority in the context of a national urban growth policy, or more appropriately, a national metropolitan development policy. Such a policy should include, among many others, such goals as:

—the elimination of transportation barriers to and between employment and housing;

—the development of patterns of land use which can be efficiently served by public transportation;

—the development of integrated multi-mode transportation systems which fit travel mode to type and intensity of land use, and permit convenient interchange between modes;

—the provision of low- and moderate-income housing in relation to and in proportion to existing and potential opportunities for employment throughout the region.

None of the goals is in any way inconsistent with the policies set forth in existing metropolitan development and transportation legislation. Goals such

---

42. The more recent moratorium on spending imposed by President Nixon has halted the construction of all subsidized housing for which funds had not already been committed.

as these merely express a recognition of the widespread effects of narrowly conceived programs that have been implemented in isolation from each other. Based on that recognition, they incorporate other dependent goals such as equal access to opportunity.

Such a metropolitan development policy should, of course, also include goals directed at guiding growth, protecting and improving the environment, preserving open space and natural resources, providing public services and facilities, conserving energy, and so on. All of these goals will then require the development of reasonable standards against which the performance of regional planning agencies and local governments can be evaluated.

We have both the experience and the expertise to formulate such standards. It is as feasible to establish standards for equal access to employment and housing as it is to establish standards for clean air or pupil-teacher ratios. And, more important, we have learned through experience that the injunction to "consider," in the absence of standards, is an empty gesture which has failed to provide any protection at all. The following are a few examples of equal-opportunity related standards which might be employed in evaluating transportation proposals:

—New transportation systems and facilities should improve accessibility between centers of low-income minority-group housing and centers of employment.

—New transportation facilities must support and reinforce existing public transit systems.

—Access points to new transportation facilities may not be located within jurisdictions which do not provide housing opportunities for minority-groups and low-income households.

In order to implement such policies, regional planning agencies must be granted powers which they do not now have:

1. The planning function must be made both meaningful and objective by transference of all responsibility for the planning of transportation facilities from state highway and transportation departments to the regional planning agencies. The state agencies would of course continue to build, maintain, and operate transportation facilities and conduct research in the areas of technology and safety.

2. The regional planning agencies must be granted limited development control powers in order to meet clearly defined regional goals: as in areas where it is necessary to control growth stimulated by transportation facilities, or where low- or moderate-cost housing sites are not provided by local authorities; or to protect regional resources, provide regional facilities, or prevent the development of land not suitable for development.[43]

3. The regional planning agencies should function as regional relocation agencies to provide the maximum opportunities for access to employment-related housing for households displaced by public actions.

4. The regional planning agencies should be authorized to function as a housing authority of last resort—that is, to be able to condemn land and

43. The political feasibility of delegating development control authority to a regional planning agency is considerably enhanced by the continuing trend toward increased revenue sharing. As long as municipal services must be financed primarily out of property taxes, local zoning prerogatives will be jealously guarded. However, the transfer of any substantial portion of this burden to revenue sharing funds should reduce the opposition of those who view local zoning control primarily as a tool for controlling municipal expenditures.

construct low- and moderate-income housing where local authorities fail to meet regional needs for such housing.

An early step toward the provision of improved access should be the extension of control over the locations of regional employment growth. This could be accomplished by selecting several existing suburban commercial or industrial centers which have the best access to inner-city residential areas, and which have room for expansion, and directing new development to them by restricting the further growth of the other less accessible centers. In this way a concentration of employment could be achieved which would warrant the provision of bus service to inner-city residential areas.

An analysis should also be made of land adjoining the rights-of-way of all existing rail lines in the region to identify sets of locations that have the potential for development capable of supporting rail transit service. Relocating or adding stations along existing commuter rail lines may serve to reinforce improved access policies and generate additional transit patronage. There may be instances in which relatively short extensions to these lines may serve to establish viable links between inner-city housing and suburban employment.

Current policies should be challenged

In the interim, to the continuing attacks on restrictive zoning should be added widespread efforts to bring the federal highway and transit programs into compliance with the wishes of Congress as expressed in the highway, transit, and metropolitan development legislation of the past ten years and the provisions of existing civil rights legislation. The restriction of funds to financing a single mode of transportation which discriminates against low-income persons should be challenged. Similarly, challenges should be raised to any project, road or rail, that has the effect of reducing the travel options of low-income persons—depriving them of access to basic necessities such as employment and housing. The construction of pending projects should be halted until a determination is made that each proposal meets equal opportunity criteria and "most effectively serves both national transportation needs and the comprehensively planned development of urban areas."

It is no longer reasonable to accept the conclusion that:

We are not yet able to plan transportation networks or systems with adequate sophistication to preclude the occurrence of unanticipated results in land use patterns, nor have we been able to plan and implement land use patterns that assure the continuing adequacy of installed or proposed transportation facilities.[44]

We have been able to do so for a long time. We have, however, been unwilling, perhaps because "the state of the public mind" has attached insufficient importance to the burdens imposed on some by decisions made by those who experience only the benefits of public programs.

44. Max L. Feldman, "Transportation: An Equal Opportunity for Access," in *Environment and Policy*, W. R. Ewald, Jr., ed. (Bloomington: Indiana University Press, 1968), p. 186.

James Gordon Bennett, the *New York Herald*, and the Development of Newspaper Sensationalism

By JAMES L. CROUTHAMEL

Sensational journalism had its origins much earlier than is generally believed. James Gordon Bennett, a pioneer sensationalist, utilized the art to create a truly popular press and to underwrite new journalistic techniques. Dr. Crouthamel, author of a biography of James Watson Webb, is a member of the history department at Hobart and William Smith Colleges.

HISTORIANS of newspaper sensationalism have concentrated on the "yellow journalism" of the 1890s, especially the colorful circulation war between Joseph Pulitzer's New York *World* and William Randolph Hearst's *Journal*, to the neglect of the earlier nineteenth century origins of this journalistic technique. In fact, sensationalism had been developed and its effectiveness proved in New York half a century earlier, and it had been carried to other metropolitan centers before the Civil War. Most of the credit for this must go to James Gordon Bennett and his New York *Herald*.

What is sensationalism? A standard dictionary defines it as "sensational subject matter or treatment of subject matter or use of such matter or treatment."[1] It is, then, both content and style. George Juergens has provided a more useful extended definition in his study of Joseph Pulitzer. He describes sensationalism as the "strategy of attracting a large audience by

An earlier version of this study was read at the annual meeting of the Organization of American Historians at Dallas, Texas, in April, 1968. The author wishes to thank Julian Rammelkamp and David C. Smith for their helpful criticisms and suggestions.

[1] *Webster's Third New International Dictionary of the English Language Unabridged* (Springfield, Mass., 1966), p. 2067.

"The Newsboys." Frank Leslie's Illustrated Newspaper, *August 2, 1856.*

concentrating on stories of timeless appeal—sex, crime, tragedy. . . ." Juergens continues:

> . . . the goal of sensational journalism is to catch the interest, even to titillate, the vast body of men and women who for one reason or another are unconcerned with happenings in government, business, or the arts. . . . The first rule is that a different standard must apply in determining what articles will be printed. Sensational newspapers expanded the meaning of the human interest story to report what had hitherto been regarded as private, the gossip and scandals about individuals, and discovered a rich source of news in crime and everyday tragedy. A corollary of the same point, they began to pay as much attention to personalities as to local or national events. . . .
>
> The second rule follows directly from the first. Sensational newspapers have their own idea of the relative importance of different stories. They can usually be counted upon to evict matters of statecraft from the front page if a provocative item of scandal is available to take its place. . . .
>
> Finally, the form requires a unique prose style. The argument applies with different force to different papers, but in general the language in a sensational newspaper tends to be slangy, colloquial, personal. In such a way does the sensational journal express its identity with the masses of people who patronize it.[2]

What is the significance of sensationalism? It attracted the readers, and then the advertisers, to create a mass audience.

[2] George Juergens, *Joseph Pulitzer and the New York World* (Princeton, 1966), pp. viii–ix.

James Gordon Bennett. Harper's Weekly, *June 22, 1872.*

The revenue, in turn, financed improvements in news-gathering and transmission, reporting, and technology that widened the appeal of the popular papers to their middle-class readers. No longer were newspapers bound to the fortunes of a political faction and the requirements of a small group of subscribers. The stultifying grip of the "special audience" or "mercantile" newspapers on circulation was broken. Until this happened, during the 1830s, most metropolitan newspapers survived on political patronage and a small mercantile audience interested in business news. A daily with a circulation of one or two thousand could manage nicely; its readers subscribed by the year and its advertisers took a yearly space. The cost, about six cents for a single copy, which was available only at the newspaper office, was prohibitive for most people, who had little interest in the partisan editorial or the detailed mercantile listings. The bulk of the remaining space was filled with "features," most of which were clipped from other newspapers and very few of which would meet modern standards of news. Almost uniformly the mercantile papers were dull and heavy in appearance and tone. Typical of those in New York City were the *Daily Advertiser,* the *Mercantile Advertiser,* and the *Commercial Advertiser.* The titles were accurate. These papers were, as Bernard Weisberger has remarked, "expensive bulletin boards for a small trading clientele."[3]

When James Gordon Bennett issued the first number of his *New York Herald* on May 6, 1835, there were already two penny papers in New York, the *Sun* and the *Daily Transcript.* They were not only cheap, designed to appeal to the audience neglected by the larger six cent commercial papers; they were also frankly sensationalist.

Benjamin H. Day's *Sun* demonstrated that a sensational paper could succeed in America as the *Penny Magazine* had done in England earlier.[4] He announced in the *Sun's* first issue, September 3, 1833, that his object was "to lay before the public, at a price within the means of everyone, all the news of the day." He was more explicit two days later:

We newspaper people thrive best on the calamities of others. Give us one of your real Moscow fires, or your Waterloo battlefields; let a Napoleon be dashing with his legions throughout the world, overturn-

[3] For background information on the mercantile press see Bernard Weisberger, *The American Newspaperman* (Chicago, 1961), pp. 70–84; Frank Luther Mott, *American Journalism* (rev. ed., New York, 1950), pp. 193–198, 257–262; James L. Crouthamel, "The Newspaper Revolution in New York, 1830–1860," *New York History* XLV (1964), 91–113, and *James Watson Webb. A Biography* (Middletown, Conn., 1969), *passim.*

[4] The *Sun* was not the first cheap popular paper in America. Several New England sheets, Seba Smith's Portland (Maine) *Daily Courier,* and the Boston *Transcript, Morning Post,* and *Mercantile Journal* cost only half the price of a mercantile paper, were smaller in size, and devoted much of their space to non-business affairs such as literature, the theatre, and humor. Mott, *American Journalism,* pp. 216–218; Mary Alice Wyman, "Seba Smith" in Allen Johnson and Dumas Malone, eds., *Dictionary of American Biography* (22 vols.; New York, 1928–1944), XVII, 345–346. In New York in December, 1831, William T. Porter started publishing *The Spirit of the Times,* a weekly sporting journal which attained a large circulation and demonstrated that there was a market for such a publication. One of Porter's typesetters, Horace Greeley, in January, 1833, tried a cheap two cent paper, the New York *Morning Post,* sold by newsboys on the street, but it survived for less than a month. Nelson F. Adkins, "William Trotter Porter," *Dictionary of American Biography,* XV, 107–108; Norris W. Yates, *William T. Porter and the Spirit of the Times: A Study of the "Big Bear" School of Humor* (Baton Rouge, 1957); William Harlan Hale, *Horace Greeley, Voice of the People* (New York, 1950), p. 40.

ing the thrones of a thousand years and deluging the world with blood and tears; and then we of the types are in our glory.

The *Sun* cost only a penny, cash, and it revolutionized circulation methods by being sold by newsboys on the street instead of by subscription, as all the mercantile papers were. Its small pages were filled with interesting but trivial bits of news and gossip, much of it humorous. Facetious police court reports were its specialty, but anything that would interest its readers found a place in the *Sun*—murders, suicides, and crime, stories about animals, uplifting essays if they were brief and pointed. Economic and political news, the specialty of the mercantile sheets, was deliberately neglected.[5]

The *Daily Transcript,* established soon after the *Sun,* was almost as successful with a similar formula, with even more emphasis on crime. Both penny papers outstripped their rivals in circulation by 1835.[6]

James Gordon Bennett learned much from the experience of the *Sun* and the *Transcript.* But he believed that they had not begun to tap the potential mass audience of literate middle-class New Yorkers. The population of New York City in 1835 was over a quarter of a million; its eleven mercantile dailies had an average circulation of 1,700 each; the *Sun's* circulation was about 10,000 and the *Transcript's* slightly less.[7] Bennett proposed to attract his share of readers not by duplicating the field of his penny rivals, but by expanding his coverage to other areas, hitherto neglected, and by covering economic and political news as well. He would challenge both the penny and the mercantile papers by doing what they were both doing, but doing it better, and by doing even more.

Bennett was almost forty years old when he launched his new venture. He had failed in several attempts to establish his own paper, but he had a wealth of experience, much of it gained on the nation's largest mercantile sheet, the New York *Courier and Enquirer.* Here Bennett had pioneered as a Washington and Albany correspondent and as an analyst of eco-

[5] New York *Sun,* Sept. 3, 5, 1833. There is a good summary of the *Sun's* early years in Frank M. O'Brien, *The Story of the Sun* (New York, 1918), p. 5ff.

[6] Dorothy MacGill Hughes, *News and the Human Interest Story* (Chicago, 1940), p. 10; Frank Luther Mott, "Facetious News Writing," *Mississippi Valley Historical Review,* XXIX (1942–43), 39.

[7] Frederic Hudson, *Journalism in the United States from 1690 to 1872* (New York, 1873), pp. 430–431. Circulation claims were notoriously inaccurate.

nomic news. He was ready in the spring of 1835, with $500 he had scraped together, to take a plunge born of desperation.[8]

From a basement room furnished with a few packing cases Bennett issued the first number of his *Herald* on Wednesday morning, May 6, 1835. Like the *Sun* and the *Transcript,* the paper was to be sold by newsboys for a penny. Most of the four small columns on page one were filled with a biographical sketch of Matthias the Prophet, the famous fraud. In the editorial column of page two Bennett discussed the *Herald's* debut. He disavowed any party affiliation.

We shall endeavor to record facts, on every public and proper subject, stripped of verbiage and coloring, with comments when suitable, just, independent, fearless, and good tempered. If the HERALD wants the mere expansion which many journals possess, we shall try to make it up in industry, good taste, brevity, variety, point, piquancy and cheapness. It is equally intended for the great masses of the community—the merchant, mechanic, working people—the private family as well as the public hotel—the journeyman and his employer—the clerk and his principal.

The first issue indicated what Bennett intended the *Herald* to be. There were numerous short, spicy tidbits: a murder and a death by suffocation, the explosion of a steamboat, a balloon ascension, horse racing, romantic poetry, and police court reports. A long editorial affirmed the *Herald's* identity with "The Mechanics," and on page four there were two columns of advertisements. Unlike the *Sun,* the *Herald* tried from the beginning to cover world and national news and economic developments as completely as its limited space would allow. In this first issue there were almost two columns of "LATE AND IMPORTANT FROM EUROPE," tersely discussing the defeat of the Peel ministry and other foreign items.[9]

From the beginning the *Herald's* readers were fed a steady diet of violence, crime, murder, suicide, seduction, and rape, both by straight news reporting and by gossip. In its first two weeks of publication there were accounts of three suicides,

[8] For biographical information on Bennett see Allan Nevins, "James Gordon Bennett," *Dictionary of American Biography,* II, 195–199; Oliver Carlson, *The Man Who Made News, James Gordon Bennett* (New York, 1942); Frederick B. Marbut, "Early Washington Correspondents: Some Neglected Pioneers," *Journalism Quarterly,* XXV (1948), 369–374; James Parton, "The New York Herald," *North American Review,* CII (1866), 373–419.

[9] New York *Herald,* May 6, 1835.

three murders, a fire which killed five persons, an accident in
which a man blew off his head, descriptions of guillotine ex-
ecutions in France, a riot in Philadelphia, kangaroo hunting,
and the execution of Major André half-a-century earlier.[10]

There was no let up. If no immediate murder was at hand,
a grisly one was revived from the past and described again.
During this first year there were stories of a wolf hunt, of the
"Causes of Infanticide Among Savages," of drownings and
duels and more murders, of Vice President Richard M. Johnson's relations with his mulatto mistress, of the veracity of
Maria Monk's *Awful Disclosures*, of a juicy breach of promise
suit, of the trial of an accused slave trader, and of the last
hours and execution of a murderer.[11] This was standard fare.

During its first year, 1835-1836, the *Herald* developed all
of the traits of sensationalism. In the process Bennett also was
making the *Herald* into a great newspaper in its completeness
of news coverage, its zeal for rapid news-gathering and transmission, its economic analysis, its technical innovations, and
its political independence. This parallel development is beyond
the scope of this study, and it has been described elsewhere.[12]

The *Herald* was not merely salacious; it appealed to the
interests of its readers. It took a booster attitude toward local
developments and reforms, opposing corruption at City Hall
and in the Customs House and fighting for retrenchment.[13]
And Bennett shared the patriotic, nationalistic sentiments of
his audience. The *Herald* supported the independence of Texas
in mid-1835 and predicted the emigration of thousands of
Americans "to aid their brethren against the rapacious and
bloody Spaniard."[14] During the difficulties over the French
Spoliation Claims, the *Herald* backed Jackson's firmness and
supported a war, if necessary, to force payment and safeguard
American national honor.[15]

The *Herald's* most important continuing news story was its
own editor, Bennett, and its own meteoric success. It was as-

[10] *Herald*, May 11-15, 19, 22, 1835.
[11] *Herald*, May 28, June 17, July 11, 13, 15, Aug. 4, Nov. 19, 20, 1835,
Feb. 6, 11, 12, 13, 1836.
[12] Crouthamel, "Newspaper Revolution in New York," pp. 91–113 and
James Watson Webb, pp. 67–94; Elwyn B. Robinson, "The Dynamics of
American Journalism from 1787 to 1865," *Pennsylvania Magazine of History and Biography*, LXI (1937), 435–445.
[13] *Herald*, May 11, July 22, Sept. 16, 25, Oct. 6, 1835.
[14] *Herald*, July 28, 1835.
[15] *Herald*, Oct.–Dec. 1835, e.g., Nov. 10, Dec. 11, 12, 24.

sumed that the readers would have a sense of identity with the *Herald;* it was their paper and Bennett was their spokesman. Its rapid increase in circulation was noted often, and when it was two months old it claimed to have passed the *Sun* and the *Transcript.*[16] After three months it was, Bennett said, the only New York paper operating on a cash basis, able to pay all its expenses out of its cash receipts. Burned out in a fire a few days later it bounced back "larger, livelier, better, prettier, saucier, and more important than ever." About once a month some new item of growth was noted: rising circulation, a doubling of advertising, a move into larger quarters.[17]

Bennett gave himself all the credit. In October, 1835, he printed the first of many autobiographical pieces, boasting that he had started a "vast and important revolution" in newspapers and predicting an ever upward rise for the *Herald.* "I am no novice in the business, and I cannot make a mistake in public feeling."[18]

The *Herald* deprecated its rivals while puffing itself. No newspaper in the city escaped the *Herald's* barbs, and one day Bennett managed to attack in a single issue seven newspapers and their editors.[19] The *Sun* was its favorite target, but all the mercantile press felt Bennett's vitriolic pen,[20] especially M. M. Noah's *Evening Star.* The abuse was too much for Peter Townsend, one of the *Star's* editors. Townsend administered a public beating to Bennett in October, all duly reported in the *Herald* as a news story (as were later beatings by James Watson Webb and William Leggett).[21]

The subject matter of the *Herald,* as described, was of a piece with the *Sun* and the *Transcript.* Bennett went further by devoting some attention to society news. He appreciated the interest of the common people in the doings of the rich, and he made a marketable commodity out of what had been back fence gossip. For example, in its first year the *Herald* carried long accounts of the masked balls in New Orleans, the social season at Saratoga Springs, and President Jackson's annual Christmas ball in Washington.[22]

---

[16] *Herald,* June 18, 28, July 9, 1835. The claims were false.

[17] *Herald,* Aug. 7, 31, Sept. 7, Oct. 19, Dec. 28, 1835, Mar. 10, 31, Apr. 9, 1836.

[18] *Herald,* Oct. 10, 1835, Mar. 2, 1836.

[19] *Herald,* Sept. 16, 1835.

[20] *Herald,* May 12, 16, Sept. 10, Oct. 17, 1835, Jan. 5, 1836.

[21] *Herald,* May 23, 30, June 22, 25, Oct. 8, 9, 1835.

There were other innovations. Bennett tried to move the news from the inside of the paper onto page one, and for several months he did this until he abandoned the experiment and returned to conventional practice.[23] Illustrations were first used in December, 1835, in the *Herald's* extensive treatment of New York's great fire.[24]

The most amusing innovations were the parodies Bennett wrote of Jackson's messages. The first one appeared on December 9, 1835, purporting to be the original of Jackson's annual message, smuggled out by a secret connection in the Kitchen Cabinet. The early paragraphs surveyed the state of the union.

In performing my duty at the opening of another session, [Bennett had Jackson begin] I have great pleasure in stating to you that the country is going ahead as fast as ever, and much faster than any country ever did under heaven. For this we ought to thank God Almighty and the democratic party—not forgetting the little aid I have given in helping to keep things in the right trim.

The Country is in a flourishing condition. Cotton fetches a good price—corn is abundant—beef and pork plenty—the dews of heaven fall as richly as ever—gold and silver overflows the land—the United States Bank is down and I have got my foot on the monster's neck.

Our foreign relations wear the most favorable and peaceful aspect imaginable. You all know how fond I am of peace, love, union, and harmony. Through a long life it has been my invariable practice to try mild measures before picking up pistols or shouldering the rifle. This peaceful disposition has invariably been successful with the solitary exception of two or three wars and about forty or fifty rows. The Honorable Senator from Missouri can tell you how calm and considerate I talked to him in the streets of Nashville about twenty years ago before I cocked my pistols and let fly. That gentleman has long since regretted his conduct on that occasion, and in consideration of his helping me to put down the monster, I have ordered him a plate of pork steaks at the Kitchen Table.

I have merely alluded to these glorious days of my past life as an illustration of my course towards Louis Phillipe and the French nation.

And so on, at length, until the conclusion, where Jackson was used to puff the *Herald*.

I cannot leave you, however, without one word more. For some time past, I have been receiving a little neat penny paper from New York called the HERALD.—It is uncommonly smart, and tells its whole mind on every subject. This is exactly what I have always done myself. It is the only honest method of getting through the world. The

---

[22] *Herald*, May 19, Aug. 25, Sept. 2, Dec. 29, 1835.
[23] *Herald*, Feb. 9, 1836. By April the news was back on the inside.
[24] *Herald*, Dec. 19, 21, 1835.

editor, I believe, is some fellow by the name of Bennett, who once wrote me a letter blowing up Amos Kendall. I could not attend to him on that time, but on further inquiry, I find that he was once Editor of the New York Enquirer, when it first came out for me, and generally wrote two-thirds of that paper, for which Noah claimed and got an office. Bennett never came to Washington seeking an office, for which my Kitchen Cabinet called him 'a d - - - - d fool.' I recommend every member of Congress to order the Herald during the season. They will be pleased with it—for it has, daily, a world of information on politics, stock, fashion, police, accidents, news, markets, and every thing in every line.

Trusting that you will attend, one and all, to the hints I have hereby given you, I subscribe myself

ANDREW JACKSON.[25]

In the next few months there were three more parodies of Jackson's messages and one of Governor William L. Marcy's. They are all similar in their skillful plays on Jackson's plain talk and idiosyncracies, and they are all funny.[26]

The famous Robinson-Jewett murder case confirmed the sensational trend already apparent in the *Herald* and the penny press. The facts of the case can be stated simply. Early Sunday morning, April 10, 1836, someone murdered a beautiful young prostitute, Ellen Jewett, with a hatchet, and then set fire to her body and bed at Rosina Townsend's luxurious house of ill fame. Richard P. Robinson, a wealthy, handsome young man about town, and a frequent visitor of Ellen's, was arrested for the crime on the basis of circumstantial evidence.[27] The penny press seized on the case and created an excitement never known before in popular journalism.

On April 11 the lead editorial in the *Herald* was headed "Most Atrocious Murder." All the ingredients were present for a human interest story of rare dimensions. The *Herald's* account was a column and a half long and presented every detail then known, even the most trivial. Bennett, like almost everyone else, assumed that Robinson was guilty.

Bennett had visited the scene of the crime himself, and he described it vividly for his readers. He entered Ellen's room:

What a sight burst upon me! There stood an elegant double mahogany

25 *Herald*, Dec. 9, 1835. It was reprinted Dec. 19.
26 *Herald*, Jan. 6, 7, Feb. 20, Mar. 23, 1836.
27 There is a long account of the Robinson-Jewett case in Carlson, *The Man Who Made News*, pp. 146–167. Bennett had earlier covered the famous Crowninshield murder case in Salem, Mass. as a reporter for the *Courier and Enquirer*. Wallace B. Eberhard, "Mr. Bennett Covers a Murder Trial," *Journalism Quarterly*, XLVII (1970), 457–463.

bed all covered with burnt pieces of linen, blankets, pillows, black as cinders. I looked around for the object of my curiosity. On the carpet I saw a piece of linen sheet covering something as if carelessly flung over it.

'Here,' said the Police Officer, 'here is the poor creature.' He half uncovered the ghastly corpse. I could scarcely look at it for a second or two. Slowly I began to discover the lineaments of the corpse as one would the beauties of a statue of marble. It was the most remarkable sight I ever beheld. I never have and never expect to see such another. 'My god,' I exclaimed. 'How like a statue! I can scarcely conceive that form to be a corpse.' Not a vein was to be seen. The body looked as white, as full, as polished, as the purest Parisian marble. The perfect figure, the exquisite limbs, the fine face, the full arms, the beautiful bust, all, all surpassed in every respect the Venus de Medici. . . .

The editor went on to describe the room in detail—Ellen's clothes, her desk, her books and magazines, the decorations on the wall.[28]

His role in the case was that of a gossip, telling as much of the story behind the news as possible. As Helen Hughes said: "He was just the person his readers would have loved to have a long talk with—one who had seen everything and was ready to tell all about it."[29]

[28] *Herald*, Apr. 11, 1836. This was reprinted on April 12, and all the stories about the case were reprinted April 13.
[29] Hughes, *News and the Human Interest Story*, pp. 197, 11–12.

Illustration from The Truly Remarkable Life of the Beautiful Helen Jewett *(1878). New York State Historical Association.*

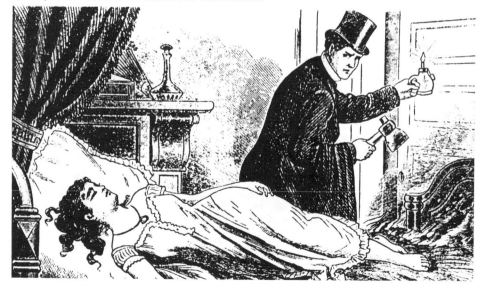

Front cover of a contemporary murder pamphlet. New York State Historical Association.

The next day the *Herald* carried a mass of detail and speculation about Ellen's background and the circumstantial evidence responsible for Robinson's arrest. There were lapses into moralizing: the reasons that Ellen had become a fallen woman, the men responsible, her passion for Byron's *Don Juan* which "has no doubt produced more wretchedness in the

world, than all the other moral writers of the age can check."
Other news, such as the Texas revolution, was shunted aside
to satisfy the public demand for information about the murder
case.[30]

Bennett gratified his readers with further details on April
13. He returned to Ellen's room and again described its con-
tents, along with the letters, scrapbook, and miscellany con-
fiscated by the police. The article concluded by raising doubts
about Robinson's guilt.[31] These questions were repeated the
next day, and each day thereafter. "How could a young man
perpetrate so brutal an act? Is it not more likely the work of
a woman? Are not the whole chain of circumstances within
the ingenuity of a female, abandoned and desperate?" One of
the paintings in the Townsend house was of a beautiful woman
on her knees, about to be scalped by two Indians (probably
Vanderlyn's "The Murder of Jane McCrea.") "What a remarkable type—or hint—or foregone conclusion of the awful
tragedy perpetrated upstairs." Why were none of the other
girls held for further questioning? Why did the police believe
Madam Townsend rather than Robinson?[32]

Obviously the case had more appeal if there was an element
of mystery, if there was some doubt of the murderer's identity.
The *Herald* the next day suggested a conspiracy at the Townsend house involving the police and the deliberate suppression
of evidence. This remained its line until the trial.[33]

On April 16 Bennett reported an interview with Rosina
Townsend in verbatim dialogue, probably the first formal interview published in an American newspaper. She discussed
her recollections of the night of the murder and of Robinson's
visits to her house, with Bennett interjecting questions.[34]

The *Herald's* circulation increased markedly with the great
interest in the case, as did that of the *Sun* and the *Transcript*.
Bennett boasted daily of the rise, to 15,000 copies, which he
predicted would reach 30,000 before long; some issues were
so scarce that newsboys sold them for a shilling each.[35] His

[30] *Herald*, Apr. 12, 1836.
[31] *Herald*, Apr. 13, 1836.
[32] *Herald*, Apr. 14, 1836.
[33] *Herald*, Apr. 15, 1836.
[34] *Herald*, Apr. 16, 1836. That Bennett pioneered the interview in American newspapers is verified in Nils Gunnar Nilsson, "The Origin of the
Interview," *Journalism Quarterly*, XLVIII (1971), 707–713.
[35] *Herald*, Apr. 11–16, 1836; Hughes, *News and the Human Interest
Story*, pp. 158–159.

defense of Robinson embroiled him in a bitter controversy with the *Sun* and the *Transcript,* which he had accused of conspiring with the police to cover up the true story.[36] Other papers attacked Bennett for devoting so much space to the case, but he was able to justify his coverage.

Instead of relating the recent awful tragedy of Ellen Jewett as a dull police report, we made it the starting point to open up a full view upon the morals of society—the hinge of a course of mental action calculated to benefit the age—the opening scene of a great domestic drama that will, if properly conducted, bring about a reformation—a revolution—a total revolution in the present diseased state of society and morals.[37]

As interest in the case sagged while Robinson was awaiting trial, Bennett would revive it with a long analysis of how and when Ellen Jewett lost her virtue. Other stories of similar crimes capitalized on the craze. One, "Another Awful Consequence of Seduction," told how a New Jersey girl of seemingly good character had strangled her illegitimate infant and then taken poison.[38]

The trial itself received extensive coverage in the *Herald.* On a typical day, June 7, almost all the front page was devoted to the trial transcript, while the lead editorial analysed critical testimony. Bennett was delighted when the jury agreed with him and acquitted Robinson. The mystery was not solved —it never was—and he still intimated that a conspiracy existed with police collusion. He moralized again:

The evidence in this trial and the remarkable disclosure of the manners and morals of New York is one of those events that must make philosophy pause, religion stand aghast, morals weep in the dust, and female virtue droop her head in sorrow. A number of young men, clerks in fashionable stores, are dragged up to the witness stand, but where are the married men, where the rich merchants, where the devoted church members who were caught in their shirts and drawers on that awful night? The publication and perusal of the evidence in this trial will kindle up fires that nothing can quench.[39]

The case was not dropped, finally, until mid-July, with Bennett continuing to speculate about the real murderer. Journalism

[36] *Herald,* Apr. 13–19, 1836; *Sun,* Apr. 16, 18, 1836; New York *Daily Transcript,* Apr. 14, 16, 1836.
[37] *Herald,* Apr. 30, 1836.
[38] *Herald,* Apr. 25, 30, 1836.
[39] *Herald,* June 2–10, 1836. Quote from June 8.

would never be the same again. The sensational press, Hughes observes, "by focusing general interest on a topic to which the demos was, it seemed, spontaneously responsive, brought it about that murder trials became occasions of great popular excitement, just as hangings were in Tyburn and Newgate."[40]

In fairness it must be said that the *Sun* and the *Transcript* gave the case almost as much attention as the *Herald* did, and of a similar kind. Neither was as inventive as the *Herald* in manufacturing human interest features about the peripheral aspects of the case, and both tended to drop the case after a week, until the trial, to concentrate on other news, which the *Herald* would not do.[41]

The mercantile press treated the Robinson-Jewett case like a traditional news item, or else neglected it completely. Most of these sheets printed a news story about the murder on April 11, and then nothing more until the trial.[42] They did present a complete trial transcript, but few went beyond this except for a moralizing editorial on the day of Robinson's acquittal.[43] The penny press, in contrast, provided analysis of the testimony, local color at the trial, and speculation about the verdict.[44] And the *Sun*, like the *Herald*, refused to drop the case. It was convinced of Robinson's guilt and decried the miscarriage of justice for another month.[45]

Bennett learned a great deal from the success of his coverage of the case. For the remainder of its first decade the *Herald* became more sensational, continuing along lines it had already marked out, and adding new touches. Murder and misfortune remained the *Herald's* specialty. The same gossipy, probing treatment was given to other murders, suicides, seductions, and accidental deaths.[46] In at least eight other mur-

---

[40] *Herald*, July 1, 11, 12, 13, 1836; Hughes, *News and the Human Interest Story*, p. 159.

[41] *Ibid.*, pp. 158–159; *Sun*, Apr. 11–21, 26, 1836; *Daily Transcript*, Apr. 11–16, 1836.

[42] New York *Courier and Enquirer*, Apr. 11, 13, 1836; New York *Evening Post*, Apr. 11–19, 1836; New York *Evening Star*, Apr. 11, 13, 18, 1836; New York *Commercial Advertiser*, Apr. 11–30, 1836; New York *Journal of Commerce*, Apr. 11–30, 1836; New York *American*, Apr. 11, 1836.

[43] *American*, June 8, 1836, said that it was not covering the trial for "imperative reasons" which it did not explain. The papers immediately cited above covered the trial in their editions of June 2–11, 1836.

[44] *Herald*, *Sun*, and *Transcript*, June 2–11, 1836.

[45] *Sun*, June 14, 20, 21, 22, 23, 24, July 13, 14, 19, 1836.

[46] *Herald*, July 14, 1836, July 13, 1843, July 10, 1838, Jan. 5, 1842.

der cases in the next eight years the *Herald* provided coverage as extensive as the Robinson-Jewett case, or more so, and usually included an engraving of the victim, the accused, or the scene of the crime.[47]

Other human interest stories were featured. When one prize fighter bludgeoned another to death, there was a long, moralizing account. Fires were always news, and shipwrecks, and trans-Atlantic steamboat races exciting. The death of Aaron Burr, certainly an interesting and tragic personality, was the occasion for a serial sketch. Nationally important stories such as the *L'Amistad* affair, the arrest and trial of Alexander McLeod, and the *Somers* mutiny were treated in all their human interest aspects as sensational items.[48]

Bennett had an unerring sense of what news would appeal to his readers. Not only violence and gore, but the interesting Joseph Smith and his Mormons, the capture of that strange animal the giraffe, the discovery of the ruins of Maya civilization in Central America, and Millerite camp meetings which were covered by the *Herald's* own correspondent—all these human interest stories received extended treatment in the *Herald*.[49]

Bennett still catered to the preconceptions of his middle class audience. He continued to call for reform in the police department and to expose corruption in banking and at the Customs House.[50] Attempts to appropriate money for the Erie Railroad were denounced as a corrupt promotion by a "clique of heartless speculators." Bennett urged reform in the quarantine laws at New York port to lessen the inconvenience to local merchants. He exposed a "landlord conspiracy" which was keeping rents high during the depression of 1837-1838. And

[47] Two of these, for example, were in the *Herald*, Oct. 7, 22–25, 1839, Mar. 25–30, Apr. 1, 1840, and *Herald*, Dec. 31, 1843, Jan. 1–11, 21, June 27–30, July 2–6, 1844.

[48] *Herald*, Sept. 15, 16, 1842; Oct. 7, 1839, had long accounts of five fires; Jan. 5–9, 1837, Jan. 16–30, Feb. 7, 1840, Jan. 8–11, 1841; July 8, 10, 17, 20, 21, 1841; Sept. 16, 19, 1836; Sept. 2, 9, 10, 13, 14, 23, 1839; Jan. 5, 6, 15, 19, 21, May 7, July 14, Oct. 1, 2,5, 6–15, 1841. The *Herald* sent a special correspondent to cover the McLeod case and it printed many dispatches from correspondents in Montreal discussing Canadian affairs; Dec. 17, 18, 20, 22, 23, 27, 29, 1842, Jan. 4–7, 10–13, 16, 17, 19, 20, 23, Feb. 2, 3, 7, 15–18, Mar. 29, Apr. 14, 1843.

[49] *Herald*, July 8, 10, 1844, Aug. 14, 1838, May 10, June 25, July 22, 1841, Nov. 4, 6, 12, 13, 14, 16, 1842.

[50] *Herald*, July 20, Sept. 9, 21, 1836, Jan. 2, 9, 1843; Nov. 13, 14, 15, Dec. 3, 7, 8, 13, 1838, Jan. 26, 28–30, Feb. 1, 6, 11, 1839, Jan. 6, 7, 13–18, Feb. 18, 21, 24, 25, 1840.

the *Herald* joined the call for a correction in the horrible conditions at the debtors' prison.[51]

The *Herald's* nationalism reached the point of chauvinism. In the late 1830s there was persistent trouble with England over a Canadian insurrection, the *Caroline* affair, Alexander McLeod's arrest, and the northeastern boundary. Bennett started with a moderate tone, lamenting American involvement in the Canadian rebellion and disavowing any American designs on Canada.[52] But the *Herald* was firm in defense of American claims to the Aroostoock region. Maine was entitled to every foot of the territory in dispute, and if England chose to resist by war, so much the worse because the United States would win the war and annex Canada in the process.[53]

On the Texas question Bennett also shifted from a moderate position to one increasingly jingoistic. Until 1841 the *Herald* denied any American interest in annexing an independent Texas.[54] Then when war between Mexico and Texas seemed imminent in 1842 Bennett became decidedly anti-Mexican and began to preach Manifest Destiny.

It would appear by all that history and our own observation teach us, that the Anglo Saxon race is intended by an overruling Providence to carry the principles of liberty, the refinements of civilization, and the advances of the mechanic arts through every land, even those now barbarous.—The prostrate savage and the benighted heathen, shall yet be imbued with Anglo Saxon intelligence and culture, and be blessed with the institutions, both civil and religious, which are now our inheritance. Mexico, too, must submit to the o'erpowering influence of the Anglo Saxon. . . .[55]

Bennett believed, with President Tyler, that England was intriguing to free the slaves in Texas and convert that area into a counterweight to the United States. By 1844 Bennett was a bellicose advocate of Texas annexation, as jingoistic an expansionist as then existed:

It is now assumed as being the destiny of this republic, that her power will be incomplete and her civilization restricted, until her institutions of all kinds be spread from the Isthmus of Panama at the South to Hudson Bay at the North, and from the shores of the Pacific to the

[51] *Herald*, Apr. 24, 26, May 4, June 1, 1839; July 25, 1839; Feb. 8–10, 13–17, 1837, Jan. 12, 1838; July 18, 1838.
[52] *Herald*, Jan. 1, 3–6, Aug. 13, 1838.
[53] *Herald*, May 2, 3, 8, 9, 13, 25, 1839.
[54] *Herald*, Oct. 11, 1837, Nov. 10, Dec. 17, 1841.
[55] *Herald*, Mar. 22, 27, 29, 1842. Quote from July 17, 1843.

Bennett's villa overlooking the Hudson at Washington Heights. Frank Leslie's Illustrated Newspaper, *October 1, 1859.*

shores of the Atlantic—thus embracing the whole of North America. . . . This is believed to be the ultimate destiny of this republic. . . .[56]

The *Herald's* own rise and the genius of its editor was still its most important news story. It continued to detail its circulation increases and general prosperity. In 1836 it raised its price to two cents and started weekly and evening editions, and in 1839 it raised its price to three cents, secure in its growing audience.[57] Bennett boasted that his genius was solely responsible. He called himself "the NAPOLEON of the newspaper press" and compared himself to Zoroaster, Confucius, Charlemagne, Homer, Herodotus, Alexander the Great, Caesar, Luther, and Byron. "Accompanied with talent, independence, genius, and success, my vanity and egotism will in two years be deemed and taken for inspiration and virtue.[58] After presenting his readers with "Sketches of My Own Life" he commented: "What Shakespeare did for the drama—what

[56] *Herald,* Nov. 10, 1843, Mar. 7, 9, 10, 18, 20, 26, 31, Apr. 2, 18, 20, 29, 30, June 14, July 28, Aug. 2, 30, Sept. 22, 25, Nov. 14, Dec. 7, 9, 11, 21, 1844. The quotation is from the *Herald,* Mar. 19, 1844.

[57] *Herald,* May 23, 25, Aug. 18, Sept. 3, Nov. 21, 1836, Aug. 14, 1837, Oct. 26, 29, 1838, June 1, 6, Sept. 27, 1839, May 12, Sept. 21, 1840, Aug. 19, Oct. 20, 1841, Apr. 26, 1842.

[58] *Herald,* Nov. 26, 1841, May 8, 1837.

Milton did for religion—what Scott did for the novel, such shall I do for the daily newspaper press." With this collosal egotism, Bennett regarded himself as one of the most newsworthy figures of the time, and he kept his name constantly before his readers.[59]

Abuse of his less talented rivals continued. The mercantile press was denounced regularly for concentrating on "the fashionable vices and profligacy of London and Paris" and "catering to speculators, hypocrites, stock-jobbers, bankers, brokers, and political and moral rascals of all kinds."[60] The *Sun,* he said in 1838, was "entirely in the abolition interest" and declining so fast as a result of the *Herald's* competition that it resembled a real newspaper only as much "as a respectable nigger generally gets to an Anglo Saxon."[61] When Horace Greeley's *Tribune* was established in 1841 Bennett immediately saw the potential competition and shifted his attacks to it. Greeley's eccentricities, especially his vegetarianism, made him easy prey. "Horace Greeley, BA and ASS," went a typical *Herald* jibe, "is probably the most unmitigated blockhead connected with the newspaper press. Galvanize a large New England squash, and it would make as capable an editor as Horace."[62]

The scant coverage of society which the *Herald* carried in its first year was expanded. Bennett's trips to Europe, Washington, Saratoga Springs, and Niagara Falls, and the dispatches of the *Herald's* Washington correspondents, gave regular and detailed reports of the vacation resorts and galleries and restaurants of the fashionable—of anything that was interesting in the way of fashion and taste.[63] By 1840 the *Herald* was sending a reporter and an artist to the most prominent fancy dress balls in New York, providing its readers with long descriptions, and usually some illustrations, of the ball and the fashions.[64]

[59] *Herald,* Oct. 11, 17, Dec. 1, 1836, June 5, 1837, Feb. 9, 14, May 6, 1838, July 17, 1839, Feb. 3, June 8, 1840.

[60] *Herald,* July 16, 1836, May 16, 1839, Oct. 22, 1841.

[61] *Herald,* Jan. 11, 1838, Aug. 3, 9, Oct. 13, 1841, Sept. 24, 1842.

[62] *Herald* Apr. 20, 29, 1841, Sept. 14, 1842. The *Herald* usually referred to Greeley as a "galvanized squash" but sometimes as a "pumpkin" or a "miserable dried vegetable."

[63] *Herald,* July 17, 18, 21, 23, 30, Aug. 16, Sept. 18, 20, 1838 for a sampling of Bennett's European society letters; Aug. 7, 1838, for an example of society in the regular Washington correspondence; Aug. 7–20, 1839 for Bennett's letters from Saratoga Springs; July 30–Aug. 22, 1840 for Bennett's

Illustrations were commonplace in the *Herald* by this time. Bennett started slowly because of the great expense of woodcut engravings. In 1838 six of them appeared; in 1839 they were used more frequently;[65] and by 1840 the month was rare in which there was not an interesting illustration or map or both—of murderers and their victims and the scenes of the crimes, fires, Harrison's inauguration and funeral, Fourth of July celebrations, fashionable balls and distinguished foreign visitors, camp meetings, a giraffe, a mermaid, and, inevitably, one called "Dancing Girls of Egypt." The maps were of almost every newsworthy location. The *Herald* was printing about twenty illustrations a year by the 'Forties, all of them either educational or entertaining.[66]

The message parodies were dropped after a short time. A bogus "original" of Jackson's last annual message appeared in December, 1836, and Bennett then tried his hand at Van Buren's Inaugural Address.[67] This was the last of them. Bennett was going to abandon the practice, he explained:

General Jackson's character was an excellent subject for such *jeux d'esprit*. It had some strong, original, excellent points about it, mixed up, cunningly enough, with some absurdity and folly. Mr. Van Buren is a tame, insipid creature, and does not deserve to go down to future ages in any original message of mine. His real Message will, no doubt, be sufficiently absurd, without any attempt to make it absurder. . . . The present state of the country is too serious a thing to make a jest of.[68]

One final innovation should be mentioned. Bennett treated religious developments as news and gave the same kind of coverage, often light and frivolous, to religious meetings and religious leaders as he did to anything else.[69] This attitude toward religion was an important cause of the "Moral War" of

---

letters from Niagara Falls; Jan. 4, 22, 29, 31, Feb. 1, 6, 22, Mar. 9, 1839 for a sampling of Bennett's own Washington letters dealing with society.

[64] *Herald*, Oct. 17, 1840, for an example.

[65] *Herald*, Apr. 12, July 17, 25, Aug. 14, Sept. 15, Nov. 5, 1838, Mar. 11, 14, Aug. 13, 15, 27, Sept. 4, 11, 12, 22 25, 26, Oct. 8, 22, 25, Nov. 25, Dec. 12, 1839.

[66] *Herald*, Mar. 26, May 5, June 6, 19, July 6, Aug. 14, 1840, Mar. 5, Nov. 29, 1841, Feb. 15, 16, July 17, Aug. 6, 1842, for examples of illustrations; for examples of maps see Mar. 19, 23, July 10, Oct. 19, 21, 1840, Dec. 6, 1841, June 4, July 13, 1842.

[67] *Herald*, Dec. 7, 1836, Mar. 4, 1837.

[68] *Herald*, Sept. 1, 1837.

[69] *Herald*, June 26, 1835, June 17, 1837, Feb. 26, 1838, May 10, 1839, Jan. 13, May 11, 14, 1840.

Detail of full front page woodcut, The Herald, *June 25, 1845.*

1840 in which the other New York City papers tried, unsuccessfully, to ruin the *Herald* by having readers and advertisers boycott it and ostracize Bennett.[70] The "Moral War" had no effect on the *Herald's* tone; it continued to treat religion like any other subject. Each May the annual meetings of the religious and moral reform societies in New York City received more extensive coverage in the *Herald* than in any other paper. Starting in 1844, each Tuesday the *Herald* would summarize the leading sermons preached in the city on Sunday.[71]

While Bennett was directing the *Herald* into these sensational paths, what of the rest of the New York press? How did other papers deal with events that appeared in the *Herald* as sensational stories? A comparison of press treatment of several such stories, selected at random from the *Herald*, suggests the answer. In 1836, for example, the New Orleans papers re-

---

[70] For the "Moral War" and its effects see Carlson, *The Man Who Made News*, pp. 173–190, and Crouthamel, "The Newspaper Revolution in New York," pp. 96–98. There was an earlier attempt, in 1837, at the same kind of ostracism of the *Herald*, and it also failed. *Herald*, Nov. 13, 14, 17, Dec. 9, 1837.

[71] Hudson, *Journalism in the United States*, p. 453.

ported a gruesome murder. All the New York papers except the *Evening Post* picked up the item and gave a short summary of the crime, about the same length as the *Herald's*. But except for the *American*, which copied its account from the *Herald*, they were all from two to five days later than the *Herald's*.[72]

Another random example was the "Shocking Suicide—Supposed Murder—Imputed Adultery" which happened on Long Island in 1843 and received two full columns in the *Herald* on July 13. The next day there was a two inch summary in the *Courier and Enquirer*. None of the other papers mentioned the story although it had all the ingredients to make it very popular reading.[73]

A different sort of story was the coverage of the Long Island racing season in the summer of 1837. The *Herald* published regular news reports written by "Old Turfman," and on June 21 carried a three column lead editorial on the famous race horse "Mingo."[74] The other papers published advertisements for the race tracks and terse race results, but none of them treated horse racing as a news story, as the *Herald* did.[75]

One of the *Herald's* most extensive society pieces was the account of the lavish grand ball on the yacht of J. C. Stevens in 1840. The *Herald's* story on October 17 was long and complete and included a three column wide woodcut of some of the belles there. None of the other New York papers covered the fete except the *Courier and Enquirer*, which merely mentioned it.[76]

Was the reverse true? Did the other newspapers publish human interest stories that the *Herald* missed? None that I have been able to discover. The *Herald* was conducted with

---

[72] *Herald*, July 14, 1836; *American*, July 14, 1836; *Journal of Commerce*, July 16, 1836; *Sun*, July 16, 1836; *Courier and Enquirer*, July 18, 1836; *Commercial Advertiser*, July 19, 1836; *Evening Post*, July, 1836.

[73] *Herald*, July 13, 1843; *Courier and Enquirer*, July 14, 1843. A search of the *Sun, Express, Tribune, Journal of Commerce, American*, and *Evening Post* yielded nothing.

[74] *Herald*, June 3, 21, 1837.

[75] For example, *Express*, June 2, 3, 1837; *Sun*, June 2, 6, 1837; *Commercial Advertiser*, June 1, 2, 1837; *Courier and Enquirer*, June 1, 2, 1837; *Evening Post*, June 1, 2, 3, 1837. The *Journal of Commerce* printed neither race advertising nor results.

[76] *Herald*, Oct. 17, 1840; *Courier and Enquirer*, Oct. 16, 1840. A search of the *American, Commercial Advertiser, Sun, Evening Post*, and *Journal of Commerce* yielded nothing.

great enterprise and energy. If a story escaped its attention it was likely to be dull and uninteresting.

This has been only half of the story of the *Herald's* success during its first decade. It was not only the most sensational of antebellum newspapers, but it was also the most newsworthy and complete. It was the technique of sensationalism that paid the bills. It attracted the mass audience, and then the advertisers to the *Herald*. Out of his cash income from advertising and circulation, Bennett could afford the expenses of improved news-gathering techniques. On the one level the *Herald* was able to surpass the *Sun,* on the other level it surpassed the mercantile press and the *Tribune.* James Parton's contemporary estimate of Bennett as "the best journalist and the worst editorialist this continent has ever known"[77] was close to the mark. Bennett had no equal in his unerring sense of what the public wanted to read.

[77] Parton, "The New York Herald," p. 418.

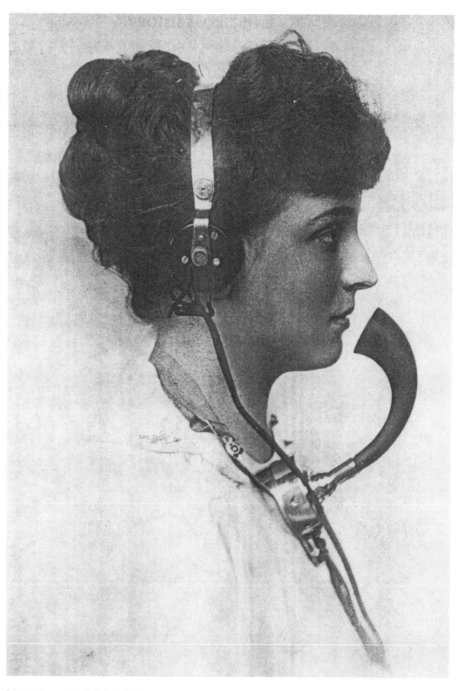

Operator Margaret Gottschalk in 1916 wearing the kind of headset
in use from the turn of the century until 1940. Illinois Bell.

The Necessary Toy:
The Telephone Comes to Chicago

BY ROBERT H. GLAUBER

"It is not too much to say that the telephone makes modern society possible, that no substitution for it would suffice." Margaret Mead

ON MARCH 10, 1876, Alexander Graham Bell made the first telephone call. With the stressful sentence, "Mr. Watson, come here, I want you!" the world shrank. Instant voice communication over distance (something men had striven after for almost a century) became a reality. The consequences of Bell's success did not, of course, become manifest at once nor did their full potential affect people immediately. But as the use of the telephone rapidly spread, it changed people's lives, attitudes, and expectations.

Now the telephone and its services are taken for granted: we build around it, count on it, use it by instinct rather than conscious planning. It has become a natural extension of our voice and style. Now, as political scientist Ithiel de Sola Pool has pointed out, it plays a pivotal role in "magnify[ing] whatever processes [are] taking place in society at a given time."

In 1877 the telegraph dominated communications in Chicago. Telegraph poles and wires made stark patterns against the downtown sky. Several companies provided the service and a small army of uniformed boys scurried through the streets delivering messages. As the number of telegrams grew, sending and delivering them became an increasing problem. Delays were frequent, errors not uncommon. Conditions were ripe for change.

A little more than a year after Bell's first call, the telephone appeared in Chicago. It was brought by Gardiner Greene Hubbard. He had six instruments and he meant to promote them. From the start Hubbard had a financial interest in the telephone. In May 1877, he and Thomas Watson issued a circular touting the instrument. It read, in part:

> Conversation can be carried on after slight practice and with the occasional repetition of a word or sentence. . . . After a few trials the ear becomes accustomed to the peculiar sound and finds little difficulty in understanding the words.
> The advantages of the Telephone over the Telegraph . . . are:
> 1st. That no skilled operator is required, but direct communication may be had by speech without the intervention of a third person.
> 2d. That the communication is much more rapid . . .
> 3d. That no expense is required either for its operation, maintenance, or repair. It needs no battery, and has no complicated machinery. It is unsurpassed for economy and simplicity.

The advertisement then gets down to figures and warrantees. "The Terms for leasing two Telephones for social purposes . . . will be $20 a year, for business purposes $40 a year, payable semiannually in advance. . . . The instruments will be kept in good working order by the lessors, free of expense, except from injuries resulting from great carelessness." So the basic appeals of the telephone were established early by Watson and Hubbard. They have not changed essentially in the ensuing century.

Senior writer and curator for Illinois Bell, Robert H. Glauber has written for encyclopedias, magazines, newspapers, and television.

Soon after Hubbard arrived in Chicago in June 1877, he called on General Anson Stager, the leading telegraph man in the city. It was a logical choice. Stager was a principal of the American District Telegraph Company, president of Western Electric Manufacturing Company, and vice president of the Western Union Telegraph Company. Stager authorized Charles Summers, an electrician with the Western Union Telegraph Company, to test the two telephones Hubbard had left with him.

Summers later testified at a patent hearing in 1881: "I believe I made the first telephone experiments with Bell instruments ever made in Chicago, using them almost every day for a month or two after I obtained them." These experiments stirred up enough interest to induce one newspaper to keep a reporter on the story. He wrote:

At 10 o'clock on yesterday morning a wonderful test was made of Bell's telephone between the Western Union telegraph office, Chicago, and the residence of Mr. C. H. Summers, in Highwood, Ill., a distance of about twenty-two miles, one of the Western Union telegraph wires being used for the purpose. At the Chicago end were Gen. Sheridan, and several of his staff, Gen. Stager . . . and other notables, Mr. Summers and family being at the remote end. . . . After some preliminary tests, Mr. F. W. Jones (without Mr. Summers' knowledge) spoke in the Chicago instrument and said: "Good morning, Summers. Do you know my voice?" Reply by Summers: "Good morning. It is Jones."
Jones: How are you all today?
Summers: All first-rate. What is your weather?
J.: Partly cloudy, and sultry.
S.: Cloudy and warm here.
J.: Gen. Sheridan is here and would like some music by telegraph [sic].
S.: Who?
J.: Gen. Sheridan.
S.: Good morning, general. What will you have?
Gen. Sheridan: Please play on your flute for us.
This request was readily granted, and every note and variation was distinctly heard by all those present.
Gen. S.: That is splendid.
Summers: We will sing you a song if desired.
Gen. S.: Thank you. Do so.

"Sweet By-and-By" was beautifully rendered . . . all the parts being readily distinguished. [*Chicago Times*, June 25, 1877]

Thus, in one of the earliest telephone calls in Chicago, a familiar use of telephone lines was established at once—information about the weather was sought and given.

But not all the early tests of the telephone in Chicago were quite so social. At about the same time Summers was singing, Leroy Firman, general manager of the American District Telegraph Company, conducted some useful tests with the Fire Patrol. Instead of telegraphing the patrol the location of a fire, he had a telephone attached to the same wire so that the location could be given verbally. It worked relatively well; the only real problem was that the fire horses made so much noise when the alarm bell sounded that it was difficult to hear on the telephone.

There were many other experiments to test the versatility and reliability of "Bell's telephone." One involved City Hall (then located at LaSalle and Quincy streets) and a system for calling messenger boys quickly. Unfortunately, it worked in only one direction. In another experiment, Firman tried to use water pipes as telephone conductors. It didn't work.

In the main, however, everyone was more than satisfied with the performance of the new device "for the transmission of articulate speech" and excited about its potential for business and for social life. The telephone was obviously going to be both popular and profitable. In the fall of 1877 Hubbard wrote to the Board of Managers of the Bell Telephone Company in Boston that he was having good success finding agents to take on telephone franchises. He noted: "Negotiations are pending with Gen. Stager of Chicago."

That phrase covered a ticklish situation. For while Stager was interested in the performance potential of Bell's instrument and held on to his option to become the Bell agent in Chicago, he was also encouraging the companies in which

The first commercial telephone. The round opening served both as transmitter and receiver. Developed by Bell in 1876 it went into service in 1877. American Telephone and Telegraph.

he already had interests to go into direct competition with Bell. Hubbard, knowing nothing of this, continued to woo Stager. Several competing telephones were eventually made by Stager companies but proved short-lived since the courts declared them infringements on Bell's patents.

But there were those who didn't care to wait for a full-time agent to be appointed. On July 2, 1877, while Summers and Firman were still experimenting, a private line telephone service was set up between the headquarters of N. K. Fairbank and Company on Dearborn Street and the firm's lard and oil factory at Blackwell and 19th streets. Business started the bells ringing.

Horace Eldred joined the Bell Telephone Company of Boston in January 1878 and worked for it in St. Louis, Louisville, and Cincinnati. On April 19 he wrote to Gardiner Hubbard to ask, "What is being done in Chicago?" The response was his transfer to Chicago in late May. He was sent to design, install, and promote the first telephone exchange there. Three men assisted him: John Taubold, a lineman; William Murray, superintendent of construction; and Edwin Poindexter, a solicitor-salesman.

Eldred worked on the exchange designs, applied for a franchise from the City of Chicago, hired local workers, and purchased the necessary materials. Poindexter sold the services of the new exchange to businesses around town. The team worked quickly, smoothly, and successfully.

Everyone involved with the early telephones recognized that their usefulness was greatly restricted because they could operate only in pairs—one instrument connected directly to the other. You spoke back and forth along a single wire. As early as May 1877 Bell had lectured on the potential importance of a "telephone exchange"—equipment that had yet to be perfected. Such an exchange would make it possible for any telephone in a town to be connected with any other one through a central office.

The first central exchange was established in Boston in May 1877 almost as Bell was talking. It served a private burglar alarm system. The first commercial exchange opened in New Haven, Connecticut, on January 28, 1878. It did not take Poindexter long to gather 33 interested subscribers to the Chicago Telephonic Exchange, "Licensed under Alex. Graham Bell Patents." In late May an advertisement was issued listing their names and inviting other businesses to join the Exchange. Its operation was described as follows: "Any subscriber desiring to communicate with any other subscriber not on their wire, notifies the Central Office to make necessary connections, and immediately the communication is established. . . . The Lines, Call-Bell and Telephone are furnished and maintained at a monthly rental." At least three of the initial subscribers are still in the directory. Their original listing read: "Chicago North-Western R.R. . . . U.S. Post Office . . . Illinois Central R.R."

Not everything went without hitch. The city council of Chicago took its time granting the franchise. Western Union strongly urged Eldred to join forces with them, as had been done in St. Louis, arguing that "if both go ahead there [would be] no money for either." But the public was enthusiastic. On June 24, 1878, Eldred published a *Notice to Subscribers* informing them that the Chicago Telephonic Exchange had

Gardiner Green Hubbard brought the first Bell telephone to Chicago in June 1877. Within a year Bell Telephone and the American District Telegraph Co. had both established exchanges in the city. Illinois Bell.

General Anson Stager was a leading figure in various communication enterprises in the city, including Western Union Telegraph Company, Western Electric Manufacturing Company, and the American Speaking Telephone Company. Illinois Bell.

secured the top floors of buildings 123 and 125 LaSalle Street, for our Central Office. Business Office, first floor, Room C, of No. 125. . . . Having now 267 subscribers, embracing all lines of trades, we have insured, beyond question, success of our enterprise in this city. Our office will be open for business, Wednesday, June 26.

So, with 267 pioneers behind them, the first Bell exchange was opened in Chicago in a small building located at what now would be just south of the alley on the east side of LaSalle Street between Washington and Madison. Money had not yet been raised locally, so financing for the exchange came from the Bell Telephone Company in Boston. In October 1878 the company issued its first proper alphabetical and classified directory. It carried 291 listings.

But this was not the first telephone exchange in Chicago. Eldred and his crew had been beaten by the American District Telegraph Company. This corporation had been chartered in Illinois in 1875 to provide telegraph and messenger services. Its financial backers were Anson Stager (who owned $30,000 worth of the initial $50,000 of capital stock), Norman Williams, John B. Drake, E. B. Chandler, and L. B. Firman. Working with a Western Union subsidiary, the American Speaking Telephone

Company, A.D.T. opened its exchange at 118 LaSalle Street with Leroy Firman as general manager on or about June 18, a week before Eldred opened the Bell exchange.

At first, Western Union had not been interested in the telephone. In the fall of 1876, Gardiner Hubbard had offered Western all of Bell's patents for $100,000 but William Orton, the company's president, turned them down. As Thomas Watson noted in his memoirs: "The Company evidently had no faith in the future of the telephone for they refused to buy the patents and wouldn't even make an offer for them." Because they needed money to continue their work, Bell and Watson were disappointed—but not for very long. Watson further recalled, "Two years later the same patents could not have been bought for twenty-five million dollars."

Within a year of Bell's offer, Western Union had changed its mind. Under the urging of men like Anson Stager, the company aggressively entered the telephone field. By 1878 its subsidiaries were producing instruments that used a transmitter developed by Thomas Edison and a receiver by Elisha Gray. Its sales force was large and hard-hitting. By early 1879 Western Union had more telephones and exchanges in service

The Mercantile Building at 118 LaSalle Street. In 1878 the American District Telephone Company established its exchange in the quarters here designated as the premises of Williams & Montgomery Insurance. CHS.

Telegraph wires dominate this view looking south on LaSalle Street from north of Washington. The first offices of the Bell Telephone Company in Chicago were located in this block at 123-125 LaSalle. Illinois Bell.

than did Bell. Neither company was about to give up what was proving to be a very lucrative market.

A patent infringement suit had been filed by Bell against Western (technically, against Peter A. Dowd, a Western Union agent) in September 1878. The case dragged on. The evidence was voluminous and complex—more than 600 pages of testimony and thousands of documents. To most the case seemed clear-cut: Bell had filed his patent application in 1876 and Western had come along later. Justice was on his side.

On November 10, 1879 a settlement was reached out of court. Western Union agreed to get out of the telephone business and to turn over to Bell a network of 56,000 telephones in 55 cities in exchange for a healthy financial settlement extending over the life of the Bell patents. Bell companies now had the right to a legal monopoly over all the telephone service in the country until Bell's initial patents ran out in 1893 and 1894.

It was "probably the greatest victory in the whole history of the world's largest corporation," wrote John Brooks in his book, *Telephone.* Thomas Watson celebrated by going to the shore at Marblehead and "declaiming to the skies all the poetry I remembered. It was an un-

dignified thing for the Chief Engineer of the Telephone Company to do . . . but I certainly felt better for it next day."

Meanwhile, back on LaSalle Street, business flourished for both the Bell and A.D.T. exchanges. The number of customers rose rapidly. The primitive equipment was pushed to its technical limits and then expanded with jerrybuilt improvements. The central offices were crowded, noisy, and inefficient. The operators making the connections and disconnects yelled back and forth at one another and shouted at the customers to make themselves heard.

In a letter of reminiscence George H. Bell, who was probably the first telephone operator in Chicago, recalled how makeshift it all was. "Early in June 1878," he wrote, "I was an A.D.T. [messenger] boy sitting in the waiting line of a dozen boys one morning when the telephone solicitor came in with Mr. Ensley, our manager. . . . Mr. Ensley asked me to step up, saying, 'This boy is George Bell. Bell invented it and Bell may as well operate it.' And so I went to work [as an operator]." Without any training, he was put on the board.

Bell was probably also the first operator to be fired. "One day [the solicitor] came in and told me I was fired because William Hibbard of Hibbard, Spencer and Bartlett had said I talked back to him. . . . Hibbard added that the telephone was a nuisance and anyway, he would have it taken out if I wasn't discharged." The boy was resourceful. He went over to Hibbard, Spencer and Bartlett, applied for a job and was hired by Mr. Hibbard "because I seemed to be a nice boy."

But boy operators did not last long. In September the first woman was hired by the Bell Company, Isabel Maunsell. She started as a disconnect operator and within four years became chief operator.

Also in September 1878, the city council finally granted the Chicago Telephone Exchange its franchise. The company was given blanket permission to use the city streets to set poles and string overhead wires save only that the work had to be done under the supervision of the Department of Public Works. For its part, the company agreed to provide free telephones in four municipal offices.

Historian Bessie Louise Pierce noted:

The council was to regret its lack of foresight. . . . [In 1881] the city became embroiled in a bitter dispute to force the wires underground. . . . Eventually in 1885, the municipality reached a compromise by which the company promised to place its wires underground, and the city accepted a delay in the accomplishment of the project.

Crude the exchange service was in 1878 and rough the equipment, but it offered something to the public that was earnestly wanted—quick, easy, and direct communication at affordable prices. Like most innovations, the telephone grew faster in Chicago than anywhere else in the country. The city rapidly became a hub of experimentation and development. There was no standardization. Exchange managers made it up as they went along.

As early as October 1878 Eldred recognized that the switching board at 125 LaSalle was hopelessly inadequate to the demands being put upon it. He ordered another one from Boston

and hoped. When a new form of board—one that he had never seen before—arrived, he installed it in one crash weekend. Many years later Eldred recounted how he did it.

In order to disconnect the wires from the old switchboard and transfer them to the new one, all the wires . . . had to be rerun. . . . Notice was given to the subscribers that from a certain hour on Saturday we should be unable to give them any service until the following Monday morning. . . . I selected 15 or 20 of the best employees we had and notified them that we should require their continuous work from Saturday morning until Monday morning without any rest; that I would furnish them with the necessary refreshments in the exchange room, but none of them would be permitted to leave unless they were sick, and a certain amount of extra pay would be allotted to them for the successful connection and working of this new switchboard on the following Monday morning. On Saturday we cut off all the wires from the old switchboard and immediately commenced running the new ones from the cupola of the building down to the switchboard on the floor below.

The space between the roof and the ceiling of the exchange being only two and a half feet, the men were required to lie on their backs and by the light of candles pull the wires through cleats down to the switchboard. Sometime during Sunday . . . the insulation of the wire took fire from a candle which was overturned and a large portion of the work was destroyed and had to be done over again.

On Monday morning about 7 o'clock we had all the wires connected, and out of the 15 or 20 employees . . . there were only two that were able to keep awake under the strain they had passed through . . . the rest of them being so worn out that they were asleep on the bare floor of the exchange, completely exhausted. [*Deposition of Horace H. Eldred*, U.S. Circuit Court]

The new switchboard went into operation on January 3, 1879. By June the company was able to issue a directory informing customers that it now had six branch exchanges scattered throughout the city to serve them. It listed more than 1,000 subscribers out of the city's population of 491,516.

OVER 15,000 IN USE.

CHICAGO TELEPHONIC EXCHANGE,

LICENSED UNDER ALEX. GRAHAM BELL PATENTS,

125 LaSalle Street,

Chicago, June 24, 1878.

NOTICE TO SUBSCRIBERS.

We have secured the top floors of building Nos. 123 and 125 La Salle Street, for our Central Office. Business Office, first floor, Room C, of No. 125.

We are building our wires and connecting stations with Central Office at the rate of six daily. Our linemen will call on you in a few days to get location of your instrument. Having now 267 subscribers, embracing all lines of trades, we have insured, beyond question, success of our enterprise in this city.

Our Office will be open for business, Wednesday, June 26, and will furnish Messengers promptly by ordering through Telephone. We are concluding arrangements to furnish Liverpool, London and New York Market Reports to our subscribers.

We would caution our subscribers against misrepresentations of other parties, and state our Company hold the only patents under which Telephones can be manufactured and used. Suits will be commenced against parties using any speaking Telephone other than the Bell Telephone, as infringers on the Bell patents, and injunctions obtained to enforce the rights of the patentees.

H. H. ELDRED,
Gen'l Agent for Bell Telephone Co.

This notice was sent to subscribers to the Chicago Telephonic Exchange to announce its opening on June 26, 1878. Illinois Bell.

Alexander Graham Bell initiates the New York-Chicago telephone line in 1892
by calling Angus S. Hibbard, a telephone company executive, in Chicago.
Illinois Bell.

78 Chicago History

Company wagons like this one, photographed in 1901, transported telephone company representatives to the districts in which they solicited subscribers. Illinois Bell.

On December 21, 1878 the Bell Telephone Company of Illinois was incorporated. Gardiner Hubbard, who had introduced the telephone to Chicago, became its first president. Over the next few years the ownership, officers, organization, and name of the company changed with bewildering frequency, for this was a period when control over the financial management or sale of corporations was non-existent. They were traded like horses. Nevertheless, the essential thread of equipment and service from the Chicago Telephonic Exchange in 1878 to Illinois Bell at present has never been broken.

The approaching settlement between Western Union and National Bell did nothing early in 1879 to stop the struggle between their two exchanges in Chicago, but it did take some of the wind out of A.D.T.'s sails. As a result, Bell swept ahead. Anson Stager was not about to accept this lightly, so he hired Horace Eldred away from Bell and made him general manager of all Western Union's nationwide telephone efforts.

Late in 1879, the telegraph and telephone companies reached their agreement in Boston, but the American District Telegraph Company continued to compete with Bell in Chicago as a separate entity until 1881. In that year there

was a merger in the public interest. "Competition between (the two companies) was exceedingly keen, and the subscribers to the two lines suffered an inconvenience which the consolidation eliminated," wrote Bessie Louise Pierce. The new organization was called the Chicago Telephone Company. Its first president was the tireless General Anson Stager.

The formation of the single company allowed the combined staff to concentrate its energies on getting new subscribers and making technical improvements rather than wasting resources on rough competition with each other.

What was first derisively called Bell's "electrical toy" proved to be extremely popular. Chicago people needed little encouragement to use it. By 1882 there were 2,610 in use within the city limits; by 1890 there were 6,518. The figures really jumped then reaching 26,661 by 1900, 575,840 by 1920, and crossing the one million mark in 1940. This had doubled by 1963.

Exchanges spread throughout the rest of Illinois quickly. The field was wide open to anyone with capital and ambition. Bell and Western Union subsidiaries started many of the early offices but by no means all. The Bloomington exchange opened in June 1879, the Danville and Decatur exchanges in October. Joliet started its exchange in March 1880, and Freeport in April. In 1881 Champaign opened in July and Canton in December. Aurora followed in early 1882 and Evanston later that year, and so on into the twentieth century. In 1882 there were only 392 telephones in Illinois outside of Chicago. By the late 1960s there were more telephones in Illinois outside of Chicago than in it.

There were occasional setbacks. A representative of the Bell Company demonstrated telephones in and around Kankakee during October 1880. By November 9 he had obtained only 23 subscribers—2 short of the 25 needed to start an exchange. So he left town. The exchange in Kankakee was finally opened in June 1881.

From the beginning, nature and/or man's carelessness contrived to make exchange life a succession of crises. The Gold and Stock Telegraph Company, a Western Union subsidiary, opened the first exchange in Springfield in September 1879. Bell interests quickly took over in November. On the night of November 12, customers and staff alike were treated to the kind of emergency that became a regular feature of telephone life. The local paper, the *State Register*, reported the next day:

There was a confusion of telephonic voices in the city last night. A derrick fell across the wires and mixed them, resulting in the right calls for the wrong person. . . . The boy at the central office was driven to the verge of insanity. He couldn't get the wires untangled, and finally he fled the office in despair. His hair turned white in a single night.

The claim in the last sentence may not be absolutely plausible, but it makes its point.

The telephone exchanges which spread out to blanket the state were by no means exclusively owned and operated by Bell companies. There have always been independent companies in Illinois providing service in their own franchised territories. True, there has been a steady trend toward consolidation; but, as the number of independent companies has diminished, the number of telephones they control has increased sharply. In 1940 there were still 415 independent telephone companies in Illinois serving 323,205 telephones. By 1950 those figures were 360 companies and 502,918 telephones; by 1960 there were 198 companies and 708,376 telephones; and by 1970 there were 68 of the former and 1,121,183 of the latter. At present the companies range from General Telephone Company in Bloomington with 825,000 telephones to the Adams Telephone Cooperative in Golden with 6,000, and the Alhambra-Grantfork Company of Alhambra with about 1,000 telephones.

At the close of 1976 the businesses and homes of Illinois had 8,994,110 telephones—7,448,692 Illinois Bell and 1,545,418 independent. The state population was 11,191,000, giving a density of slightly more than 80 telephones per 100 people. On an average day approximately forty million calls were made in the state. So much for statistics!

As popular and intriguing as it may have been in the early days, the flooding of the telephone across the plains of Illinois was due in great part to technical advances that made the use of the telephone convenient and cheap.

Switchboards made it possible for two subscribers who were not on the same line to talk together. Starting out as manually operated "switching" boards, they eventually became automated electronic installations capable of an enormous variety of functions.

Loading coils boosted the distance it was possible to transmit a voice along a wire. In their various forms and refinements they made long distance transmission possible, up to and including the use of underseas cable and microwave.

The dial made it possible to complete a call without the intercession of an operator. Originally called "automatic telephone exchange," the first dial in Illinois was used at Fort Sheridan in 1892. Without such an automatic device (now expanded to include push buttons and various automatic dialing devices), it would not be possible to assemble enough operators to handle manually the volume of calls made each day or to construct buildings large enough to house the equipment they would require. This invention by Almon B. Strowger was manufactured in Chicago by the Automatic Electric Company, and the first system was introduced in La Porte, Indiana, in 1892. The company that installed the automatic system in Chicago was the Illinois Telephone and Telegraph Company, which operated from 1903 to 1917, when the Bell system took it over.

The transistor made miniaturization possible and vastly increased the speed at which telephonic equipment could operate. Both are pivotal for handling the mass volume that made telephone talk what Marshall McLuhan termed "an absolute right."

All of these technical improvements had important effects on the financial aspects of telephoning and were essentially responsible for the fact that the long distance call from Chicago to New York which cost $9 in 1892, when the service was initiated, now costs less than 50¢.

The proliferation of patents, statistics, and corporate financial reports tells only the impersonal side of the story. For more than the first half-century of its life, the telephone was primarily a humanistic device—both for the people who worked for the company and for its customers. There are numerous documents from the old days that convey some of the special flavor that working for a telephone company must have had. William Boyd, who started with the company in 1881 as a night operator and retired as treasurer in 1930, described his first job this way:

After 9 PM it was our job to clean out the office, clean and fill the lamps with kerosene, fill the stove with coal . . . blow the dust out of the switchboard with a bellows. . . . After all our chores were done, we would connect up with a few nearby towns. They would furnish singing or music on an organ.

(*The Old Oaken Bucket* was the favorite.) We had our own quartet: the manager, myself, the janitor of the building and his friend.

The instructions for personal conduct issued in 1911 to troublemen—now called installation/repair technicians—included such exhortations as:

Put up a "good front." It is not necessary to advertise any tailor shop, neither is it necessary to go about your work looking like a coal heaver. Overalls can look as respectable as anything else, but they must at least show that they are on speaking terms with the laundryman.

If requested to go around to the back door, don't consider yourself insulted, but try to realize that the lady of the house may not have a maid and is only trying to save work for herself.

If you ever believe that a subscriber is a crank, forget it. All of them are wise enough to tell when a telephone is not working right. Not every troubleman can do this.

Treat everybody as you like to be treated, not forgetting your horse. If you want to know the horse's side of it, just take off your coat and hat some zero day, hitch yourself to the post with your belt, and stand there about two hours.

Don't go pellmell through the streets regardless of pedestrians as though you were going after your salary check.

If you don't like your job, resign.

That was hard-bitten candor for the working man. For the working girl, all was sweetness and light. *The Beauty of Service*, a 1918 recruitment booklet, recounts in rather purple prose the tale of Joy Miller who comes to Chicago to help earn enough money to care for her ailing Mother. After several disheartening experiences in looking for suitable work, Joy sees an advertisement for telephone operators: "Life, death, personal happiness, business, the fate of nations—all rest with the Girl at the Board. She never fails." Joy resolves to apply. This "was not merely an advertisement to her; it was a call from Mother." The rest of the story deals with virtue rewarded. Joy gets the job . . . loves her work . . . saves a life by telephone . . . is reunited with her

Advertisements such as this in the *Chicago Daily Tribune* of November 1, 1878 contributed to a patent infringement suit by Bell against Western Union (of which the American District Telegraph Company was a subsidiary). The suit was settled out of court in favor of Bell in 1879. CHS.

Mother. "It was the Chicago Telephone Company that did it all!" she cries in triumph.

Heady stuff, no doubt. But it helped produce the kind of operators who, as "Central," helped knit communities together for generations. In the popular mind, "Central" *was* telephone service.

For the public the telephone quickly became more than just a device to make running a business easier. In a decade or two, use of the telephone took on very special social overtones. A 1903 report indicated:

Ingenious householders have found innumerable other uses for the telephone never originally thought of. The operator many times gets the request, "Please ring my bell at 6 o'clock tomorrow morning"; 5,000 times every day in Chicago she is asked the time of day; election and prizefight results, football and baseball scores are asked for and reported.

Many attempts have been made to allow telephone patrons to listen to operas, concerts and church services, but without any permanent success. I have heard of people taking French and German lessons over the telephone.

There was social prestige in it. If you had a telephone *and* a victrola, it was fashionable to tie up the lines for hours entertaining your friends with records by Caruso, Tetrazzini, Journet, or the John Philip Sousa Band. A telephone also brought a certain tone of respectability to any establishment. The 1905 book carried a listing for Ada Everleigh (South 412), but one can only speculate about its uses.

Songs about the telephone were very popular. *Hello! Ma Baby; Drop a Nickel, Please; Hello Central, Give Me Heaven; Call Me Up Some Rainy Afternoon,* and the real winner, *All Alone by the Telephone* were sung everywhere. There was a special kind of joshing telephone humor that made it feasible in 1903 for the Illinois Telephone and Telegraph Company to advertise its new dial as "the cussless, waitless, out-of-order-less, Girl-less telephone."

Not all uses of the telephone were acceptable.

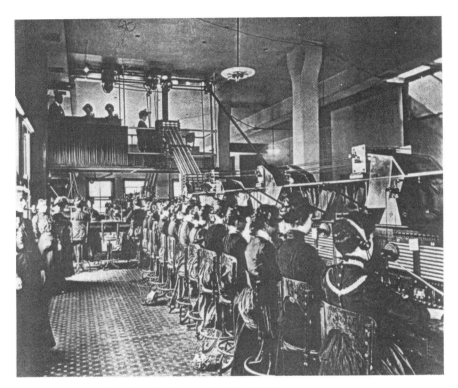

Operators at the toll call switchboard in the main Bell office in Chicago, 1902. The women standing by the wall are supervisors. Illinois Bell.

There were quick legal restrictions on the use of the telephone for "the dissemination of race-track information"—in a word, on bookies. The battle against them started early and hasn't ended. In a sample period, from 1939 to 1950, Illinois Bell took out more than 13,500 illegal bookie lines in Chicago.

Since the early 1960s, many companies have used the telephone as a tool for mass sales solicitation. This was not practical until dials were widely available (Chicago was not "all-dial" until 1960), but once they were, the practice spread quickly. While some customers now regard telephone sales solicitation as an invasion of privacy, the sales rationale seems to be that we can easily throw out junk mail unopened, but there are few people who can resist answering a ringing telephone.

In recent years telephone lines have been used to put forth all sorts of messages. In Chicago one can, or recently could, dial a prayer, a joke, a nature story, a tip on losing weight, a hint on kicking the smokers' habit. In December 1977 almost 8½ million calls were made to Santa

Claus. Obviously such messages fill a public craving.

Telephone lines have also been used to transmit the propaganda of racist and anti-semitic groups. Many of these hate-filled messages have stirred up great public protest and have been debated in the courts. The consistent ruling has been that the law gives equal access to telephone service for all—the sound and the sordid.

In times of national joy or crisis people turn to the telephone for confirmation and comfort. On the night of November 11, 1918, bedlam broke out all over Chicago when news of the armistice swept through the city. In one hour, more than 370,000 calls were made by people anxious to share the news. A night operator in Woodstock recalled that she first heard the news from a lineman in Chicago. "Shortly after, a long distance call came in for the superintendent of the town factory. Then the plant's whistle—normally used for fire alerts—began to sound. The board lit up like a Christmas tree with almost everybody in town calling to find out where the fire was. She remembered that "We answered the calls with: 'The war is over. Number, please?' "

The attack on Pearl Harbor caught everyone unprepared. A usually quiet December Sunday in the central offices was turned into a frantic rounding up of operators to handle the avalanche of calls. Long distance was especially busy. Some 42,000 calls were placed in a very short time causing long delays for some circuits. In contrast, at about 11:30 AM on December 8, 1941 all telephone traffic virtually ceased as President Roosevelt went on the air to assure Americans that "with the confidence in our armed forces, with the unbounding determination of our people, we will gain the inevitable triumph."

But that was only one instance of volume calling. That memorable early fall night in 1959 when the White Sox clinched the pennant, every civil defense siren in the city screamed out in celebration. Within moments, central offices all over the city were overwhelmed by callers trying to find out from each other and the operators if Chicago were under attack.

The entire telephone system was jammed again just after noon on November 22, 1963, when the country was stunned by the assassination of President Kennedy. In the hour between 12:30 and 1:30, more than 90 percent of all local callers in Chicago had to wait for a dial tone.

The biggest calling day in the city's history, however, was January 26, 1967—the day Chicago was buried under two feet of snow. But Chicagoans were only technically snowbound. They stayed at home and made 16.7 million telephone calls.

Anthropologist Margaret Mead has expressed the view that:

> It is not too much to say that the telephone makes modern society possible, that no substitution for it would suffice, but that all the applications [of the telephone] are all necessary in a world in which people need to be both far apart and very close.

There is no way of knowing how much, if any, of this idea may have been in Alexander Graham Bell's mind when he invented the telephone or in Gardiner Greene Hubbard's when he brought it to Chicago. But, at present, it would be virtually impossible to disengage from our day-to-day lives the telephone and the innumerable services based on its worldwide switching network that we use without thought.

The growth of that network was an effective factor in the growth of American business, both the multiplant, multinational corporation and the smaller or isolated company. Instant voice-to-voice communication wiped out the problems that distance and time had always posed for business. In a similar way, our densely packed, vertical cities are practical only because of the elevator (which makes internal movement feasible) and the telephone. The ubiquity of the telephone plays almost as important a role in modern mobility as does the airplane. People

Shopping by Telephone is a Most Convenient Way

"Private Exchange-One"

MARSHALL FIELD & COMPANY'S TELEPHONE SWITCHBOARD, THE LARGEST PRIVATE TELEPHONE SYSTEM IN THE WORLD. OVER 250 BRANCH LINES. REQUIRES FOURTEEN DAY OPERATORS TO MAKE CONNECTIONS. A BRANCH LINE RUNS TO EVERY SECTION OF THE STORE

There is a growing satisfaction in shopping by telephone with us. Every effort is being made to make it so. Every order, inquiry, or request will be quickly and intelligently cared for. Every section of this store is at your service.

Call for "Private Exchange-One," and then ask the operator for the section you wish to speak with. If in doubt, explain briefly to the operator, who will give proper connection. For directory of sections, see page 90.

Marshall Field & Company
State, Washington, Randolph and Wabash Ave.

are willing to move from city to city because distance is now measured in the cost of a long distance call or the length of a jet flight rather than in miles.

Bell's simple black box contained seeds that have grown into an immense variety of services all transmitted over the telecommunications network. Television and radio broadcasts of major newsbreaks, of sporting events, presidential elections and inaugurations, wars and civil disturbances, our favorite entertainment, and the Academy Awards—all go via the network. The stock quotations in brokers' offices, the wire photos printed in newspapers, the remote alarm and security systems that watch over closed or distant homes, offices, banks, museums, and factories, the computer in one place that feeds or gathers information from another computer across the city, state, or country, the credit and identity checks made in stores and banks, the medical data sent from patients remote from a hospital or doctor's office—all travel over the network. An electronic blackboard is just around the corner. A little further along in the future, the network will help a person away from home turn off the lawn sprinkler, turn on the oven, and put out the cat. These and other services are all grandchildren of that first "Mr. Watson, come here, I want you!" They make possible, in Daniel Boorstin's phrase, that "mass-production of the moment" that is probably the most characteristic feature of contemporary society. Bell invented it. Clever men promoted it. As one social commentator has observed:

our lives [are] shaped by this innocent and unobtrusive invention, which began as a toy, acquired status as a luxury, became a comfort, and is now such a necessity that welfare families can have their telephones paid for by the state. We shaped it to us, and now it shapes our lives in ways to which we give no thought. [Ithiel de Sola Pool in *The Social Impact of the Telephone*.]

Selected Sources

Illinois Bell Historical Collection. ADVERTISEMENTS: May 1877, May 1878, June 24, 1878, in *Illinois Telephone and Telegraph Company Directory*, 1903. LETTERS: from Gardiner Hubbard, September 1877 (?); from Horace H. Eldred, April 19, 1878; from A. O. Morgan to Thomas Sanders, June 2, 1878; from George H. Bell, August 27, 1940. INTERVIEW: taped, with Dorothy Teeple Suhr, November 23, 1976. DEPOSITIONS: Charles H. Summers, U.S. Patent Office, *Firman vs Shaw*, Chicago, 1880; Horace H. Eldred, U.S. Circuit Court, *Western Electric Co. vs Citizens Telephone Co.*, Chicago, May 31, 1898. PUBLICATIONS: *Rules for Troublemen*, 1911; *The Beauty of Service*, ca. 1918; *Pioneer Reminiscences*, by William J. Boyd, ca. 1930.

Boettinger, H. M. *The Telephone Book*. Croton-on-Hudson: Riverwood Publishers Ltd., 1977.

Brooks, John, *Telephone*. New York: Harper & Row, 1976.

Kingsbury, J. E. *The Telephone and Telephone Exchanges*. London: Longmans, Green and Co., 1915.

Larned, S. J. "Telephone Service." *Western Electric*, March 20, 1903.

Mackenzie, Catherine. *Alexander Graham Bell*. Boston & New York: Houghton Mifflin, 1915.

Mead, Margaret. "Looking at the Telephone a Little Differently." *Bell Telephone Magazine*, January-February, 1976.

Pierce, Bessie Louise. *A History of Chicago*. III. New York: Alfred A. Knopf, 1957.

Pool, Ithiel de Sola, ed. *The Social Impact of the Telephone*. Cambridge: The M.I.T. Press, 1977.

Watson, Thomas. *Exploring Life*. New York: Appleton and Company, 1926.

By John Cameron Sim

19th Century Applications of Suburban Newspaper Concepts

Some "modern" suburban press concepts such as central plant publishing began a century ago.

► As pointed out in a previous paper,[1] the name "suburban" was applied to some newspapers as early as the 1870s, although of course not in precisely the same sense as it is used today.

Further research indicates that the concepts of central plant publishing, of a group of newspapers all substantially the same except for one or more local news pages, and of promotion of the special merits of a suburb as an audience for advertising, now so closely identified with the suburban press, also came into extensive use during the '70s.

Similarly, the sometimes disconcerting (to directory compilers and historians) practice of a publisher starting, and then a short time later killing, papers in order to test markets and formats occurred then as it continues to occur now. This certainly was one of the practices which led to the inflated statistics of numbers of weekly newspapers

published in the last decade of the 19th century and the first 15 years of the 20th.[2]

That the name "suburban" had been applied to a newspaper in some form at least as early as 1875 is attested to by inclusion of the name in a list of "The Newspaper Names of America."[3] The list did not show when or where the name was used. Some possibility exists that its use may have been suggested by publication in 1870 of the popular work by William Dean Howells, "Suburban Sketches."

In 1877 the former West Roxbury (Mass.) *Gazette* was re-named the Boston *Suburban News.* It was a weekly published Saturdays, identifying itself as Republican and claiming a circulation of 1,000. (Then, as later, such rounded-off circulation figures were suspect.) The next year the paper changed its name to Jamaica Plain *Suburban News,*[4] thereby more closely identifying itself with a suburb. Publisher E. Porter Dyer was very much like a present-day suburban chain promotion director in saying that the *News* "only paper published in the 23rd ward of Boston (formerly the town of West Roxbury and lately annexed) circulates among a population of 13,000, comprising many of the wealthiest and most aristocratic families of Boston. The paper is devoted to local interests."[5] This is essentially the same kind of presentation on demographics a suburban newspaper publisher might use to sell national advertising. In another suburb, Dorchester, the *Beacon & Newsgatherer*

---

[1] John Cameron Sim, "Toward a Definition of 'Suburban Newspaper,'" paper read at the 1973 convention of the Association for Education in Journalism, and printed in *Grassroots Editor,* Vol. 14, No. 8, Nov.-Dec., 1973. Reference is made (page 27) to a New England Suburban Press Association existing at least as early as 1883, therefore probably some years earlier.

[2] Sim, *The Grassroots Press: America's Weekly Newspapers* (Ames, Iowa: Iowa State University Press, 1969), pp. 43-7.

[3] *Geo. P. Rowell & Co.'s American Newspaper Directory for 1876* (New York: Geo. P. Rowell & Co., publishers), p. 12. A search of the directory did not turn up a paper using any form of the title "Suburban," so it must have been a short-lived sheet published prior to 1875.

[4] *Pettengill's Newspaper Directory for 1878* (New York: S.M. Pettengill & Co., 37 Park Row, publisher.) Pettengill was one of the first advertising agents.

[5] *Ibid.*

► Prof. Sim is a member of the journalism faculty at the University of Minnesota.

627

used another timeless selling point, describing itself as "essentially a family and local paper of eight pages...more carefully read and preserved than dailies."[6]

By the 1870s, too, the practice of printing newspapers in one plant for several communities had taken hold, most notably in Massachusetts and Illinois. At Palatine, Ill., the firm of E.J. Dougherty and Benedict published the Palatine Sun, the Barrington News Enterprise, and the Wanaconda (Wauconda?) News. The firm's directory blurb described these as "...all thoroughly independent, form-(ing) a journalistic phalanx which ought to be powerful. Each of the papers constitutes a magazine of home news, and is freely used by the local advertisers, an earnest that they have a circulation of consequence."[7] But Messrs. Dougherty and Benedict refrained from saying exactly what that circulation was.

A considerably more ambitious undertaking, one very much resembling a modern suburban group down to popularity of the name "Sun" for units of the group, was that of H.L. Goodall & Co., whose base was the daily and weekly editions of the Drovers Journal, serving the Chicago Union Stockyards and the livestock and meat-packing industries of the Midwest. Goodall moved into a new plant, erected just outside the main entrance to the Stockyards, in 1875.[8] This new production capacity, unquestionably augmented by the surge in popularity of readyprints (also called patent insides and cooperatives) evidently led Harvey Goodall by 1877 to try serving a number of Chicago suburbs and neighborhoods with new community weeklies.

The firm claimed the following circulations for these units: Hyde Park Sun, 2,500; Jefferson Sun, 500; Lake Sun, 3,000; Lake View Sun, 750; Maine Sun at Des Plaines, 500; Blue Island Sun, 500; Cicero Sun (said to be at Austin in the Town of Cicero), 750; Cook County Sun, at the Stockyards, 1,000; South Chicago Enterprise, 1,000; and Thornton Sun (published at Dalton and Riverdale) 500.[9] The papers at Hyde Park, Lake,

Lake View, Des Plaines, Blue Island, Cicero and Thornton were not in the next year's directory, and it may be that they failed against competition. On the other hand, the omission could be due to the publisher's failure to send in directory data.

Although publication days of these units were scattered throughout the week— Monday, Tuesday, Wednesday and Saturday, leaving Friday for the weekly Drovers Journal—it is certain that the firm could not have accomplished the production job without the aid of the readyprint system. But just because that made it comparatively easy to start new papers without substantial capital investment or greatly expanded work force, just so it was easy to abandon the papers if the enterprise did not prove to be as profitable as expected.

Prof. Elmo Scott Watson has credited the original idea of readyprints to English printers "early in the Nineteenth Century," and said the first experiment in the United States was made by T.L. Terry, editor of the Berlin (Wis.) Courant early in 1862:

He conceived the idea of forming an association of publishers for printing inside sheets at a central office where material to meet the common wants of the various newspapers could be set in type and advertisements be included in the readyprint sheets as a source of revenue for the enterprise ... nothing came of it.[10]

Thus the idea of central plant publishing, now seen as the savior of the rural community weekly, was tried a century before its flowering in the late 1950s or early 1960s. It is true that the intent was to produce "insides" or "outsides"

[6] Ibid.

[7] Pettengill's Directory for 1877.

[8] A.T. Andreas, History of Chicago, Vol. III, 1871-1885, p. 35. (Chicago: A.T. Andreas Publishing Co., 1886).

[9] Pettengill's Directory for 1877.

[10] Elmo Scott Watson, History of Auxiliary Newspaper Service in the United States (Champaign, Ill.: Illinois Publishing Co., 1923). Watson wrote three versions of this monograph, one of them as a master's thesis. Of interest is his final sentence: "The final estimate of its (readyprint) importance in American journalism is that, in giving service in the truest sense of the word, it will live when other more superficial developments have long since passed away." (P. 46) The service was discontinued in 1952.

instead of finished newspapers, and the participating publishers needed their own printing plants to complete the product, whereas the concept is applied today to eliminate the heavy capital investment needed for modern offset presses and plate-making facilities. Those who have scanned files of smaller 19th century papers will be aware that "outsides" consisting of rewrites of world and national news could be ordered instead of the more usual feature time copy used on inside pages. That is, the readyprint house printed pages one and eight, or one, four, five, and eight instead of the inside pages two, three, six and seven. In either case, the customer received his order of flat newsprint sheets printed entirely or partially on one side and then ran these through his flatbed press to print the other side off forms containing the local news and ads set in his own shop.

As the man who picked up the idea and made it work, Ansel N. Kellogg came to be known as "the father of the readyprint industry." Others[11] would give Andrew Jackson Aikens of Milwaukee no less than an equal share of the credit. Both men were unquestionably aggressive leaders in the rapid expansion of the service, but Aikens did seem to have moved the fastest and the farthest. In 1870, under the firm name of Cramer, Aikens & Cramer he established the Chicago Newspaper Union and shortly thereafter opened a New York Newspaper Union (not to be confused with printers' trade unions).

Kellogg founded the Chicago *Evening Lamp,* a weekly, in 1862 and three years later began his first experimental efforts with readyprints. He also founded the *Publishers' Auxiliary*[12] as a monthly aimed primarily at the typographical

trade. After building his extensive readyprint service he gave it up because of ill health and died March 23, 1886, at the age of 54.[13]

A directory published in 1871 noted that the number of papers joining a readyprint service had doubled in the previous year and that "more than 1,000 new papers have been established since January 1, 1870."[14]

Readyprints made it relatively simple and inexpensive for newcomers to start new papers. But clearly quite a number of established publishers saw that they were in an even better position to do the starting. By 1880 it was a fairly common practice in northeastern states to issue papers for several towns from one printing plant. The Marlboro (Mass.) *Mirror-Journal,* for instance, printed papers for Hudson, Northboro, Shrewsbury and Southboro. The papers at South Acton and South Framingham in Massachusetts each printed for three other nearby towns, and Clinton and Lynn printed for one other town. The El Paso (Ill.) *Journal* printed papers under the nameplate of *Home Journal* for Benson, Gridby, Minonk, Rutland, and Secor. Competition was provided by the Minonk *Journal* which printed the Benson *Argus,* Dana *Herald,* Ransom *Times,* Roanoke *News* and Rutland *Post.* The White Hall (Ill.) *Register* printed papers for Greenfield, Kane and Roodhouse in the Chicago area.

It is probable that the peak effort in these suburban projects was that undertaken in 1889-90 when Charles R. King & Co., operating at 415 Dearborn St., Chicago, with the Chicago *Record* as its cornerstone publication, reorganized (or established a subsidiary) as the Suburban Newspaper Co. It seems more than coincidental that the office of the Western Newspaper Union, ultimately the dominant firm in the readyprint industry, was next door at 417 Dearborn, and it is probable that Suburban Newspaper Co. was a subsidiary or joint enterprise with WNU.

The firm established 51 newspapers in neighborhood communities and sub-

[11] See the monograph "Centennial Newspaper Exhibition," prepared by Geo. P. Rowell & Co. It was distributed at the Philadelphia Exposition of 1876, which put on display bound volumes of newspapers contributed by the New York Newspaper Union, Chicago Newspaper Union, Milwaukee Newspaper Union, St. Paul Newspaper Union, Aikens Newspaper Union, and Southern Newspaper Union.

[12] *Pettengill's Directory for 1878.*

[13] *Inland Printer,* Vol. III, No. 7, April, 1886, p. 435.

[14] *Geo. P. Rowell & Co. Newspaper Directory for 1871,* p. 4.

urbs of Chicago,[15] some of them lying seemingly quite far out for the transportation of that day. It must be remembered, however, that by the 1890s, dozens of trains daily came into Chicago along routes passing through or by those suburbs.

Many of these papers were facing well-entrenched opposing publications, which by that time had operated in the communities 25 or 30 years. The experiment failed, and most of the new sheets disappeared within a year or two, undoubtedly because they did not solve the problems imposed by the required suburban formula—local news and comment.[16] Simply printing a nameplate over the time copy that the readyprints offered was not enough; similar content could be found in the metropolitan dailies, which offered also timely news of the world. Yet to provide local editors and reporters would be impractical. Obviously the publishers hoped to make their profit from the national advertising placed in the readyprints, almost all of it for patent medicines and nostrums (hence "patent insides"). But that's where the readyprint industry had to get its profit, and at the low rates and relatively low volume of the times there simply couldn't be enough revenue to go around.

The best hope for profit in a multiple operation of this kind was in legal notice printing, but established publishers faced with competition from the multiple operations had been quick to persuade state legislatures to pass statutes containing restrictive definitions of what constituted "a legal newspaper." Before all the loopholes were closed, some enterprising printers scored coups, as the following rather plaintive letter to the editor of *Inland Printer* attests:

Manchester, N.H.—A recent enterprise in this city is the Kendall Newspaper Company, which publishes a paper for 40 different towns in this vicinity, the papers being all practically the same, with the heading changed for each town. This company had a "fat take" in publishing the laws passed at the last session of the legislature, each paper publishing the same receiving between $60 and $70 regardless

of the number of copies circulated, the combined circulation of the 40 papers being much less than that of many single papers.—F.T.Z.[17]

As could be expected, sharp differences of opinion on the desirability of readyprints existed. The Chicago *Daily Times* of March 11, 1876, is quoted as saying:

Within the past 12 years there has been a vast improvement made in the country newspaper—particularly in respect to the care in which all the news of the day is gathered and edited and the literary ability displayed in its columns, which relieves the provinicial press from the few objections that have heretofore been urged against it, and has accorded to it higher dignity, extended its influence, and greatly augmented its power for good. The typographical appearance of the country journal of today is perfect; the display and classification of its advertisements tasty; the quality of the paper used and the printing far superior to that of its city contemporaries."[18]

The article further described the great advantages for national advertisers:

... Ridiculously cheap. *One cent per line* in each paper covers the expense... But one cut is necessary for the whole 1,200 papers. If he were to advertise direct with 1,200 different papers it would require an

[15] *Pettengill's Newspaper Directory for 1878* listed the 51 papers: Almira *Democrat*, Auburn Park *Monitor*, Austin *Times*, Blue Island *Chronicle*, Bowmanville *Democrat*, Central Park *Herald*, Chicago Lawn *News-Record*, Colehour *Journal*, Cragun *Democrat*, Cummings *Sun*, Downers Grove *Press*, Elmhurst *Gazette*, Elsdon *Herald*, Fernwood *News*, Glencoe *Gleaner*, Grand Crossing *News*, Grayland *News*, Gray's Lake *Enterprise*, Harlem *News*, Highland Park *Herald*, Hinsdale *Herald*, Irving (Park) *27th Ward Democrat*, Jefferson Park *Democrat*, Hyde Park *Times-Sun* (and *Times* for Kensington), Lake Forest *Era*, Lake Zurich *News*, Libertyville *Lake County Times*, McCaffrey *Sun*, Maywood *News*, Melrose *Times*, Moreland *Herald*, Morgan Park *Reflector*, Naperville *Home News*, New Bremen *Journal*, Normal Park *Collegian*, Oakdale *Enterprise*, Oak Park *News*, Pullman *Journal*, River Forest *Glen*, Roseland *News*, Shermerville *Weekly*, South Chicago *Star*, South Englewood *Mirror*, South Evanston *News*, Washington Heights *Post*, Wauconda *Journal*, West McHenry *News*, Wheeling *News*, Winnetka *Times*.

[16] Margaret V. Cosse, *The Suburban Weekly* (New York: Columbia University Press, 1928). Miss Cosse said "the suburban weekly... has the function of explaining the community to itself; the obligation of editorial interpretation." (P. 5)

[17] *Inland Printer*, Vol. VII, No. 2, November, 1889, p. 145.

[18] From a sketch entitled, "The Co-operative Newspapers," in a booklet published for the Centennial Newspaper Exhibition at Philadelphia, 1876, by Geo. P. Rowell & Co. of New York.

equal number of cuts...and the cost for cuts and postage would equal or exceed the cost of the advertising, as many advertisers have found to their sorrow heretofore.

Despite such glowing praise from the *Times*, then one of the nation's major dailies, readyprint service was disparaged by many, especially papers with readyprint competitors. This was an attitude which was to become much more pronounced as the 20th century advanced, and linecasting composition equipment replaced handsetting of type on even the smallest weeklies. For instance, the Boston *Suburban News* in 1877 advertised that it was "Not cooperative."[19] This rather odd-sounding disclaimer did not mean that the paper would pay little attention to the needs or wants of advertisers or subscribers, but rather that the *Suburban News* did not use readyprint service, thus was not part of a Union or "cooperative." These firms were not, of course, cooperatives in the sense of being controlled by member-shareholders as we understand the term today; they were individual proprietorships or corporations.

The Chelsea (Mass.) *Telegraph and Pioneer*, founded in 1845 and claiming a circulation of 2,000 in 1877, said, "The publishers wish it distinctly understood that they are not in need of, nor do they have the help of patent outsides to issue the paper, doing everything connected with its publication in their own office."[20]

The experimenting in this fashion grew to such proportions that the N.W. Ayer Directory for 1890 felt compelled to print a prefatory note under the title "Sub-editions." It said:

> From the very first we have endeavored to distinguish between independent and distinct publications and those which are simply reprints of some other paper with little and often with no change beyond the

name. This practice has been increasing year by year until it has assumed proportions that in our estimation call for more decisive action. A concern[21] in one of the New England States last year proposed to publish not less than 40 papers, each of which apparently was published in the town for which it was dated, and, except the name of the paper, not a line was different in any one of the 40. Another in a Western State publishes about 60 papers on very much the same plan, the only difference being that, instead of all being exact copies of one paper they are prepared in groups, consisting of from three or four up to, perhaps, a dozen; the papers of each group being identical and the groups themselves varying only a little in the makeup or in a local column; there are numerous other instances of this in various parts of the country to a lesser extent.[22]

As previously noted, 1889-90 probably saw the crest of this effort to serve suburbs with newspapers bearing at least the community's name. Ayer's Directory for 1892 credited the Suburban Newspaper Co. of Chicago with just 21 such papers instead of 51, and Marlboro, Mass., was down to 19. The 1895 Directory did not even list a Suburban Newspaper Co. A more modest enterprise in the Chicago area, the *Suburban Times*, with five papers in 1890, still had five in 1895. By 1897 it had moved to Des Plaines and became the *News*, but it still published newspapers for Edison Park, Norwood Park, Palatine, and Park Ridge. Marlboro, Mass., now had a newspaper named the *Enterprise*, publishing editions under the name of *Enterprise* for Acton, Berlin, Concord, Hudson, Maynard, Northboro, Southboro and Sudbury.

The flowering of the trade press, strengthening of state editorial associations, and the rise of the National Editorial Association all gave impetus to a greater feeling of professionalism and the importance of community news and comment among weekly newspaper editors. The increased availability of mechanical typesetting equipment strengthened the trend toward individual locally based newspapers for suburbs. Distinctive suburban group arrangements and centralized publishing were not to re-emerge on the American scene in any number until after World War I.

[19] *Pettengill's Newspaper Directory for 1877*, p. 12.

[20] *Ibid.*, p. 15.

[21] May have referred to the Marlboro, Mass., *Mirror*, or could have referred to the Kendall Newspaper Co. of Manchester, N.H.

[22] *N.W. Ayer & Son Directory of Newspapers and Periodicals for 1890*, p. 13.

Telecommunication and the City

By FORREST WARTHMAN

ABSTRACT: Telecommunication has historically been interrelated with transportation. Since railroads first used the telegraph as a scheduling aid, this interrelationship has affected almost every mode of point-to-point and broadcast telecommunication. For example, commercial radio and television depend on advertising which, in turn, is dependent on the transportation of goods to and from urban market centers. Telecommunication also serves as a substitute for travel, greatly increasing the speed of information consumption and processing and greatly broadening the availability of information and entertainment to individual homes and moving vehicles. As long distance communication continues to decrease in cost, major urban centers will become more international, since they are the focal points of travel. Telecommunication will also assist the outward spread of metropolitan areas, but social and transportation factors will constrain this outward movement.

Forrest Warthman is an urban planner in Berkeley, California. He holds Masters Degrees in Architecture and City Planning from the University of California. He has planned physical developments, open spaces, transportation systems and information networks. He is the author of Cable Television: Its Urban Context and Programming.

THROUGHOUT history the transmission of information over distances has depended primarily on transportation. Messages, whether written in letters or carried in the memory of humans, have required a mode of physical transportation to unite the sender with the receiver. Roads and waterways formed the basis of these networks: in America, as in other parts of the world, the post office was an early and major builder of roads. Any method which could improve the speed or reliability of transportation had a similar effect on communications.

A BRIEF LOOK AT THE PAST

Transportation was not the only means of sending messages. In ancient times simple messages were sent over long distances by signaling. Sound from the human voice, gongs or battle trumpets could carry through any weather, but they offered no privacy and traveled relatively short distances. Visible signals by smoke, light or flags could travel much longer distances with varying degrees of privacy.

During the French Revolution Claude Chappe developed a systematic network of optical telegraph which incorporated more advanced codes for signaling.[1] His system evolved into a signaling network for railroads which used it to monitor and schedule the movement of trains. By mid-nineteenth century Samuel Morse's electric telegraph was in wide use, often running its wires along the same railroad lines it helped to control. Not only did it

1. Ben Dibner, "Communications," in *Technology in Western Civilization,* ed. Melvin Kranzberg (New York: Oxford University Press, 1967), vol. 1, pp. 452–468.

improve the speed and reliability of the railroad system—which at this time was beginning to take long distance postal delivery away from stagecoaches and steamboats—but it also carried messages in itself, becoming important in the Civil War. While telecommunications helped railroads run better, railroads were instrumental in bringing communications, rural population and industrial raw materials to cities.

Alexander Graham Bell won his patent for the telephone in 1876; within a decade the basic organization of the Bell Telephone system was established. City telephone companies were brought together by region and, then, across states by American Telephone and Telegraph (AT and T). The first open-wire line between New York and Boston was dedicated in 1885. Unlike telegraph, the telephone allowed an untrained public to use their natural voice and language without conversion into machine code.

Somewhat later another technology of equal importance began its impact on urban life: the automobile and its relative, the truck. By the 1920s city streets were being equipped with electric traffic signals to improve the flow of automobile traffic, just as they were used in conjunction with telegraph to improve the control of railroads. It may be reasonable to speculate that the convenient and independent access to distant people afforded by telephones influenced public expectations with respect to physical travel, hence, the demand for individual automobile travel. At any rate, the telephone and the automobile came to be used in an interrelated fashion. Potential destinations of travel could be reached more quickly and cheaply by telephone than by actual travel. Moreover, telephone could help travelers and truckers predict the consequences of their trip: they could arrange and cancel appointments quickly, check on the availability of services or goods and communicate with their families while on the road. Once telecommunications had improved the predictability of travel, everyone started traveling more. However, the inertia of travel time and cost worked harder against transportation than against telecommunications.

Just before 1900 Guglielmo Marconi patented his wireless telegraph system. By 1901 he succeeded in transmitting a message across the Atlantic, some forty years after the first transatlantic telegraph cable had been laid. His invention opened communications with ships, land vehicles and, eventually, aircraft; it also led to commercial radio and television broadcasting. With others, Marconi developed reflectors to focus short-wave radio signals in narrow beams that traveled great distances with little energy, forming the basis for many of today's microwave systems.

Scientific discoveries and technological improvements accelerated the growth of telecommunications. By 1915 vacuum tube amplification made transcontinental telephone service possible. By 1920 the first commercial radio station was built by Westinghouse, and vacuum tube radio receivers were successfully sold to the public. The cathode ray tube was widely used in radar systems by 1936 and in commercial television by the 1940s. Transistors began replacing vacuum tubes in all types of electronic systems by the late 1950s. These led to large scale integration circuits for today's com-

puters and consumer electronics. Magnetic tape was introduced in the 1950s for storage of voice, data and television information. Storage media have since become an integral part of telecommunications, one that still depends on transportation-oriented delivery networks, such as the postal system.

Telecommunications—the electronic transmission of information—has special importance for the development of cities. It is closely related to transportation, one of the major shaping factors of urbanization throughout history. The vast majority of information or programming carried over telecommunication networks originates, terminates or passes through urban centers. Telecommunications has become a nerve-like web in our urban economic, political and social lives.

TODAY'S TELECOMMUNICATION NETWORKS

Some telecommunication networks are used so frequently that they cannot escape our attention. Today, the average individual receives more telephone calls than pieces of mail, and the average household has a television turned on more than six hours each day. Other networks are hidden from view, supporting our daily or occasional activities in many ways. Few telecommunication networks rely on only one technology. The telephone network stands above others by making use of almost every technology in the field.

Telegraph

In the postwar era Western Union has experienced a dramatic decline in domestic message traffic. Many of its personal message offices have been closed in efforts to further concentrate services on teletype and computer-related business applications. Telegraph lines and offices are concentrated in urban business districts; furthermore, many clients not located near a telegraph cable use standard telephone lines to reach the main telegraph switching office.

Western Union's automatic teletypewriter exchange service (TELEX) offers international teleprinter and data communications through the international carriers: International Telephone and Telegraph (IT and T), RCA, AT and T and Western Union International —now unrelated to the domestic Western Union company. The domestic teletypewriter exchange service (TWX) allows machine conferencing and automatic receive-answer capabilities over direct distance dialing lines. Both the TELEX and TWX services employ message-switching at their exchanges; this allows an accumulation of messages at busy switching centers, with automatic forwarding to final destinations as long distance trunk lines become free. Western Union was the first company to be granted rights by the Federal Communications Commission (FCC) to own and operate a domestic satellite.

It is possible that the telegraph network may evolve into a more pervasive electronic postal network.[2] Western Union has recently begun its mailgram service in cooperation with the United States Postal Service. Under this system messages delivered to the local telegraph office are transmitted to the destination city, where a teleprinter in the post office records the

message for delivery as regular first class mail.

Telephone

Originally begun as a voice communications network, the interstate telephone trunk lines now carry a higher proportion of nonvoice— that is, data, facsimile and television—traffic, although local telephone lines still carry primarily voice communications. The telephone network is experiencing a shift from local to long distance traffic; annual rates of increase related to the postal system are listed in table 1.

TABLE 1

TELEPHONE TRAFFIC RELATED TO THE POSTAL SYSTEM

TYPE OF COMMUNICATION	ANNUAL INCREASE, PERCENT
Pieces of mail	3.5
Local telephone calls	5.0
Toll telephone calls	10.0
Overseas calls	25.0

SOURCE: John R. Pierce, "Communication," Scientific American 227, no. 2 (September 1972), p. 37.

Interstate trunk lines use coaxial cables and microwave networks. Overseas calls are carried on shortwave radio, submarine cables and satellites. New solid state switching devices—actually, special purpose computers—and digital transmission techniques are coming into use by local switching offices and major domestic trunk lines. This new equipment can provide faster, more accurate switching and transmissions.

New terminals are being introduced by telephone companies and private entrepreneurs. Dataphone is a telephone company device for in-

2. President's Task Force on Communications Policy, Final Report (Washington, D.C.: Government Printing Office, 1968), p. VI–43.

terfacing computers with the telephone network. The touchtone telephone and other pushbutton devices can be used to communicate with distant computers; the user enters information in coded form on the pushbuttons, and the computer responds in simulated human voice.[3] Home devices which receive computerized information over telephone lines can display this in print form on standard color television receivers. Combined video-keyboard terminals for commercial data systems are operated regularly over the telephone network. They are heavily used by the airlines and other transportation industries, financial industries, government and law enforcement agencies and, more recently, newspaper publishers. Electrostatic copying machines and voice scramblers for privacy are also available for acoustic coupling with any standard telephone headset. In a few major cities two-way personal television communication, using small viewing screens, is offered on the picturephone system.

Many telephone companies operate landmobile telephone exchanges which link automobile, maritime and private aircraft telephones with the landline telephone network. AT and T has proposed an ambitious landmobile telephone plan, offering improved reception characteristics, for large urban areas which would link a matrix of low power landmobile broadcast stations with the local telephone network. Portable walkie-talkie telephones for use on sidewalks of the largest urban centers will also be available in the very near future.

New areawide calling rates are

3. James Martin, *Future Developments in Telecommunications* (Englewood Cliffs, N.J.: Prentice-Hall, 1971), pp. 135 f.

being introduced by telepone companies. Optional residential telephone service (ORTS) expands customers' local call area over a somewhat larger toll call region for a fixed monthly cost. Wide area telephone service (WATS) allows users to place or accept calls from major regions of the country for a fixed rate. Among the heavier users of incoming WATS lines are credit card companies and hotel reservation systems; sales organizations and federal government agencies have been major users of outgoing WATS lines.

Mobile service

Most land, maritime and aeronautical radio systems utilize pairs of frequencies: one frequency carries communications from the control or dispatching station to all vehicles; the other frequency is used by all vehicles, one at a time, and is received only by the control station. Common carrier radiotelephone systems are equipped to utilize several frequency pairs, depending on the geographic area in which the vehicle is located. Advanced mobile data and teleprinter systems can send coded messages from the control station to selected vehicles, greatly increasing privacy. In some cases, such as radiotelephone, highway patrol and trucking company applications, several mobile broadcast stations are interconnected by microwave or cable to service large geographic areas.

Aeronautical and maritime voice communications are supplemented with radiolocation and radionavigation systems which transmit voice, data or other electronic signals. Every major airport is equipped with instrument landing systems which allow aircraft to be homed onto electronic guide beams and navigation signals within a few miles of their runways. Advanced radar systems for tracking aircraft can identify particular aircraft, their location and speed and display it in alphanumeric print on a video screen in airport control towers. Spacecraft are special users of telemetry equipment; space radio control information can be input to computers which remotely operate many mechanical systems aboard the spacecraft or satellite.

Radio broadcasting

Radio systems have been operated for many years over telephone or special wire networks in various parts of Europe and Japan, much as have closed-circuit public address systems. The Muzak system of subscription radio in the United States is still carried in this manner. Broadcast radio, however, has the advantages of reaching more people in densely populated urban areas at lower cost and of diversifying the capital investment in equipment by having listeners purchase their own radio receivers. More recently, the advantage of reaching automobiles has been a major factor in radio's survival after the appearance of television broadcasting.

Amplitude modulation (AM) radio stations cover widely varying geographic areas, depending on the power of their transmitter, atmospheric conditions and topographic and conductivity features of the earth's surface. A 50,000 watt AM station may reach several hundred miles on a summer night, bouncing its low-frequency signals off ionospheric layers above the earth's surface, while the same signal on a winter's day may reach less than one hundred miles.

Frequency modulation (FM) stations are higher in frequency than AM stations, and they rely primarily on line-of-sight signal transmission. Because their signals cannot follow the curvature of the earth, they reach smaller geographic areas. However, the fidelity of FM is much better than that of AM and has led to increasing use of stereophonic broadcasting. A few FM stations have also attempted noncommercial subscription broadcasting for business and industrial background music which requires special receivers.

Only AM stations are nationally interconnected over the telephone networks and not to the extent they once were. National radio networks were nearly eliminated by the advent of national television networks. Nevertheless, today's AM and FM radio stations are interconnected by an international program network based on the postal distribution of disc records and audiotape cassettes. Radio has become a vital marketing link for the music recording industry.

Television broadcasting

More United States households have television receivers than have telephones—96 to 92 percent in 1970. While radio took over two decades—1920 to 1940—to penetrate 80 percent of United States households, television accomplished this same penetration in half the time—1950 to 1960. It was a development which seriously affected the broadcast radio and the motion picture industries.

Very high frequency (VHF) channels are much more desirable to broadcasters than the ultra-high frequency (UHF) channels, partly because VHF channels have better natural propogation and reception characteristics and partly because televison set manufacturers have not provided UHF tuning devices equal in quality or convenience to the tuning devices for the twelve VHF channels.

Occasionally, television stations will elect to broaden their viewership by reaching cities beyond their normal reception area of fifty to seventy-five miles. This is done with translator stations placed at the fringes of broadcast reception; the translator antenna receives the weak signal, amplifies it and rebroadcasts it. Cable television systems perform a similar, although competitive, service.

Broadcast television and radio are well suited to urban area program distribution, because large concentrations of the population within geographic ranges can be reached by these media. Advertising support of the broadcast industry eliminates the need for privacy in communications: the more people who can be reached, the better. The large audiences in urban areas earn higher than average incomes and consequently consume advertised products heavily. Foods, home and hygiene products, automobiles and gasoline are major advertising interests. All of these products are commonly available in urban and suburban shopping centers, reachable by automobile. Suburban shopping centers, in particular, have specialized in convenient access by automobile; this has made them a focus of economic activity which commercial broadcast media have promoted.

Subscription television—or pay television—provides programming on a user-fee basis rather than an advertising-sponsored basis. It has the potential of being more respon-sive to market demands for specialized programming not presently available on commercial television.[4] These systems generally use coded broadcast signals and small decoding devices at each subscriber's television set to ensure subscription payment. Public television, supported by federal taxes, foundations, industrial sponsors and individual viewer donations, still remains in an uncertain state of development, searching for reliable funds to support programming of broad cultural and social interest.[5]

Cable television

Cable television systems were first built in the 1950s when broadcast television was growing rapidly, and the growth of both media has been closely related. As were broadcast translator stations, cable television systems were built primarily around the reception fringes of major urban broadcast stations. The cable companies pick up the weak broadcast signals with high quality antennas placed on high points of the earth's surface, then amplify and retransmit them through their network of coaxial cables which run down city streets on telephone or utility poles.

The cable systems are franchised by local cities to operate within the boundaries of the city. Unlike most television translator stations, cable systems can finance the importation of signals directly through monthly subscriber fees. Cable systems are also more versatile than broadcast

4. Roger G. Noll, Merton J. Peck and John J. McGowan, *Economic Aspects of Television Regulation* (Washington, D.C.: Brookings Institution, 1973), pp. 136–150.

5. Les Brown, *Television: The Business Behind the Box* (New York: Harcourt Brace, Jovanovich, 1971), pp. 314–346.

translator stations. They can be economically viable on very small subscriberships: the average United States cable system had less than 3,000 subscribers in 1973. Furthermore, they can import and carry many signals on their coaxial cable, while translator stations carry only one channel per station.

Two major factors have influenced the growth of cable television. First, people in small towns have become increasingly aware, through travel and traveling friends, of the entertainment programming available on major commercial broadcast television stations. Second, the gradual, but decisive, movement of United States populations away from the city center into suburban residential areas has brought larger television markets into outlying areas.

While the average capacity of United States cable systems is below twelve channels, those presently being built in the largest urban areas have up to twenty activated channels, with capacity exceeding this number. Two-way data transmission on cable systems can provide automatic billing for special subscription programs by monitoring the channel to which each subscriber is tuned at any point in time. The same system can be used to monitor burglar or fire sensors built into subscribers' homes. With the addition of small keyboards—ten to twenty pushbuttons—subscribers would be able to make specific requests for subscription television programs, services or merchandise displayed on television advertising programs or in printed catalogs distributed through the postal system. By adding small strip printers to home terminals, verification of purchases can be made. By adding electronic image storage devices in homes, still pictures or print information can be displayed on home television sets from an information library.[6]

There are advantages of speed to the computer-controlled system of data transmission and monitoring services two-way cable systems can provide. Local data transmission over the telephone and telegraph networks has been limited by the slow speed of electromechanical switching equipment used in local offices. Only as this older switching equipment is replaced with new computerized equipment will local data transmission by telephone and telegraph approach the speeds obtainable on the polling networks proposed for two-way cable systems. Yet, the telephone and telegraph networks have two present advantages for data communications and for telemetry or monitoring services. First, they are pervasively accessible in downtown business districts where data users are concentrated, whereas cable television systems have been built primarily in the suburbs to carry entertainment programming. Second, the telephone and telegraph networks are regionally and nationally interconnected for two-way service, allowing software services to reach geographically dispersed locations economically, whereas cable television systems are not presently interconnected in this manner.

Microwave networks

Point-to-point narrow beam microwave systems were developed

6. Walter S. Baer, *Interactive Television: Prospects for Two-way Services on Cable* (Santa Monica, Cal.: Rand Corporation, 1971), pp. 18–19.

most vigorously by AT and T to carry their interstate television and voice traffic. These systems operate at frequencies above one billion cycles per second and utilize repeater stations spaced about thirty miles apart. The ability to carry all types of electronic information on microwave systems has been improving both with technological advances in FM, AM and digital transmission techniques and with advances in channel multiplexing and amplification circuitry.

Simple and multiple microwave links are used for interoffice communications in business, industry and government. Cable television companies use them to import distant broadcast signals. Utility companies use them to control and monitor remote dams and reservoirs. Transportation and communication companies use them to link landmobile radio stations and other control centers. More recently, new common carrier microwave links have been built between major urban centers to carry commercial traffic of many kinds. These common carrier links operate as trunk lines to which the individual users gain access by their own telephone, private cable, landmobile radio and private microwave links. Eventually, these independent microwave networks will span the country to compete with the business service networks operated by the Bell system and Western Union.

New technological developments and new frequency space allocations by the FCC are expanding the potential applications of broadcast microwave stations. Instructional television fixed service (ITFS) was established for use by educational institutions in reaching many school buildings or other receiving sites from a single broadcast studio.

ITFS systems can carry up to four television channels; also, they can be equipped to carry voice communications from individual viewers back to the central broadcast studio, much like landmobile radio. Multipoint distribution service (MDS) can carry one-way television, voice or data over a radius of approximately twenty-five miles. MDS will operate as a common carrier for many applications.

Satellites

International communication satellites have been in commercial operation by the International Telecommunications Satellite Corporation (INTELSAT) since 1965, carrying voice, data and television signals. In 1970 telephone voice transmission accounted for almost 80 percent of the total INTELSAT channel capacity and revenues.[7] Other satellites are operated for meteorological, telemetry and communications research. Experiments in satellite communication with conventional aircraft and maritime vehicles are in progress.[8] All satellite communications are dependent on the transportation technology of rocketry.

In 1972 Canada became the first country to launch a geostationary satellite for domestic use in long distance communications. Several corporations are seeking authorization from the FCC to operate domestic satellites in the United States. The FCC has indicated that it will grant approval to more than one applicant.

7. Communication Satellite Corporation, *Eighth Annual Report* (Washington, D.C., 1971), p. 6.
8. International Telecommunication Union, *Tenth Report by the ITU on Telecommunication and the Peaceful Uses of Outer Space* (Geneva, 1971), pp. 58 f.

Any number of earth stations, located in widely dispersed areas of the country, can be reached using wide-beam broadcast antennas on-board the satellite. Unlike land lines of cable or microwave, satellite systems do not require hardware interconnection of the various receiving sites. This eliminates terrestrial distances as a factor in the cost of communication. It is a method of transmission for delivery of many television signals from central program centers to large urban cable systems throughout the country presently under study by the cable television industry. As the cost of earth stations declines in the future, satellite broadcasting to more remote areas may become economically feasible.

TELECOMMUNICATION AND THE FUTURE CITY

Life in the future city will be carried on at a faster pace of communicating, learning, producing, consuming and moving on to different experiences. Our printed communications will continue to be replaced by telecommunicated messages over teleprinters, facsimile devices and television screens, although print media will continue to grow. The translation of languages on overseas communications will occupy our attention more. We will obtain more information over telephone and microwave networks from regionally or nationally centered data banks for specialized business purposes. There will be many more private closed-circuit telecommunication networks built with underground cable and over-the-air microwave or laser. If cities provide underground communication conduits in the streets of urban centers in which private companies or individuals lease space, this proliferation of specialized communication networks will grow even faster.

In our homes we will receive more information over devices linked to the telephone network and cable television systems. Magnetic tape cassettes and magnetic discs will grow in our libraries; most will be produced nationally, but some will be produced and exchanged between friends and business associates, as are letters. Together with expanded broadcast television, cable television and radio, these storage media will give us more choice in programs in a manner approaching real-time, interactive access to central libraries. Access to entertainment and basic information will be supported by user fees, in addition to advertising. We will be on more television programs and will use closed-circuit or common two-way television systems, first in business and then in the home.

The transactions and events which support human activity and mobility will grow at an even faster pace as our expectations of convenience, safety and dependability increase. Financial transactions, investments and commodity prices will respond much more rapidly to news. Computing and telecommunicating power will become more interrelated with monetary wealth. Access to data collection and recording will continue to raise questions of privacy. Our physical movements in homes, automobiles and airplanes will be monitored more by telecommunication and telemetry systems. We will have telecommunicating ability literally following us down the streets with walkie-talkie telephones and paging devices.

Mobile communications from automobiles and airplanes will broaden our sense of work and leisure. We will spend more work time more efficiently in travel, communicating with clients and home offices as we move. We will expand our reach to clienteles in more remote cities and suburbs, making more overnight sojourns. Both mobile communications and landline telephone networks will provide voice and data contact with central information centers. Broadcast news and entertainment to mobile vehicles will be greater and more specialized, adding comforts to travel. Broadcast frequencies will concentrate more on reaching moving receivers rather than stationary ones.

As microwave and satellites expand their networks, the relative costs of long distance communication will decrease. This will initiate and support more long distance travel. The airports necessary for this will expand near urban centers, bringing more hotels and international business offices with them, until they cannot grow further due to traffic congestion and other environmental side effects. They will gradually move out from urban centers to larger regional airports served by small aircraft with additional reservation, control and general communication support systems.

Metropolitan areas will continue their expansion outward to form megalopolitan areas. The very largest urban centers of the country will be linked by satellite to many lesser urban centers; these lesser centers will use microwave to relay information around their suburbs.

Industrial and office employment will continue their decentralization, linked to urban centers by these telecommunication links over which distance plays a lesser role in communication cost. However, the outward move will continue to be gradual: the ties to urban centers are still dependent on physical travel at less frequent intervals. The need for face-to-face contact within the metropolitan area will continue, and the mainstream of international business, social and political life will continue to revolve around urban centers. The desires for privacy, security and convenient transportation motivate the outward movement of populations from urban centers; yet, other social factors influencing our lives hold us in denser living patterns.[9]

It is possible that electronic communications will be used in urban centers to create a sense of privacy and security where it cannot be maintained by social institutions. Urban residents may find themselves behind electronic surveillance devices and remote communication systems which they neither understand nor control directly. To the extent that these devices and networks improve the convenience and comforts of urban life, leaving more time for social interaction, they will be beneficial. But to the extent that they isolate us socially, they will impair our well-being.

9. Christopher Alexander, *The City as a Mechanism for Sustaining Human Contact* (Berkeley, Cal.: Institute of Urban and Regional Development, University of California, 1966), pp. 19–32; and Richard L. Meier, *A Communications Theory of Urban Growth* (Cambridge, Mass.: Massachusetts Institute of Technology Press, 1962), pp. 20–44.

By Andy McCue

Evolving Chinese Language Dailies Serve Immigrants in New York City

Two new publications are among the seven dailies now serving the steady influx of Chinese into the metropolitan area.

► While New York's once flourishing ethnic press is generally dying out, the Chinese language press is experiencing a period of rebirth and change. Two newspapers have been started in the past year and a half, raising the number of Chinese dailies to seven. Renewed competition and improved technical quality have made the papers almost unrecognizable to people who read them 10 years ago.

The Chinese press has been spared the fate of most of the ethnic press by the McClellan Immigration Act of 1965, which insured a steady flow of Chinese immigrants to New York. The Act set an upper limit of no more than 20,000 immigrants per year from any one country, but it also perpetuated special categories outside the quota system for relatives of people already in the United States, and for special skills needed in America.

As a result, more than 20,000 Chinese a year are entering the United States, according to figures compiled by Professor Betty Lee Sung of City College of New York. Almost all of the immigrants are from Hong Kong. In addition to the immigrants, said Professor Sung, approximately 8,000 Chinese a year, most-

ly from the Republic of China on Taiwan, enter the United States on student visas. Professor Sung estimated that only 2% of these students returned to Taiwan after completing their education.

Professor Sung estimated that 25 to 30% of the Chinese immigrants come to New York. Thus, 5-7,000 Chinese a year, the vast majority of whom cannot read English, are dropped into the laps of New York's Chinese newspapers every year.

The number of Chinese immigrants explains the continuing existence of the Chinese press, but it does not completely explain the changes taking place within the papers. These changes reflect the nature of these immigrants.

The first Chinese immigrants to the United States were almost all laborers from the rural districts of Kwangtung Province. They were often illiterate, and their sole interest in coming to the United States was to make money to take back to China. The new wave of immigrants comes from a modern, urban environment. They have been educated in the Hong Kong schools.

The differences between the two groups of immigrants are reflected in Professor Sung's statistics. Only 3% of Chinese immigrants before 1934 fell into the category of "Professional and Technical, Managers, Administrators," but 30% of the immigrants in the period from 1965 to 1970 fell into this category.

The urbanized, better-educated wave of immigrants demanded new and different things from the Chinese press. The immigrants were accustomed to reading the lively Hong Kong press. The New

► The author has a master's degree in Chinese Studies from the University of Washington in Seattle. This article was based on his master's project at Columbia's Graduate School of Journalism where he received a second master's degree last May.

York Chinese language press often did not measure up to their standards.

The older Chinese papers were reflections of the intellectuals who ran them, men who were more worried about the future of China than the future of their countrymen in the United States. Some were northerners who could not speak the Kwangtung dialects which over 90% of the Chinese in America speak exclusively.

Papers were started in New York to support the various factions which were struggling for power back in China. There was a Communist paper, *The China Daily News,* now a bi-weekly, and several Kuomintang papers, notably *The China Tribune* and *The Chinese Journal.*

Fifty years of competition for political power in China and the support of the American Chinese communities insured that the focus of the American Chinese language press remained China. While news of international and national importance, like the energy crisis and the Arab-Israeli war, could grab headlines, the emphasis remained on Asia and particularly on China.

As a result, there was little news of importance about Chinatown. Community news was usually confined to coverage of the official activities of the major establishment organizations, such as the Chinese Consolidated Benevolent Association.

But the old-style newspaper with its Asian emphasis, intellectual concerns and lack of community news could not continue to exist when faced with the new immigrants. The gap that was appearing between the papers and their readers was brought home to the New York Chinese newspapers by the arrival of *Sing Tao Jih Pao* in 1969.

Sing Tao is a Hong Kong newspaper. When it arrived in New York, it had already been tempered in the fires of the highly competitive Hong Kong newspaper market. Technically and editorially, it was much better than the older Chinese press.

Sing Tao used smaller characters than the established papers, and thus got twice as much news on a page. *Sing Tao* also was produced by offset printing, a great improvement in efficiency and clarity over the old letterpress.

But the biggest change that *Sing Tao* brought to New York was the way its growing circulation emphasized the importance of the new immigrants to the Chinese newspaper community. Then, as today, with roughly 5-7,000 people arriving every year (an average of 15 to 20 new people in Chinatown every day), new readers were constantly appearing.

And the new readers, whether due to habits brought from Hong Kong, or to critical appraisal of *Sing Tao's* competitors, usually turned to *Sing Tao* for news.

But *Sing Tao* was still a Hong Kong newspaper. Only half a page of news was put into the paper in New York. All the rest was flown here daily from Hong Kong. By the time it gets to New York, it's a day old. Some of the older papers, notably the *China Tribune* and the *Chinese Journal,* reacted slowly to the lessons of *Sing Tao's* success, but others in Chinatown sensed the opportunities. They established a new paper, the *China Post,* in the fall of 1972.

In the year and a half since its establishment, the *Post* has succeeded in gaining a circulation as large as those of its two largest competitors, *Sing Tao* and the *United Journal,* roughly 10,000. It has succeeded both by imitating *Sing Tao* and by attempting to fit itself to the New York Chinatown community in a way no Hong Kong newspaper could. In doing so, it has become a catalyst of further changes in the Chinatown press.

Some changes are indicated in a content analysis of the seven Chinatown dailies during the week of March 4 to March 9, 1974. (See Table 1)

For comparison in reading the tables, the *China Post* and *Sing Tao* are the leaders of change, with the *United Journal* and the *China Times* as imitators. The *China Tribune* and the *Chinese Journal* represent the old style Chinese press. The *Youth Daily* is a new paper whose days seem numbered already. It is heav-

TABLE 1

*International News in Chinese Language Daily Newspapers in
New York for Week of March 4-9, 1974, in Square Inches*

	People's Republic	Hong Kong	Taiwan	Other Int'l News	Total[A] Space
China Post	168½	1,028½	771	867	**19,440**
China Times	217½	871½	1,506½	765	21,168
China Tribune[1]	283	613	518	1,021½	15,360
Chinese Journal[1]	481	335	572	892	15,360
Sing Tao	584½	1,742	280	1,781½	17,640
United Journal	492½	1,077	1,032½	1,190½	17,640
Youth Daily	224	306	481½	581½	15,360

[1] The *China Tribune* and the *Chinese Journal* use larger characters. Their pages hold only half as much news as those of the other papers.

[A] The total number of square inches in the newspaper for the week under study.

ily overloaded with advertising and its news judgment and printing quality are nothing short of atrocious.

TABLE 2

U.S. News in Chinese Dailies,
March 4-9, 1974

	American National[A]	Watergate	Other
	Inches	Inches	Inches
China Post	660½	212½	448
China Times	539½	227	312½
China Tribune	438	235½	202½
Chinese Journal	796	173½	622½
Sing Tao	685	240	445
United Journal	582	230½	351½
Youth Daily	256½	126½	130

[A] American-National is the main category, the others are subdivisions of it.

The first major difference between the old and the new style of newspapers is the amount of space devoted to features. While the new style papers generally print as much serialized stories and other fiction as the older papers, they print substantially more factual features, i.e. sports, household hints, movie and entertainment news.

The *Post* devotes a third more of its space to features than its closest competitor in this category, *Sing Tao* (Table 4). But it is the amount of factual features which constitute the biggest difference. The *Post* is 20.2% factual features; its closest competitor, the *China Tribune,* is 8.8%. It must be noted that the type size of the *China Tribune* and the *Chinese Journal* allows them to print only half as many characters per page as the other papers.

The *Post*'s editors emphasize features because they feel that their papers must be interesting as well as informative. "Papers, for the Chinese people, are a form of entertainment, and we have everything; sports, movies, stories," said editor H.S. Kung. General Manager Robert Tam said, "If they have a dime, we think they should buy our paper, because we cover everything."

The editors of the *Post* claim that one of their major innovations was to bring more American news to the Chinatown reading public. But the figures don't bear this out. The *Post* devotes 3.4% of its space to American national news. The average for the seven papers, not allowing for the type size differential, is 3.3%.

While the *Post*'s claim for a new emphasis in news coverage is not upheld for American national news, the paper has substantially upgraded and expanded

TABLE 3

Local News in Chinese Dailies
March 4-9, 1974

	Local[A]	*New York*	*Chinatown*	*Other Chinese Communities*
	Inches	Inches	Inches	Inches
China Post	2000½	503	1475	22½
China Times	1444½	283	1086	75½
China Tribune	818½	176½	609½	32½
Chinese Journal	959½	93	717½	149
Sing Tao	347½	318½	29	0
United Journal	1431½	223	861½	347
Youth Daily	513½	224½	281	8

[A] Local is the major category, the others are subdivisions of it.

its local coverage. The *Post* prints more news about Chinatown (7.6%) than any of its competitors. It also runs half again as much news about New York. This has been one of the major reasons for the *Post's* rapid rise. The production process of its closest competitor, *Sing Tao*, prevents it from producing a great deal of up-to-date local news.

The urbanized immigrants from Hong

TABLE 4

Features and Advertising in
Chinese Dailies, March 4-9, 1974

	Fictional Features	*Factual Features*	*Advertising*
	Inches	Inches	Inches
China Post	1618	3918	8282
China Times	2784	1105	11802½
China Tribune	1330½	1366½	8785
Chinese Journal	1260	51½	9791
Sing Tao	2582	1141½	8388
United Journal	1328	1466	8866
Youth Daily	1010	1020	10832

Kong are interested in and expect that kind of news. Danny Yung of the Asian-American Field Study Project, which is sponsored by HEW, said, "The new immigrants are urbanized people from Hong Kong. They are more familiar with modern things than the old rural immigrants, who tended to come from more conservative areas. The new immigrants expect more from the papers and they're starting to get it. The *Post* has more local news, more community news, than any of the other papers."

At times, the local news is played up to attract even more readers. This is especially true of crime stories. "We always have crime news," said H.S. Kung, "just like the *Daily News*, we always have something that happened in the community. I think that people, all the people in the world, want to read something sensational."

One major reason for the success of the new style papers is their ratio of copy to advertising. The *Post* is only 42.9% ads, with *Sing Tao* at 47.7%. The older papers check in at 57.4%, for the *China Tribune*, and 63.8%, for *Chinese Journal*.

But statistics cannot reflect all of the important changes occurring in the Chinatown press. The arrival of the politically moderate *Sing Tao,* and President Nixon's visit to the People's Republic of China in February of 1972, had created considerably more room in the middle of the political spectrum for the Chinese press. The new style papers have set out to move into that space, to appeal to a Hong-Kong-born audience which was often neutral in the debate between the two Chinese governments.

H.S. Kung said, "We want to be independent." George M.F. Law, a member of the *Post's* Board of Directors, expanded: "The old papers are controlled by the government of Formosa (Kuomintang). The immigrants from Hong Kong don't belong to the Communists and they don't belong to the Formosa government. They are individuals. They want to know what is going on in Chinatown. They don't want to read government announcements like some papers run."

But the *Post* still reads like a Kuomintang newspaper, for the Chinese language provides its own clues to a paper's loyalties. One is reflected in the varied English spellings of the capital of the People's Republic of China: Peking or Peiping.

Peking means "Northern Capital," and has been the name of that city for over 500 years. In 1927, however, when the Kuomintang came to power in China, it moved the capital to Nanking, "Southern Capital," and changed Peking to Peiping, "Northern Peace." When the Communists defeated the Kuomintang in 1949, they switched the name back.

Today, the government on Taiwan still refers to the city as Peiping while the Communists call it Peking. And, editors of Chinese newspapers must make the politically loaded choice to use the character for "capital" or for "peace" when they refer to Peking. Of the seven dailies, only *Sing Tao,* a neutral paper, uses Peking rather than Peiping.

In addition, most papers use a dateline based on the number of years since the establishment of the Republic of China by the followers of Dr. Sun Yat-sen, founder of the KMT. Thus, the year in the *Post's* dateline is not 1974, but the 63rd year of the Republic of China.

But, there are signs of movement. The *Post* has recently taken to directly translating their sources rather than editing them. Thus, on the same front page, one may find reference to Chairman Mao Tse-tung, translated from the New York *Times,* and "Mao-kung," a derogatory phrase commonly used by the Nationalists referring to the mainland. It roughly translates as Mao-Communists. But the *Post* is still the first New York-based Chinese daily to take the step of referring to Mao Tse-tung as chairman, his official title.

This is admittedly a small change, but it is a start. One reason there have not been more such changes is the influence of advertisers. One Chinatown editor, who asked to remain anonymous, asserted that the only thing which kept his paper from making the change from Peiping to Peking was the knowledge that it would draw protests from advertisers.

But the bonds of loyalty to Taiwan are definitely loosening, especially since the President's visit to Peking. T.Y. Hang, editor-in-chief of the pro-KMT *China Tribune,* noted that the most profound effect of Nixon's visit had been in the Chinese business community. Chinese businessmen can once again buy goods from the mainland and shops are opening up which carry nothing but mainland products. Because of this development, the Chinese business community is more amenable to better relations between the United States and the People's Republic. "We don't feel any impact now," said Hang, "but if the U.S. government recognizes Peking, things may be different."

In Chinese, the phrase "wo kuo," which literally means my country, has always meant China. The older papers in Chinatown still often use it that way. But, the new wave of immigrants who are coming to this country are looking to set up a new life. "My country" to them no longer will mean China but the United States. The Chinese papers are only beginning to reflect this shift in Chinatown attitudes.

ACKNOWLEDGMENTS

Van Trump, James D. "A Trinity of Bridges: The Smithfield Street Bridge over the Monongahela River at Pittsburgh." *Western Pennsylvania Historical Magazine* 58, No.4 (1975): 439–70. This publication is now titled *Pittsburgh History*. The former quarterly is no longer published under the WPHM title. Reprinted with the permission of the Historical Society of Western Pennsylvania.

Dornfeld, A.A. "Steamships: A Hundred Years Ago." *Chicago History* 4, No.3 (1975): 148–56. Reprinted with the permission of the Chicago Historical Society.

Hijiya, James A. "Making a Railroad: The Political Economy of the Ithaca and Owego, 1828–1842." *New York History* 54, No.2 (1973): 145–73. Reprinted with the permission of the author and the New York State Historical Association.

Rosenberg, Leon J. and Grant M. Davis. "Dallas and Its First Railroad." *Railroad History* 135 (1976): 34–42. Reprinted with the permission of the Railway & Locomotive Historical Society.

Halma, Sidney. "Railroad Promotion and Economic Expansion at Council Bluffs, Iowa, 1857–1869." *Annals of Iowa* 42, No.5 (1974): 371–89. Copyright (1974) State Historical Society of Iowa. Reprinted by permission of the publisher.

Rhoda, Richard. "Urban Transport and the Expansion of Cincinnati 1858 to 1920." *Cincinnati Historical Society Bulletin* 35, No.2 (1977): 130–43. Reprinted with the permission of the author and the Cincinnati Historical Society.

Kuhm, Herbert W. "When Milwaukee Streetcars Were Horse-Drawn." *Milwaukee History* 2, No.2 (1979): 30–37. Reprinted with the permission of the Milwaukee County Historical Society.

Saylor, Larry J. "Street Railroads in Columbus, Ohio, 1862–1920." *Old Northwest* 1, No.3 (1975): 291–315. Reprinted with the permission of Miami University.

Wikstrom, Debbie. "The Horse-Drawn Street Railway: The Beginning of Public Transportation in Shreveport, 1870–1872." *North Louisiana Historical Association Journal* 7, No.3 (1976): 83–90. Reprinted with the permission of the North Louisiana Historical Association.

Johnson, Frank E. "Eight Minutes to New York: The Story of the Hudson and Manhattan Tubes." *American History Illustrated* 9, No.5 (1974): 12–23. Reprinted with the permission of Cowles Magazines, Inc.

Granger, Denise. "The Horse Distemper of 1872 and Its Effect on Urban Transportation." *Historical Journal of Massachusetts* 2, No.1 (1973): 43–52. Reprinted with the permission of the Institute for Massachusetts Studies.

Handley, Lawrence R. "Settlement Across Northern Arkansas as Influenced by the Missouri & North Arkansas Railroad." *Arkansas Historical Quarterly* 33, No.4 (1974): 273–92. Reprinted with the permission of the Arkansas Historical Association.

Oihus, Colleen A. "Street Railways in Grand Forks, North Dakota: 1887–1935." *North Dakota History* 44, No.2 (1977): 12–21. Reprinted with the permission of the State Historical Society of North Dakota.

Hovinen, Gary R. "Lancaster's Streetcar Suburbs, 1890–1920." *Journal of the Lancaster County Historical Society* 82, No.1 (1978): 49–59. Reprinted with the permission of the Lancaster County Historical Society.

Mallach, Stanley. "The Origins of the Decline of Urban Mass Transportation in the United States, 1890–1930." *Urbanism Past and Present* 8 (Summer, 1979): 1–17. Reprinted with the permission of the University of Wisconsin-Milwaukee.

Hofsommer, Donovan L. "Townsite Development on the Wichita Falls and Northwestern Railway." *Great Plains Journal* 16, No.2 (1977): 107–22. Reprinted with the permission of the Institute of the Great Plains.

Siegert, Wilmer H. "Spokane's Interurban Era." *Pacific Northwesterner* 17, No.2 (1973): 17–28. Reprinted with the permission of the Westerners.

Barrett, Paul. "Public Policy and Private Choice: Mass Transit and the Automobile in Chicago between the Wars." *Business History Review* 49, No.4 (1975): 473–97. Reprinted with the permission of the Harvard Business School.

Owen, John E. "A Sociologist Looks at America's City Traffic Problem." *Social Science* 48, No.2 (1973): 87–92. Reprinted with the permission of *Social Science*.

Piper, Robert R. "Transit Strategies for Suburban Communities." *Journal of the American Institute of Planners* 43, No.4 (1977): 380–85. Reprinted with the permission of the *Journal of the American Institute of Planners*.

Rabin, Yale. "Highways as a Barrier to Equal Access." *Annals of the American Academy of Political and Social Science* 407 (1973): 63–77. Reprinted by permission of Sage Publications, Inc. Copyright 1973 Sage Publications, Inc.

Crouthamel, James L. "James Gordon Bennett, the *New York Herald*, and the Development of Newspaper Sensationalism." *New York History* 54, No.3 (1973): 294–316. Reprinted with the permission of the author and the New York State Historical Association.

Glauber, Robert H. "The Necessary Toy: The Telephone Comes to Chicago." *Chicago History* 7, No.2 (1978): 70–86. Reprinted with the permission of the Chicago Historical Society.

Sim, John Cameron. "19th Century Applications of Suburban Newspaper Concepts." *Journalism Quarterly* 52, No.4 (1975): 627–31. Reprinted with the permission of the Association for Education in Journalism and Mass Communication.

Warthman, Forrest. "Telecommunication and the City." *Annals of the American Academy of Political and Social Science* 412 (1974): 127–37. Reprinted by permission of Sage Publications, Inc. Copyright 1974 Sage Publications, Inc.

McCue, Andy. "Evolving Chinese Language Dailies Serve Immigrants in New York City." *Journalism Quarterly* 52, No.2 (1975): 272–76. Reprinted with the permission of the Association for Education in Journalism and Mass Communication.